Introduction to Forest Genetics

Jonathan W. Wright

Department of Forestry
Michigan State University
East Lansing, Michigan

ACADEMIC PRESS   New York San Francisco London

A Subsidiary of Harcourt Brace Jovanovich, Publishers

Cover illustration: Grafted seed orchard of Japanese larch.
See Fig. 10.4, p. 198.

ACADEMIC PRESS, INC.
111 Fifth Avenue, New York, New York 10003

United Kingdom Edition published by
ACADEMIC PRESS, INC. (LONDON) LTD.
24/28 Oval Road, London NW1

Library of Congress Cataloging in Publication Data
Wright, Jonathan W
        Introduction to forest genetics.

        Includes bibliographies and index.
        1.   Forest genetics.   2.   Tree breeding.   I.   Title.
SD399.5.W72   634.9'56   75-19686
ISBN 0−12−765250−7

Contents

PREFACE   xi

ACKNOWLEDGMENTS   xv

1 Introduction
Problems Peculiar to Forest Genetics   2
The Possibilities of Tree Improvement   4
General Sources of Information   4

2 Concepts from Basic Genetics
Chromosome Structure and Function   7
DNA and RNA   12
Genes   12
Nongenic Inheritance   20
Genotype versus Phenotype   21
Cell Division and the Opportunity for New Gene Combinations   22
Study Questions   26
Suggested Reading   27

3 Additional Basic Genetic Concepts
Inbreeding   29
Hybrid Vigor   41
Recovery Ratios in the F_2 Generation   44
Study Questions   45

4 Population Genetics—Selection
Definitions   47
Hardy–Weinberg Equilibrium   48

Complete Elimination of Homozygous Recessive Trees   48
Partial Selection against Recessives   50
Selection Favoring Recessives (against Dominants)   53
Selection for Genes with Additive Effects   53
"Fitness" and Fisher's "Fundamental Theorem"   55
Selection for and against Heterozygotes   56
Selection in Small Populations   59
How to Increase Selection Pressure   60
A Word Picture of Selection   60
Recapitulation   62
Study Questions   62

5 Population Genetics—Mutation, Migration, Isolation
Mutation   65
Migration   66
Isolation   77
Study Questions   82

6 Useful Techniques—Controlled Pollination
Selection of Parent Trees   86
Climbing   86
Bagging to Prevent Outside Pollination   91
Pollen Handling   95
Pollination Procedures   100
Indoor Controlled Pollination   103
Fruit Collection and Seed Handling   105
Success Expected from Controlled Pollination   108
Study Questions   109

7 Useful Techniques—Vegetative Propagation
Introduction   111
Grafting and Budding   111
Propagation by Stem Cuttings   121
Propagation by Root and Root–Stem Cuttings   127
Propagation by Leaf Cuttings   127
Root Suckers   127
Other Methods of Occasional Interest   128
Apomixis   128
Topophysis or Persistent Nongenetic Effects   129
Study Questions   133

8 Useful Techniques—Establishment and Measurement of Test Plantations
Introduction   135
Definitions   135

Commonly Used Experimental Designs 136
Nonstatistical Considerations 138
Statistical Considerations 146
Sample Schedules for Experiment Establishment 151
Hints on Measurement and Data Recording 154
Study Questions 157

9 Selective Breeding—General Principles and Methods

Role of Selective Breeding in Tree Improvement 159
General Principles 160
Breeding Methods Described 165
Study Questions 184

10 Choice of Breeding Method and Type of Seed Orchard

Factors Affecting Choice 187
Thinned Natural Stands versus Planted Seed Orchards 188
Mass versus Family Selection 189
Half-Sib versus Full-Sib Progeny Testing 191
Half-Sib versus Full-Sib Family Selection 191
Progeny Test Seed Orchards versus Clonal Seed Orchards 194
Earliness of Flowering on Grafts and Seedlings 201
Study Questions 202

11 Examples of Progress through Selective Breeding and Seed Orchard Management

Introduction 205
Examples of Simple Inheritance 205
Variation and Improvement in Quantitative Traits 207
Seed Orchard Management in Loblolly and Slash Pines 223
Management of Seedling Seed Orchards of Red Pine 232
Special Cases 235
Study Questions 237

12 Estimation of Heritability and Genetic Gain

Experimental Procedures and Applicability of Estimates 239
Parent–Progeny Regression 240
Variance Analysis—Half-Sib Progeny Test 242
Variance Analysis—Full-Sib Progeny Test 246
Conversion of Family to Single Tree Heritabilities 249
Variance Analysis—Clonal Test 250
Genotype–Environment Interaction—Effect on Breeding Plans 252
Study Questions 252

13 Provenance Testing

Role of Provenance Tests in Tree Improvement   253
Factors Influencing Amount of Geographic Variability   254
General Geographic Trends   255
Knowns and Unknowns about Natural Selection   259
Design of Provenance Tests   261
Practical Use of Provenance Data   264
Study Questions   265

14 Geographic Variation in Scotch Pine

Introduction   267
Phenotypic versus Genetic Variation Patterns   268
Amount and Pattern of Genetic Diversity   269
Description of Varieties   274
Practical Use of Scotch Pine Provenance Data   278
Typical and Atypical Aspects of Scotch Pine   280
Study Questions   281

15 Geographic Variation in American Trees

The Pacific Coast, California to Alaska   283
Western United States—A Region of Great Diversity   286
Boreal America—A Region of Clines and Great Genetic Diversity   292
Slight to Moderate Variation in Northeastern Species   295
Geographic Variation in the Central Hardwoods   300
Southeastern United States—A Region of Moderate Genetic Diversity   304
Study Questions   311

16 Species and Racial Hybridization

Use of Hybridization in Agriculture   313
Goals of Hybridization in Trees   314
Crossability Patterns   316
Hybridization and Evolution   320
Crossability Patterns by Genus   322
Study Questions   342

17 Species and Racial Hybridization—Practical Application

Historical Background   343
Mass Produced F_1 Hybrids   344
Utilization of F_2 Generation Hybrids   349
Hybridization and Subsequent Selection in Later Generations   351
Study Questions   357

18 Introduction of Exotic Species

Introduction   359
Reasons for Expecting Success from Exotics   360

Factors Governing Successful Introduction   364
Exchange Relationships among Regions   371
Methods of Introducing and Testing New Species   372
Examples of Successful and Unsuccessful Exotics   376
Regional Summaries of the Use of Exotics   391
Study Questions   397

19 Polyploidy, Aneuploidy, and Haploidy

Definitions   399
Techniques of Chromosome Study   400
Karyotypes of Pines   402
Rareness of Polyploidy in Gymnosperms   403
Importance of Polyploidy in Angiosperms   404
Breeding Behavior of Polyploids and Aneuploids   412
Induction and Recognition of Polyploidy   415
Haploid Breeding   418
Study Questions   420

Appendix: Common and Scientific Names of Trees   423

Bibliography   427

GLOSSARY   439

INDEX   457

Preface

As the world's population grows, so does the need for forest products. However, the land available for growing them has been shrinking because of agricultural demands, city expansion, and road development. Thus there is a need to increase the productivity of our available forest land.

Foresters once relied mainly on natural regeneration; now they often find it expedient to harvest and plant immediately after. Indeed, in many regions "forestry" is synonymous with "plantation culture." In Europe, where planting has been practiced the longest, foresters have tried to obtain the proper seed. Elsewhere, planting has often been done the cheapest possible way, with little concern for genetic quality of the planting stock. That situation is rapidly changing as data concerning the advantages of using genetically superior trees become available. Today all large planting programs include some breeding research, and improved species of trees are being used.

In spite of the fact that trees are long-lived and less convenient test organisms than herbaceous plants, progress in tree improvement has been rapid. Even when measured in terms of years rather than generations, the percent increase in productivity has been as great for some trees as for some annuals. Each success engenders further work so that we should expect to see a constantly growing number of new varieties. The methods used to develop these varieties vary as do the ways in which they can be used for maximum effectiveness. To a large extent, the way in which a variety is developed governs the way in which it can be mass produced and put to practical use.

Most principles involved in forest tree breeding are extensions of the basic principles of Mendelian and population genetics. Their application varies considerably among species and with economic conditions. Wherever possible I have attempted to give examples, based on actual experience with trees, which illustrate a diversity of situations and responses. To aid those practitioners who desire more information on any one subject, I have included a list of general sources of information at the end of Chapter 1.

This book is intended as an introductory text for future professional tree breeders; those, primarily foresters, who will have occasion during their careers to help in some phase of tree breeding activity such as selection of plus trees, choosing a site for a seed orchard, or establishment of a test plantation; and for those interested in planting trees and wishing to evaluate for themselves new varieties as they become available. It can be used with or without previous training in either genetics or forestry. A few basic genetic concepts most often used in tree improvement studies are summarized briefly in the early chapters. They are presented descriptively and without proof of their validity. For those interested in delving further into genetic concepts, a textbook of fundamental genetics should be consulted.

Some experimental techniques can be explained in simple, nontechnical language, but the validity of many results depends on the quality of the experimental work. To the extent that both conditions are true, I have included descriptions of experimental techniques. In doing this, I have been guided primarily by the needs of the nontree breeder who wishes only a general understanding of the work involved. Future, professional tree breeders should use the descriptions only as starting points and consult recent technical literature for additional detail.

Chapters 2–5 are devoted to the basic concepts of Mendelian and population genetics which are of greatest concern in tree improvement. These chapters are necessarily concentrated, as a complete treatment of any one of the concepts requires many pages.

Chapters 6–8 are devoted to techniques which are generally useful in tree improvement work. The descriptions are necessarily general. Some possible variations are mentioned but the chapters are not meant to comprise a complete procedural manual for any individual species.

Chapters 9–18 are devoted to four of the most commonly used approaches to tree improvements: individual tree selection and breeding (Chapters 9–12), provenance testing (Chapters 13–15), species and racial hybridization (Chapters 16–17), and introduction of exotics (Chapter 18). This sequence is not necessarily the one to be followed in an actual improvement program. I have included theoretical considerations and actual examples in the development of each topic.

Chapter 19 is devoted to chromosome studies and to two special approaches (polyploidy and haploidy breeding) based on manipulation of chromosome complements. These approaches represent a merger of cytology and genetics disciplines.

A glossary of technical terms most often used in forest genetics and tree improvement work concludes the work.

I would like to dedicate this book to two United States pioneer tree breeders: E. J. ("Ernie") Schreiner and F. I. ("Pete") Righter. Both started work in 1924. At that time there was relatively little forest planting and almost no interest in improved planting stock. They were among the few who even believed that genetic tree improvement was worthwhile. Their work opened new vistas, and their enthusiasm was responsible for others entering the field. Much of modern tree breeding was influenced by their work.

Jonathan W. Wright

Acknowledgments

Among the people to whom I am indebted in the preparation of this book, I would like to acknowledge the following who gave helpful reviews of portions of the manuscript: Drs. Calvin Bey and David Funk of the U.S. Forest Service at Carbondale, Illinois; Dr. Bruce P. Dancik of the University of Alberta; Dr. Richard A. Cunningham of the U.S. Forest Service at Lakewood, Colorado; Dr. Henry D. Gerhold of Pennsylvania State University; Dr. Ray E. Goddard of the University of Florida; Dr. Richard D. Hall of Iowa State University; Dr. Howard B. Kriebel of the Ohio Agricultural Research and Development Center; Dr. Donald Lester of the University of Wisconsin; Dr. Gene Namkoong of the U.S. Forest Service at Raleigh, North Carolina; Dr. Hans Nienstaedt of the U.S. Forest Service at Rhinelander, Wisconsin; Dr. Leroy C. Saylor of North Carolina State University at Raleigh; and Dr. Hans Van Buijtenen of the Texas Forest Service.

I am also indebted to the following who furnished photographs or drawings: Calvin Bey and David T. Funk of the U.S. Forest Service, Carbondale, Illinois; Kim K. Ching of Oregon State University; William Dvorak of the Fiji Pine Scheme, Fiji Islands; H. D. Gerhold of Pennsylvania State University; Peter W. Garrett of the U.S. Forest Service, Durham, New Hampshire; Lauri Kärki of the Finnish Forest Tree Breeding Foundation, Helsinki; Howard B. Kriebel of the Ohio Agricultural Research and Development Center; Robert P. Karrfalt and Donald T. Lester of the University of Wisconsin; Timothy La Farge of the U.S. Forest Service, Macon, Georgia; Ray E. Goddard of the University of Florida; Adolph Nanson of the Station de Recherches des Eaux et

Forets, Groenendaal, Belgium; Hans Nienstaedt of the U.S. Forest Service, Rhinelander, Wisconsin; Kathleen Nixon of the Wattle Research Institute, Pietermaritzburg, Republic of South Africa; Ralph A. Read of the U.S. Forest Service, Lincoln, Nebraska; Leroy C. Saylor and Bruce J. Zobel of North Carolina State University at Raleigh; Geoffrey B. Sweet of the Forest Research Institute, Rotorua, New Zealand; Mirko Vidakovic of the University of Zagreb, Yugoslavia; P. C. Wakeley, retired from the U.S. Forest Service; and Osborn O. Wells of the U.S. Forest Service, Gulfport, Mississippi.

1

Introduction

Forest genetics is the study of hereditary variation in forest trees. Hereditary differences are those which are caused by the genes and/or cytoplasm within the tree. They are predetermined at the time the seed is formed, and in that sense are opposed to differences which are caused by the external environment.

However, it is a mistake to speak of most tree characteristics as being strictly under either genetic or environmental control. Most commercially important traits are controlled by both internal and external factors. An example is growth rate, which can be increased by breeding for faster growth or by culture. Also, many genetically controlled traits may be more evident in one environment than another. For example, genetic differences in frost resistance will not be evident when trees are planted in a frost-free region, but may be very important when the same trees are planted in an area subject to low temperatures during the growing season.

Forest tree improvement is the application of forest genetics to practice. Usually this is accomplished by testing various wild types and determining which will grow best when planted on specific sites. In more advanced programs, this is accomplished by breeding, e.g., for increased growth rate, increased resistance to pests. The new varieties are then planted.

Basic genetic principles are the same for trees, men, and fruit flies. But inheritance patterns and methods of experimentation vary considerably among these groups. Thus, forest genetics is a distinct field of endeavor, having its own problems.

Problems Peculiar to Forest Genetics

Forest Genetics Uses Indirect Evidence

Most characteristics are under the control of genes and the environment. Genes are submicroscopic parts of a cell. Even with an electron microscope they cannot be seen and identified as "good" and "bad." Instead, they must be identified by indirect means—by growing the offspring of a tree and observing the characteristics of the offspring.

Thus progeny tests are integral parts of forest genetics research. By a progeny test is meant the growing of different species, different races, or the offspring of different trees under similar conditions in a replicated experiment. Then if one particular progeny grows faster than the others it is safe to assume that growth rate is under genetic control. Or, if one particular progeny has longer needles than the others it is safe to assume that needle length is under genetic control.

Uncertainty and the Need for Continuous Experimentation

Most forest genetics research leads to some generalizations such as "tall trees produce faster growing offspring" or "trees from southern climates grow faster." Each such generalization is usually for a single species. If the necessary research work has been done and a rule of this type has been formulated for a particular species, a grower can use it when ordering seed. The grower will be assured of obtaining a certain amount of genetic improvement.

But these rules are rarely absolute. Nearly always there is a certain amount of unexplained variation. It may be true, for example, that the offspring of the tallest trees grow 1% faster than the offspring of average trees; however, the offspring of some trees grow 5% faster than average. Thus, a grower can obtain the 1% gain by using the general rule, but he must test the offspring of each parent to obtain the full 5% gain.

In eastern white pine there may be a general rule that southern trees grow fastest when planted in southern or central locations. However, there are enough exceptions to the general rule (i.e., trees from the midsouthern state of Virginia grow relatively slowly) that trees from many places must actually be tested to determine with certainty which grow best. Because of these many uncertainties, continued experimentation is an integral part of most practical as well as theoretical tree improvement programs.

The Time Element

Trees are long-lived organisms which require several years to produce seed. A pine breeder cannot produce 8 generations in 4 years as can a maize breeder. However, time is not so much a limiting factor as might be thought at first glance. Some species produce seed in a few years, and some grow rapidly enough to be harvested in 10 years.

Tree breeders have learned to live with the problem and to adjust their procedures to compensate for the fact that they work with long-lived perennials rather than herbs. This problem can be compensated for by working on several projects simultaneously. One project may yield a worthwhile result in the tenth year, a second in the eleventh year, a third in the twelfth year, etc. Grafting may be used, since in many species grafted trees flower appreciably earlier than do seedlings. Conclusions can be drawn concerning which trees grow best when the trees are only a few years old. A great deal of work is under way to determine how well the results from young experiments apply to trees which are normally grown on rotations of 40+ years. In many of the experiments, early results are at least moderately trustworthy.

The Necessity for Seed Production

Reproduction by seed is a necessary part of any improvement work. This is almost an incidental problem in crop plant breeding because most crop plants are raised for their seeds; heavy and regular seed production is taken for granted. Most trees, however, are not grown for their seeds, and many that would be otherwise desirable fruit only occasionally or not at all. Thus, whether or not a tree breeder wants to be concerned with seed production, he must be, and he must often dilute part of his work with experiments on the stimulation of flowering or fruiting.

Scarcity of Basic Genetic Information about Trees

Research on tree improvement and tree genetics has been in progress for 150 years. But only in the past 25 years has the research been intensive. Also, trees are more difficult test organisms than corn or yeast. Thus, we lack basic genetic information about even the best studied forest tree species. We know what happens if pine is selfed for 1 generation, but we do not know what happens if it is selfed for 5 generations. We know how many chromosomes a pine tree has, but we do not know what genes are located on any particular chromosome. Nor do we know how many genes determine most tree characteristics.

This lack of basic knowledge is a fact which must be taken into account when planning improvement work. It is often necessary for a tree breeder to work on seemingly simple basic problems which have already been worked out for crops such as corn and wheat. It is always necessary to tailor tree improvement work specifically to the amount of basic information available as well as to the needs of tree planters. In a way this situation is advantageous. Crop plants have been subjected to selection for dozens or hundreds of generations so that the modern crop plant breeder must improve on an already highly improved crop. Tree breeders, on the other hand, generally start with unimproved wild types or with trees which have been selected for two or three generations at the most. Thus there is tremendous room for improvement, and many simple experiments have led to very great gains.

The Possibilities of Tree Improvement

The first forest genetic experiments were started nearly 200 years ago, but only in the past two decades has there been concentrated effort on the part of many people. Thus, it is not yet possible to forecast the total impact of forest genetic research on forestry practices in general.

It can be said, however, that many of the early results are promising. Of the many million southern pine seedlings planted each year in southern United States, most will soon be of an improved type which grows faster or is straighter than the wild trees of the same species. Tannin production in the wattle plantations of the Republic of South Africa has increased as seed orchard seed became available. From work started only two decades ago, improved second-generation pine hybrids are being planted in the Republic of Korea. In a few subtropical countries the introduction of new species has caused radical changes in planting practice in the past few years. The promise from these projects indicates that genetic improvement will play an increasing role in maintaining and increasing forest productivity.

General Sources of Information

The following books or articles give general information on forest tree breeding. They are arranged by date of publication.

Kalela, A. (1937). "Zur Synthese der experimentellen Untersuchungen über Klimarassen der Holzarten. Publ. 26. Forest Research Institute of Finland, Helsinki. (Detailed numerical data and abstracts of pre-1937 provenance tests.)

Richens, R. H. (1945). Forest tree breeding and genetics. *Imp. Agr. Bur. Joint Publ.* *No. 8*, pp. 1–79. (Compilation of work done prior to World War II.)

Lindquist, B. (1948). "Genetics in Swedish Forestry Practice." Chronica Botanica, Waltham, Massachusetts. (Available in Swedish, German and English.) Description of mass selection in Swedish pines and spruces.

Sato, K. (1950). "Forest Tree Breeding," 2 vols. Asakura Press, Tokyo. (General text. Available in Japanese only.)

Larsen, C. S. (1956). "Genetics in Silviculture." Essential Books, Fairlawn, New Jersey. (Devoted mainly to Danish experimental work.)

Rohmeder, E., and Schönbach, H. (1959). "Genetik und Züchtung der Waldbäume." Parey, Berlin. (General text.)

Wright, J. W. (1962). The genetics of forest tree improvement. *FAO Forest. Forest Prod. Stud.* **16**, 1–399. [General text. Available in English, French, Spanish, Rumanian and (some parts) Japanese.]

Gerhold, H. D., Schreiner, E. J., McDermott, R. E., and Winieski, J. A. (eds.). (1966). "Breeding Pest-Resistant Trees." Pergamon, Oxford. (Text of papers presented at a symposium held in 1964 at Pennsylvania State University, Philadelphia.)

Vidaković, M. (1969). "Genetics and Forest Tree Breeding," Mimeo. UNDP-FAO, Pakistan National Forestry Research and Training Project. (General text.)

FAO-North Carolina State University. (1969). Lecture notes. Forest Tree Improvement Training Centre, North Carolina State University at Raleigh, School of Forest Resources. (Manual for practicing tree breeders.)

Van Buijtenen, J. P., Donovan, G. A., Long, E. M., Robinson, J. F., and Woessner, R. A. (1971). Introduction to practical forest tree improvement. *Tex. Forest Serv., Circ. No. 207*, pp. 1–17.

Zobel, B. J. (1971). The genetic improvement of southern pines. *Sci. Amer.* **225**, 94–103. (Popular account of southern pine improvement.)

Enescu, V. (1972). "Ameliorarea arborilor." Editura CERES, Bucharest, Romania. (General text written in Romanian.)

Namkoong, G. (1972). "Foundations of Quantitative Forest Genetics." Government Forest Experimental Station of Japan, Meguro, Tokyo, Japan. (Text used for a short course on quantitative genetics.)

Burley, J., and Nikles, D. G. (eds.). (1973). "Selection and Breeding to Improve Some Tropical Conifers," 2 vols. Commonwealth Forestry Institute, Oxford, England and Queensland Dept. of Forestry, Queensland, Australia. (Papers presented at a 1971 symposium which was part of the 15th IUFRO Congress at Gainesville, Florida.)

Stern, K., and Roche, L. (1974). "Genetics of Forest Ecosystems," Ecol. Stud. No. 6. Springer-Verlag, Berlin and New York. (Primarily a study of natural vegetation, with heavy emphasis on ecology.)

Silvae Genetica, published by J. D. Sauerlander's Verlag, Frankfurt am Main, West Germany, is the only journal devoted solely to forest genetics. The papers are published in German, English, or French but cover work in many countries. There is also a lengthy section devoted to abstracts of bulletins and papers published in other journals.

Articles on forest genetics commonly appear in forestry journals and in experiment station bulletins, but rarely appear in genetics journals.

In Canada the Canadian Tree Improvement Committee and in the United States the Northeastern, Southeastern, Lake States, Central States and Western Tree Improvement Committees hold annual or biennial meetings. These meetings are open to the public and usually consist of field trips and a number of papers. Such papers are usually published in a proceedings, often in mimeographed form.

Almost every year there is a national or international conference concerning some phase of forest tree improvement. Such conferences are sponsored by various organizations such as IUFRO (International Union of Forest Research Organizations), FAO (Food and Agriculture Organization of the United Nations). Papers presented at such conferences are usually published in a proceedings.

2

Concepts from Basic Genetics

Certain basic principles of genetics are similar for all organisms, whether plants or animals. While it is beyond the scope of this book to cover all the basic principles in detail, it is desirable to describe a few basic concepts having most to do with inheritance in trees.

Chromosome Structure and Function

Trees are composed of cells, some living and some dead. Each living cell consists of an outer "cell wall," a fluid called the "cytoplasm," and a "nucleus" surrounded by the cytoplasm. Among other things the nucleus contains "chromosomes."

The chromosomes are of special interest genetically because they carry the bulk of the genetic information and transmit that information from one cell generation to the next. They are nearly constant in number throughout the vegetative cells of a tree or of an entire species. If it were possible to see all the details inside the chromosomes of a tree, one could understand nearly everything about that tree's genetic potential. That, however, is a difficult matter.

During a cell generation, chromosomes change much in appearance. At most times they appear as long threads, intertwined with each other and very difficult to study. At one time just prior to cell division (the "metaphase" stage) they are much contracted and easily observed. At that time, one can count them and determine certain general features of their structure (Fig. 2.1).

7

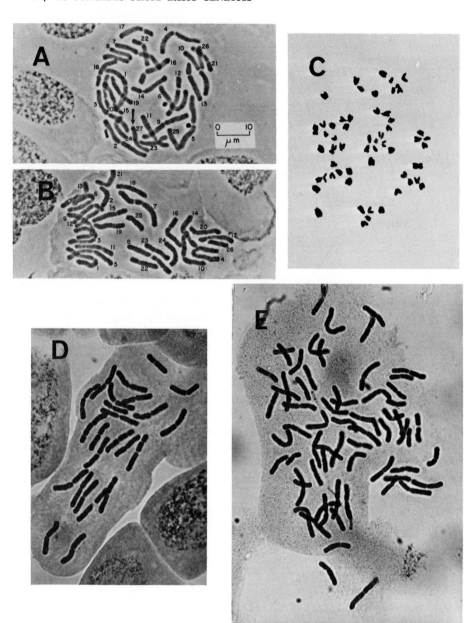

FIG. 2.1. Metaphase mitotic chromosomes of (A) dwarf aneuploid branch of Douglas-fir ($2n = 27$), (B) normal Douglas-fir ($2n = 26$), (C) normal American elm (tetraploid, $4n = 56$), (D) normal Engelmann pine (diploid, $2n = 24$), and (E) normal redwood (hexaploid, $6n = 66$). The presence of an extra chromosome

Chromosomes vary in size among species of trees. In poplar, willow, eucalypt, and many other angiosperms they are only a few microns long [1 micron (μm) = 0.001 mm], little larger than the limits of resolution of good quality light-conducting microscopes. Hence in those species, cytological work is usually confined to studies of chromosome number and of pairing at meiosis. Chromosomes are considerably large in most gymnosperms, where it is possible to make more detailed studies.

The features of metaphase chromosomes that can be observed under a light-conducting microscope are illustrated in Fig. 2.1. Note that each chromosome appears as a sausagelike organ with a near-central constriction or "centromere." In some pine species there are also secondary constrictions closer to the ends of the chromosomes. By very detailed measurements, Saylor (1972) was able to recognize each of the 12 pairs of chromosomes in a vegetative pine cell and to find differences in centromere position among corresponding chromosomes of different pine species.

Our knowledge of detailed chromosome structure and function comes mostly from work on lower plants, such as yeast and bacteria. Much of that work has been summarized in the book "Molecular Genetics" by DeBusk (1968). Research, which is still unfolding and giving us additional insights, shows that a chromosome is a long, thread-like structure consisting of deoxyribonucleic acid (DNA) and a protein sheath. The DNA is the active genetic material and is a very long molecule composed of two spiral strands. Each strand is composed of four organic bases (cytosine, C; guanine, G; adenine, A; and thymine, T) and attached sugar radicals. One such base and its attached sugar radical is called a "nucleotide." The nucleotides in a single strand are tightly joined together by phosphate radicals. The two spiral strands are held together more loosely by hydrogen bonds.

Ability to replicate itself is one of the secrets of DNA which makes it possible for the chromosomes to transmit genetic information from one cell generation to the next. This ability for self-replication is a consequence of the double-strand nature of the molecule and of the properties of the four bases. Adenine and thymine are joined together by 2 hydrogen bonds (A=T) and have such a molecular structure that only

---

(No. 27 in A) resulted in slow growth of Douglas-fir. The large constrictions near the chromosome centers are called centromeres. All chromosomes are magnified to approximately the same extent. The differences in chromosome density are mainly a matter of staining technique. [Photogaphs (A) and (B) courtesy of Kim K. Ching, Oregon State University; (C) courtesy of Robert Karrfalt and Donald T. Lester, University of Wisconsin; and (D) and (E) courtesy of Leroy C. Saylor, North Carolina State University at Raleigh.]

A and T can be joined together. Cytosine and guanine are linked by 3 hydrogen bonds (C≡G) and have such a molecular structure that only C and G can be joined together. At the time of chromosome division, the two DNA strands unravel, every base along each strand attracts unto itself its complementary base, and each single strand quickly becomes a new double strand exactly like the original. This can be illustrated as shown in Fig. 2.2 and the tabulation on p. 11.

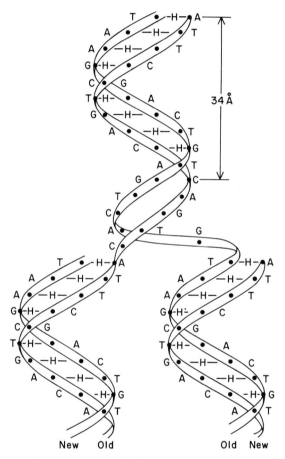

FIG. 2.2. A replicating DNA molecule showing the double helix. The distance between the two strands is 20 Å and the distance between nucleotides on one strand is 3.4 Å. In a single strand there is one complete spiral every 10 nucleotides or every 34 Å. The nucleotide pairs are shown as A-T and C-G. The strands are joined by H bonds, and nucleotides within a strand are joined by phosphate bonds (not shown). As the parental molecule (top) unravels the two daughter strands (bottom) quickly replicate themselves to form exact duplicates of the parental molecule.

Original double strand	Single strands during process of unraveling		Two new double strands, each a duplicate of the original	
CG	C	G	CG	CG
AT	A	T	AT	AT
GC	G	C	GC	GC
AT	A	T	AT	AT
TA	T	A	TA	TA
TA	T	A	TA	TA

Along a single strand the bases A, T, C, and G are arranged linearly and are organized in groups of three, called triplets. These triplets can also be considered as three-letter words, e.g., AAT, AAA, CCC. These three-letter words are the genetic code and thus are the other secret of DNA—the ability to regulate life functions within a cell and within an organism.

With a four-letter alphabet, only 64 different three-letter "words" are possible. How can only 64 "words" constitute the genetic code responsible for the myriads of differences found in the plant and animal kingdoms? This is possible because the triplets or "words" are in turn organized into special sequences or "paragraphs" which direct growth processes. The sequences responsible for the formation of particular amino acids and for the activity of certain enzymes governing protein synthesis are known for some one-celled plants. For trees, however, only the relative amounts of DNA per cell and the amounts of unique and nonunique sequences are known. For example, Richard D. Hall, working with jack pine, estimated that 24% of the DNA in a cell is in unique sequences, 56% is in sequences repeated 40–3500 times per cell, and 20% is in sequences repeated a million times per cell.

From the above it can be surmised that DNA molecule has a very large molecular weight. That is true. Hall estimated that the 12 pairs of chromosomes in a single pine cell contain 50,000,000,000 nucleotide pairs. If each nucleotide is considered a letter of the alphabet, that is enough letters to write 50,000 books the size of this one. Even allowing for the repetition of DNA sequences within a cell, that seems enough to provide a code for 13,000,000 genes.

DNA is small in actual terms, however. Each nucleotide contains approximately 30 atoms of C, H, O, N, and P and is approximately 2.5×10^{-9} cm in diameter. Completely stretched out, a human DNA molecule is approximately 6 cm long and 2×10^{-6} cm across. An enterprising mathematician calculated that the DNA from the egg cells which gave rise to the 3 billion people now living could be contained in a box 3 mm on a side.

DNA and RNA

While DNA is the self-perpetuating genetic code within a cell, the actual control of growth processes in a cell is accomplished by a related substance called ribonucleic acid (RNA). The principal differences in structure between DNA and RNA are that RNA is single-stranded, RNA sugars have one more oxygen atom than DNA sugars, and RNA has uracil (U) in place of thymine (T). Very briefly, one strand of a DNA molecule acts as a template from which a complementary strand of RNA is formed in the same manner by which a complementary strand of DNA is formed from a single strand. If, for example, the DNA sequence is CGGATTCCG, the RNA sequence becomes GCCUAAGGC.

According to its location and function within a cell, RNA can be categorized as belonging to one of three types. In very simplistic terms the functions of these three types are to carry messages from the DNA to other parts of the cell, to regulate and participate in the synthesis of amino acids, and to regulate and participate in the synthesis of proteins from amino acids. The RNA which regulates amino acid and protein synthesis is composed of relatively small molecules, some containing only 75–80 nucleotides.

Genes

Structure and Function

From a structural standpoint, a gene is defined as a sequence of triplets along a DNA molecule. So far it has not been possible to isolate and study the structure of a single gene. However, virus and bacteriophage particles are believed to be similar to or identical with genes, and they can be isolated and studied in mass. Therefore, much of our knowledge of gene structure comes from the study of viruses and phages.

A gene can also be defined functionally as that part of a chromosome responsible for the development of a particular trait in an organism. Throughout this book we shall use this functional concept of the gene, without knowing its structural details. Thus we shall speak of genes for rapid growth, for cold hardiness, for long leaf length, etc. In so doing, we shall consider the gene as the ultimate hereditary unit (although still of large molecular size) transmitted from one cell generation to the next and from one tree generation to the next. From one standpoint, practical tree breeding is a manipulation of gene frequencies.

The exact number and size of the genes in an individual cell is unknown. R. D. Hall estimates that at most there are 13,000,000 genes in the cell of a jack pine tree. In some intensively studied crop plants, different plants of the same species are known to differ by 500 or more genes. This of course does not take into account the much larger number that might differ between two closely related species, or the very much larger number that might differ between a pine and an oak tree. It is likely that the simplest breeding problem involves the manipulation of a few dozen genes.

Because a gene is much too small to be seen and identified with a light-transmitting or even an electron microscope, it must be identified by indirect means. That is accomplished by growing trees and their offspring under similar conditions and noting the characteristics controlled by genes. For example, if the offspring of tree A grow faster than the offspring of tree B, we assume that tree A contains genes for rapid growth not present in tree B. So far in trees there are few data about traits governed by individual genes; most data are applicable to traits governed by groups of genes, which have not been individually identified

Individual genes may have small or very large effects. Growth rate in trees seems in general to be under the control of genes with small individual effects so that a fast growing tree is assumed to have many "plus" genes. Brown versus blue eye color in man is under the control of a single pair of genes that have relatively large effects.

Some genes may act singly and be responsible for the development of a particular trait, no matter what other genes are present in the tree. Others may have an effect only if certain other genes are present. Thus, gene A may act independently of gene B, may act only if gene B is present, or may have only a small modifying effect on gene B.

Gene Arrangement, Linkage Groups, Chromosome Mapping

Genes are arranged linearly on the chromosomes. Genes on the same chromosome are said to be "linked" and to belong to the same "linkage group." Thus, genes on a particular chromosome tend to be transmitted as a unit from one generation to the next.

In each vegetative cell there are two sets of chromosomes, one set being considered "homologous" to the other. (Two chromosomes are "homologous" if at corresponding positions both have the same genes or genes affecting the same traits.) The gene arrangement may be *ABCD* on a chromosome and *abcd* on its homologue. Because of linkage, a pollen grain or egg cell which receives *AB* also receives *CD;* alternatively a pollen grain or egg cell which receives *ab* also receives *cd.*

Linkage is not perfect, however. At meiosis, when homologous chromosomes pair, they break and exchange parts. The original arrangement may be *ABCD* and *abcd* on homologues. If a break occurs between *B* and *C*, the new sequences are *ABcd* and *abCD*. When this happens, a "crossover" is said to occur. Crossovers usually occur with a frequency of one or more per chromosome. Obviously they occur more frequently between genes which are far apart than between genes which are close together on the same chromosome. This fact makes it possible to map genes on a chromosome. Mapping in a plant such as maize (it has not yet been attempted in a tree species) involves crossing an ABCDEFG/abcdefg plant with an abcdefg/abcdefg plant (capital and lower case letters applying to dominant and recessive genes, respectively). If a crossover occurs between the first two genes, some offspring are Abcdefg/abcdefg or aBCDEFG/abcdefg and can be identified; if a crossover occurs between the second and third genes, some offspring are ABcdefg/abcdefg or abCDEFG/abcdefg and can be identified. By common consent, the distance between genes is defined in terms of the frequency of such new combinations, 1 crossover unit (or map distance) being equal to 1% new combinations.

Linkage, although not yet measured in trees, is important in tree breeding. In the pines, for example, there are 12 pairs of chromosomes and thus 12 linkage groups. When selecting for any one gene, there are apt to be changes in the frequency of other, closely linked genes.

In trees we do not have the necessary pure lines which can be crossed in such a way as to accomplish chromosome mapping in the same way used in maize. There is, however, a way to learn something of the genes comprising different linkage groups. This is to observe the degree to which different traits are associated in very different populations of a tree species. If the genes governing two traits are closely linked, we might expect the two traits to be associated one way in one population, and in exactly the opposite way in another population.

Classification of Genes

For convenience, genes can be classified according to the traits they govern (e.g., genes for growth rate, for cold hardiness, for leaf width). They can also be classified according to desirability in a breeding program ("plus" and "minus" genes producing desirable or undesirable trees, respectively).

Genes can also be considered as "alleles" or "nonalleles." Two genes are alleles and considered as "allelic" to each other if they occupy the same "locus" (or position) on homologous chromosomes and affect the same trait but in different manners. Thus in a plant with the genetic constitution ABCDEFG/abcdefg on a particular pair of homologous chromosomes, A and a are alleles, B and b are alleles, C and c are alleles, etc. In a particular diploid plant there can be only one allele per locus (in which case the locus is homozygous) or two alleles per locus (in which case the locus is heterozygous). In a species, however, there may be several alleles per locus. In most tree breeding, the aim is to replace one allele with another, more desirable, allele.

Throughout this book, I shall denote alleles by the same letter (A and a are alleles, B and b are alleles, etc.), and shall use different letters to denote genes which occur at different loci and are therefore not allelic to each other. In more intensively studied organisms, alleles are often designated by two- or three-letter code names, the code usually being the first letter of a word descriptive of the trait which the gene governs. If there are several alleles, these may be given superscript numbers. Thus, st^1, st^2, st^3, etc., might be the designations given alleles causing self-sterility; re and mo might be alleles causing the development of red or mottled leaves, respectively.

Another useful way of classifying genes is according to their type of action, as follows.

Additive. Genes with additive effects have small effects, control the same trait, and enhance the effect of each other in a cumulative manner. Thus, trees having 1, 2, 3, 4, or 5 additive-effect genes for rapid growth might be expected to grow 1, 2, 3, 4, or 5 units faster, respectively, than normal. Many important tree characteristics are believed to be under the control of genes with small additive effects.

Dominant–recessive. Dominant and recessive genes are allelic to each other. A dominant gene is one which causes a characteristic to be expressed whether the tree is homozygous (AA) or heterozygous (Aa); the recessive gene has a visible effect only if the tree is homozygous (aa) for it. Usually, dominant genes are denoted by capital letters (A, B, C, etc.) and recessive genes by lower case letters (a, b, c, etc.). However, capital and lower case letters are also used to denote genes which are not dominant or recessive. Dominance may be partial or complete.

Epistatic. One gene is epistatic to another (which is hyopstatic) if the one is dominant over the other, and if the genes are not allelic to each other. In most practical tree improvement work, no distinction is made between dominance and epistasis.

Overdominance. This is a misnomer, since it has nothing to do with dominance or superdominance, as the word might infer. Overdominance is said to exist if heterozygosity (Aa) causes a trait to be expressed more fully than in either type of homozygous (AA or aa) parent. That is, overdominance exists if AA, Aa, and aa trees grow at relative rates of 100, 105 and 95, respectively. Overdominance is one possible explanation of hybrid vigor (superior performance by a hybrid over that of either parent).

It is important to learn about the type of gene action involved when breeding for a trait such as growth rate. Both the approach and the rate of progress differ according to whether the genes responsible for rapid growth are dominant, recessive, additive, etc. The basic experiment needed to learn these things involves crossing trees in various combinations and growing the offspring, which should be kept separate by both female (seed) and male (pollen) parent. Assume that one has some green-leaved and yellow-leaved trees and wants to learn the inheritance pattern for these traits. Then the proper procedure is to cross each of a few green trees with each of a few yellow trees, and also to cross some green trees with each other and some yellow trees with each other. The following shows how gene action can be inferred, according to the results obtained.

If green and yellow foliage colors are caused by dominant gene G and recessive gene g, respectively, the green parents have GG or Gg genotypes and the yellow parents have a gg genotype. In this case

Green × green (GG × GG, GG × Gg) will produce 100% green (GG or Gg) offspring.

Green × green (Gg × Gg) might also produce 75% green (25% GG and 50% Gg) and 25% yellow (gg) offspring.

Green × yellow (GG × gg) will produce 100% green (Gg) offspring.

Green × yellow (Gg × gg) might also produce 50% green (Gg) and 50% yellow (gg) offspring.

Yellow × yellow (gg × gg) will produce 100% yellow offspring.

If green and yellow foliage colors are caused by recessive gene g and dominant gene G, respectively, the green parents have a gg genotype and the yellow parents have GG or Gg genotypes. In this case, substitute green for yellow and vice versa in the above scheme.

If green and yellow foliage colors are caused by several pairs of genes with additive effects, no clear separation of the parental types into "green" and "yellow" will be possible; there will be many intermediate trees. Also

Green × green will produce green offspring.
Green × yellow will produce intermediate offspring.
Yellow × yellow will produce yellow offspring.

Mode of Gene Action

We commonly speak of genes in terms of their end products. Thus we speak of genes for rapid growth, long leaves, etc. It is desirable to remember, however, that genes are actually portions of a DNA molecule, exert their influence within a cell through the medium of RNA, and only finally cause a visible end product because of a long series of chemical and growth reactions. The mode of gene action in trees is unknown, so I give an example from work on the fungus *Neurospora*, which has been studied intensively. This example concerns synthesis of the amino acid arginine. It was originally thought that a single gene was responsible, and that arginine could be produced if the fungus was supplied with a suitable substrate which we shall call A. Now, arginine synthesis is known to be a seven-step process controlled by 7 genes. Substance A is converted to substance B through an enzyme controlled by gene *1*, substance B is converted to substance C through an enzyme controlled by gene *2*, etc., and finally substance F is converted to substance G (arginine) through an enzyme controlled by gene 7 (Hartman and Suskind, 1965).

From work on bacteria we get another insight into how genes act. This example involves a group of closely linked genes called an "operon." One of the genes is called an "operator," and it determines whether or not the DNA composing the other genes will be transcribed to RNA. The operator itself seems to be governed by the presence of a certain chemical within the cell. Presence or absence of this chemical controls whether or not the operator will permit the other genes to be copied as functional RNA. If not, the cell behaves as if it did not contain the other genes.

These examples are cited to give an idea of the way in which genes operate. It is still some time in the future before we shall know how particular genes operate in trees.

Mutations

The mechanism by which DNA replicates itself is nearly perfect. That is to be marvelled at considering the fact that bacteria may reproduce at a rate of one generation every several minutes. Each new generation, the double strand must untwist, the millions of nucleotide triplets must be reproduced exactly on new single strands, and the single strands must again be twisted around each other to form new double strands.

Mistakes do happen, however. An error in the replicating process is called a "mutation." Mutations arising in this manner are the ultimate source of the genetic variability upon which a tree breeder depends to produce a new variety. They are also the ultimate source of most of the genetic variability that has made possible the evolution of thousands of different tree species.

The frequency of mutation is small, variously estimated at between 1 in 10,000 and 1 in 1,000,000. That is, any one gene may mutate once in every 10,000 to 1,000,000 cells. Or, stated another way, of every 10,000 to 1,000,000 genes in a cell, one may mutate. If we remember that there may be several million genes in a cell, and several million cells in a tree, the absolute number of mutations is very large, in spite of the small mutation rate. Several million mutations may occur in a single tree.

Most new mutations have such small effects as to be unnoticeable. A seedling may contain 100 mutant genes not present in its parents, yet not be noticeably different from others.

The vast majority of new mutations are slightly deleterious and recessive. The fact that they are deleterious can be explained by the random nature of the mutation process. Presumably most trees contain a complement of genes that causes them to grow efficiently in harmony with their environment. Now a mutation occurs, quite by accident, which results in higher (or lower) production of a particular enzyme, and the result is more than likely to be detrimental to the tree.

The fact that most new mutations are recessive has been explained by the late R. A. Fisher. He postulated that most organisms are growing in harmony with their environment and that they benefit from a mechanism which preserves the *status quo*. He then postulated that over a period of thousands of generations evolution has caused most common or "wild-type" genes in any particular population to become dominant and thus repress the activity of most new mutations.

Presumably, mutations occur frequently in the vegetive parts of a tree. However, as most are recessive and have only small effects they are not noticeable. Nor are they passed on to the tree's offspring.

It is the mutations which occur in the reproductive structures— during the divisions which give rise to the pollen grains and the egg cells—in which breeders are most interested. These are passed on to the tree's offspring.

In spite of the fact that most individual mutations are deleterious, mutation as a whole is a beneficial process. It is the principal way in which variability is produced. This seeming paradox is easily explained. Because most new mutations are recessive, they tend to be retained in a population so that every tree possesses literally thousands of mutant genes incapable of expressing themselves. However, the genes which may be detrimental in one environment may be beneficial in another environment. Thus, when a species encounters a new environment, it already contains a supply of mutants which may be beneficial in the changed conditions.

Mutation rate is a very difficult thing to measure. The only practical way in trees is to confine the study to mutations with such large effects as to cause seedlings to die at an early age if present in the homozygous state, i.e., to work with lethal recessive mutants. Even this method is inexact, as it can give highly inflated estimates if there has been mating between closely related trees. The formula is

Mutation rate = frequency of trees homozygous for lethal recessive gene

So far the discussion has centered entirely around mutations in the strictest sense, that is, around "gene mutations." Sometimes the word is used in a broader sense to include any sudden change in a tree's genotype. In that sense it usually involves a gross change in chromosome structure or number, such as the deletion of a portion of a chromosome, the inversion of a portion of a chromosome, the addition or subtraction of an entire chromosome, or the doubling of chromosome number. Often these gross changes in chromosome structure are accompanied by gross abnormalities.

The low mutation rates prevailing in nature have been mentioned. These can be increased artificially by certain chemicals (called "mutagens") and by ionizing radiation (X rays, cosmic rays, γ rays). These agents cause random gene changes, as does nature, so that it is not possible to control the genes which will mutate or the direction in which they will mutate. These agents act by damaging the DNA. Therefore they cause gross chromosome abnormalities (e.g., breakage, inversion) as well as gene mutations. With too many gross abnormalities, the treated cells die. Therefore the amount by which mutation rate can be increased by increasing dosage is severely restricted. As most trees already possess a great many mutant genes, the artificial induction of mutations has been of relatively minor importance in tree improvement work.

Nongenic Inheritance

Maternal Inheritance

Most tree characteristics are believed to be controlled by genes borne on the chromosomes. As the chromosomes in a fertilized ovule are contributed to equally by the male and female parents, both the seed and pollen parents usually contribute equally to the genotype of the offspring.

There are, however, cases known in herbaceous plants (none definitely proved in trees as yet) of characteristics controlled by cytoplasmic factors. As the cytoplasm is contributed entirely by the female parent, this type of inheritance is called cytoplasmic or maternal. It is detected by making reciprocal crosses, A ♀ ✕ B ♂ and B ♀ ✕ A ♂. If the offspring of the first cross are A-like and the offspring of the second cross are B-like, maternal inheritance is assumed.

This type of inheritance, although rare, can be of great importance in improvement work. That was demonstrated with maize in 1971. Several years earlier a single maize plant had been discovered which produced no tassel and therefore no pollen. Crossing experiments indicated that the male sterility was due to a cytoplasmic factor. The male sterility was considered to be a very desirable trait, as a male sterile line could be interplanted with a male fertile line to produce hybrid maize (hybrid corn) without hand detasseling of the female parents; thus the labor of thousands of people engaged in hybrid seed production could be avoided. Accordingly, breeders transferred the male sterility factor to most of their selfed lines by appropriate crosses and became able to produce almost any hybrid without hand detasseling. As the seed for commercial production was always collected from female (male sterile) plants having the same cytoplasm, much of the hybrid maize being grown on millions of acres in eastern United States had identical cytoplasm. That became important in 1971 when a serious disease epidemic occurred. That particular cytoplasm proved extremely susceptible to the (possibly new) strain of the disease which caused very great damage in many parts of the country. Accordingly, many seed companies returned to older breeding practices involving varied cytoplasms as well as varied chromosome complements in different hybrids.

Paternal Inheritance

A case of paternal inheritance was discovered by Ohba *et al.* (1971) in Japan. They worked with sugi (cryptomeria), an important Japanese conifer. Their work involved an X-ray induced mutation causing the ends of twigs to become yellow. The same mutation was originally discovered in two different plants. If a yellow-tipped tree was used as a female (seed) parent and crossed with a normal tree, all offspring were green. If the cross was in the reverse direction, some offspring were yellow-tipped. This phenomenon is presumably extremely rare, as the authors could find only one previous reference to paternal inheritance in any seed plant.

Genotype versus Phenotype

A tree's genotype is its genetic constitution, usually expressed in terms of genes. Thus, we might speak of a tree's genotype as ABCDeFGHIJK/ABcdefGhIJk. Sometimes we also speak of a "genotype" as a group of trees having similar genetic constitutions with regard to certain genes.

A tree's phenotype is its external appearance, growth rate, chemical constitution, etc., as the tree is actually growing. In other words the phenotype is what one actually sees and feels. Phenotype is partially controlled by the genotype. There are always some recessive genes which do not express themselves in the presence of the corresponding dominants; the recessives are a part of the genotype but there is no hint of them in the phenotype. There may also be some small effect and modifier genes present in the genotype that are unnoticeable in the phenotype.

Phenotypes are also partially controlled by the environment. A tree may contain several hundred genes for rapid growth but grow extremely slowly because it was planted on a poor site. It may lack the genes necessary for cold hardiness, but this does not become known unless it is grown on a cold site. Also, it may have a genetic potential (genotype) for extreme susceptibility to a certain disease but have a seemingly resistant phenotype if grown where the disease is not present.

It is apparent from the above that some inferences as to a tree's genotype can be made from a careful study of its phenotype. Many more can be made, however, from study of a tree's offspring, especially from a study of the relationships between the characteristics of the parents and the offspring.

Most critical experiments show that a tree's ability to transmit desirable traits to its offspring is governed solely by its genotype. Thus, an inherently crooked tree may be trained in such a way as to grow straight, but it will still produce crooked offspring. An inherently slow growing tree may be fertilized heavily and thus caused to grow rapidly;

nevertheless, its offspring will grow slowly unless they too are fertilized. A non-winter-hardy tree may be grown rather far north in a protected situation, but will produce non-winter-hardy offspring just as if it had been grown farther south. The generalization that acquired characters are not inherited was one of the important conclusions of Charles Darwin's study of evolution.

If the genotype of an individual tree does not change as the result of environmental change, how does the genetic constitution of a population of trees change so that average growth rate or average winter hardiness may be increased? That is done through a process called "selection" (see Chapters 4 and 9–12). As no two seedling trees have quite the same genotypes, nature or man selects those with the most favorable genotypes as the ones to produce the most offspring. Thus over a period of generations the average genetic constitution of the population changes.

Cell Division and the Opportunity for New Gene Combinations

Mitosis

The cells in most parts of a tree contain two sets of $2n$ chromosomes, one set being homologous to the other. The type of cell division which takes place in the cambium, root tips, leaf primordia, etc., to produce the wood, leaves, etc., is called "mitosis."

In this type of cell division, chromosomes shorten and gather on an "equatorial plate" in the center of the mother cell (Fig. 2.1). Each chromosome splits lengthwise to form two identical new chromosomes. One member of each chromosome pair migrates to one end of the cell and the other member migrates to the opposite end. A new cell wall forms and the two daughter cells are completed. Aside from mutations which may occur during the process of cell division, each daughter cell has a genotype identical to that of the mother cell. Thus all vegetative cells throughout a tree have similar genotypes.

Meiosis

At flowering time a type of cell division called "meiosis" occurs. This is a reduction division because chromosome number is reduced from the $2n$ condition found in vegetative tissue to $1n$ in the gametes (egg cells or pollen grains). I shall describe only the most important features. A pine cell undergoing meiosis is shown in Fig. 2.3.

At meiosis, the two sets of homologous chromosomes come together in the center of the mother cell, and members of one set pair with

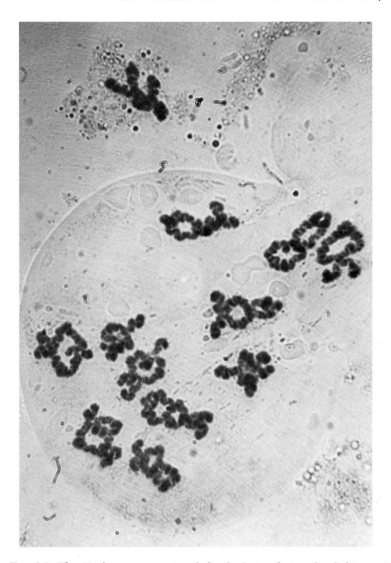

FIG. 2.3. The 12 chromosome pairs of shortleaf pine during the diakinesis (immediately preceding metaphase) stage of meiosis. The major chromosome coils are visible as are the chiasmata or points where crossing over occurs between homologous chromosomes. (Photograph courtesy of Leroy C. Saylor of North Carolina State University at Raleigh.)

their homologues from the other set. While the homologous chromosomes are paired, "crossovers" occur. That happens as a result of simultaneous breakage at exactly the same locus on both chromosomes with the subsequent rejoining in such a way that the chromosomes exchange segments. (Assuming that the original sequence for the pair of chromosomes was ABCDEFGH/abcdefgh and that a crossover occurred between the D and E loci, the new sequence on the two chromosomes becomes ABCDefgh/abcdEFGH.) Next, one chromosome of each pair migrates to one end of the mother cell, and its homologue migrates to the opposite end. A new cell wall forms and the two daughter cells or gametes are completed, each having only half ($1n$) the original number of chromosomes.

Assume that we are considering a pine tree with $n = 12$ ($2n = 24$) chromosomes and that the 24 chromosomes in a mother cell were derived from parents A and B so that they can be designated as $A1$–12 and $B1$–12. After pairing and crossing over they segregate independently to the two ends of the mother cell. Chromosomes $A1$ and $B1$ must migrate to opposite ends, but that has no bearing on the direction of migration of chromosomes $A2$ and $B2$, $A3$ and $B3$, etc. The daughter cell which receives $A1$ may receive either $A2$ or $B2$, either $A3$ or $B3$, etc. Also, it must be remembered that crossovers have occurred independently of each other between every pair of homologous chromosomes, and that no two homologous chromosomes are apt to have the same genes (because of the presence of a few hundred mutant genes in most cells). Thus, it can be seen that no two gametes produced by a tree are apt to have exactly the same genotype. Only very closely linked genes are apt to segregate together.

Meiotic pairing occurs because of an attraction between identical genes on homologous chromosomes, and proceeds normally only to the extent that the vast majority of genes at the same loci on homologous chromosomes are identical. This is ordinarily true in trees which are not hybrids. However, in the evolution of one species from another, there are often gene replacements, chromosome inversions, and chromosome translocations so that homologous chromosomes of the two species are apt to be dissimilar. In a hybrid between two such species, pairing is often impaired, with the result that the gametes may not receive a full complement of $1n$ chromosomes. In such a case the hybrid is apt to be sterile. This is particularly true in some angiosperms; most hybrids between pine species seem to be quite fertile.

Fertilization and Seed Development

Fertilization is the reverse process of meiosis. One essential feature is the coming together of a $1n$ male gamete with a $1n$ female gamete to produce a fertilized egg cell that then divides (by mitosis) to produce an embryo, which in turn develops into a seed, which in turn develops into a tree. During the process, chromosomes from the male gamete enter the nucleus of the egg cell and mingle with the chromosomes of that cell; the cytoplasms do not mingle, however, so that the cytoplasm of the fertilized egg cell is derived from the female parent only. Together, meiosis and fertilization are the mechanisms by which the genetic variability within a tree can segregate and recombine to produce offspring that are different from the parent. Thus, even if a tree is pollinated by itself, no two of its seedlings are apt to have identical genotypes.

Another feature, called "double fertilization" is found in angiosperms only. After meiosis in a (female) egg mother cell, a $1n$ nucleus is formed which then divides (by mitosis) to produce eight nuclei, of which three are especially important. One is called the "egg nucleus" and the other two are called the "polar nuclei"; all are $1n$. Meanwhile, as a pollen grain germinates on the style of the female flower, its $1n$ nucleus divides (by mitosis) to produce two $1n$ nuclei. At the time of fertilization a $1n$ male nucleus from the pollen grain fertilizes the $1n$ female egg nucleus to produce the fertilized egg, which then develops into an embryo and a tree. The other n male nucleus unites with the two $1n$ female polar nuclei to produce a $3n$ cell which then divides (by mitosis) to become the endosperm or storage tissue of the seed.

The seed coat surrounding the embryo and endosperm develops from maternal tissue (by mitosis), and therefore is of the same $2n$ genotype as all vegetative cells in the mother tree. Thus the entire seed is a mixture of three types of tissue which are quite different genetically:

A $2n$ seed coat of the same genotype as the female parent

A $2n$ embryo containing n chromosomes each from the male and female parents

A $3n$ endosperm with $2n$ chromosomes from the female parent and $1n$ chromosomes from the male parent

Double fertilization does not occur in the gymnosperms, in which the endosperm is n tissue, of the same genotype as the unfertilized egg cell. Thus, gymnosperm seeds, too, contain a mixture of three genetically different types of tissue:

A $2n$ seed coat of the same genotype as the female parent

A $2n$ embryo containing n chromosomes each from the male and female parents

A $1n$ endosperm, purely maternal

Whatever the type of seed, these three tissues develop in close proximity, compete for nutrients and space, and mutually influence each other. In a normal, nonhybrid tree there is usually a proper balance such that all three tissues develop and a normal seed results. However, there is often an imbalance when trying to hybridize very diverse species. This imbalance may be at the subcellular level. The embryo may fail to develop properly (as in some pine hybrids) because of an inability of the *n* chromosomes contributed by the male to function with the *n* chromosomes contributed by the female. An angiosperm endosperm may not develop properly because the *2n* chromosomes from a particular female do not function properly with the *n* chromosomes from the male. The endosperm might develop better if the reciprocal cross were made. The imbalance may be at the tissue level. In many species seed coats may develop without containing an embryo or endosperm. X-ray pictures of pine seeds have shown many cases of well-developed endosperms and poorly developed embryos. The opportunity for such tissue imbalance varies among kinds of trees. An oak seed, for example, contains a very large embryo but little or no endosperm, so that any imbalance in the endosperm is relatively unimportant.

Study Questions

1. Draw a cell showing the relative location of cell wall, cytoplasm, nuclear membrane, nucleus, chromosomes, and genes. Which of these can be seen with a high-powered light microscope?

2. Describe the structure of DNA and the way in which a DNA molecule can replicate itself.

3. Describe the differences in structure and function between DNA and RNA.

4. How does a gene relate to DNA? To an observed tree characteristic?

5. What is a gene mutation? Why are most mutations deleterious and recessive?

6. What is meant by a mutation rate of 1 in 50,000? How can mutation rates be increased and why is there a limit to the increase in mutation rate which can be put to practical use?

7. What types of nongenic inheritance are known, and how can they be demonstrated?

8. Define alleles.

9. Define linkage and tell how crossovers affect linkage.

10. Define the three types of gene action and tell which is the most important in tree breeding.

11. Describe an experiment to show that gene *a* is recessive to gene *A* if no inbred lines are available.

12. Differentiate between mitosis and meiosis.

13. Describe meiosis and tell why chromosomes segregate independently, and why genes may also segregate independently.

14. How does fertilization complement meiosis to produce a constant stream of new variation, even in the absence of mutation?

Suggested Reading

Several elementary texts in many languages describe basic genetic principles. The following references give information on special points.

DeBusk, A. (1968). "Molecular Genetics." Current Concepts in Biology Series. Macmillan, New York.

Hartman, P. E., and Suskind, S. R. (1965). "Gene Action." Prentice-Hall, Englewood Cliffs, New Jersey.

Ohba, K., Iwakawa, M., Okada, P., and Murai, M. (1971.) Paternal transmission of a plastid anomaly in some reciprocal crosses of sugi, *Cryptomeria japonica* D. Don. *Silvae Genet.* **20**, 101–107.

Saylor, L. C. (1972). Karyotype analysis of the genus *Pinus*—subgenus *Pinus*. *Silvae Genet.* **21**, 155–163.

3

Additional Basic Genetic Concepts

Inbreeding

Definition

Inbreeding is crossing between close relatives. The extreme case is selfing, or the crossing of a tree with itself. However, there are all gradations from this extreme, e.g., crossing between brother and sister, between cousins, between members of small isolated populations.

Primary Consequences

UNIFORMITY WITHIN AND VARIABILITY AMONG INBRED POPULATIONS

Assume that a tree is heterozygous (having different alleles) at a particular locus, having the genotype Aa, and assume that the tree is selfed. The offspring will be of the genetic constitution 1 AA:2 Aa:1 aa. In other words, $\frac{1}{4}$ of the offspring will be homozygous AA and $\frac{1}{4}$ will be homozygous aa. Altogether, half the offspring will be homozygous at that particular locus. Thereafter, if selfed, the homozygous AA trees will produce only AA offspring, and the homozygous aa trees will produce only aa offspring. The same thing will happen at other loci, and each generation of selfing will reduce the number of heterozygous loci by 50%.

Thus, starting with a tree with the genotype AaBbCcDdEe, it is possible with repetitive selfing to end up with pure lines of genotype AABBCCDDEE, aabbccddee, AABBccDDee, etc. Each of these lines will be uniform within itself, but the lines may vary greatly among themselves. The same thing can happen, but more slowly, if a breeding population is maintained at $N = 5$ for several successive generations, i.e., if 5 trees were crossed with each other in various combinations and only 5 offspring were saved as the parents of the next generation, etc.

LOSS OF VIGOR

Wild populations of trees usually carry a large "genetic load." That is, they are usually heterozygous for a large number of deleterious recessive genes. These genes do no damage because each tree is usually pollinated by others that do not possess the same recessive genes. Thus, in normal seedlings the deleterious recessive genes are not expressed. If, however, a tree with the genotype AaBbCc is selfed, 25% of the offspring will be homozygous aa, 25% homozygous bb and 25% homozygous cc. If the parent contained 100 deleterious recessive genes at different loci, its average offspring will be homozygous for deleterious recessives at 25% of the loci. Even though the individual genes may have small effects, the cumulative effects on growth rate can be considerable (Fig. 3.1).

Franklin (1968, 1969) selfed 132 loblolly pines and also collected open-pollinated seeds from the same trees. Four parents produced as much seed and as vigorous offspring after selfing as after crossing. The remaining 128 suffered some degree of lowered seed set (some yielding almost no selfed seed) and lowered vigor after selfing. In experiments with other species, selfed seedlings have often grown less than 50% as fast as normal seedlings.

Reduced vigor or "inbreeding depression" is not an invariable consequence of inbreeding. Tomato, wheat, and many other herbaceous plants have flower structures which promote selfing. At one time those plants may have contained large numbers of deleterious recessive genes, but they have been eliminated by thousands of generations of selfing. Therefore modern tomato and wheat plants produce just as vigorous offspring when selfed as when crossed.

Most temperate zone trees have mechanisms that inhibit selfing. Even so, close inbreeding does not invariably mean decreased growth rate. This is so because decreased vigor is caused by the accumulation of deleterious recessive genes rather than by selfing itself. Trees without such recessive genes can safely be selfed. That gives rise to the general rule, "Selfing is generally bad but may be good."

FIG. 3.1. Inbreeding effect in white spruce growing in northern Wisconsin. Hans Nienstaedt is inspecting one of four selfed seedlings in the front 4-tree plot. Four open-pollinated white spruce comprise the rear 4-tree plot. The average inbreeding depression was 40–50% in the nursery and increased as the trees grew older. The trees are 12 years old from seed (8 years from planting). (Photograph courtesy of Hans Nienstaedt of the United States Forest Service at Rhinelander, Wisconsin.)

LOWERED SEED SET

Deleterious recessive genes may act during the period of seed development. If so, selfed pollen can result in fertilization, but the fertilized embryos may abort at an early stage, resulting in primarily empty seed. This is the probable explanation for Franklin's (1968) lowered seed set after selfing for 128 of his 132 loblolly pine trees and for the low seed set in selfed Douglas-fir (Orr-Ewing, 1957).

There is another explanation in some angiosperms which carry self-sterility alleles s_1, s_2, s_3, s_4, etc. These act in such a manner that neither s_1 or s_2 pollen can germinate in a style with the genotype s_1s_2, nor can s_1 or s_3 pollen germinate in a style with the genotype s_1s_3.

TABLE 3.1

Seed Set and Seedling Vigor following Selfing of Forest Trees

---

Seed set much reduced, offspring usually weak
  Alnus glutinosa
  Castanea mollissima
  Eucalyptus alba, citriodora, grandis, maculata, punctata, robusta, umbellata
  Larix decidua, leptolepis
  Liriodendron tulipifera
  Picea jezoensis, sitchensis
  Pinus resinosa, strobus, sylvestris, thunbergiana, rigida
  Pseudotsuga menziesii
  Ulmus americana, parvifolia
Many viable seeds produced
  Alnus incana
  *Eucalyptus blakelyi, caesia, cinerea, macrorryncha, maideni, polyanthemos, pulveru-
    lenta, rubida*
Variable results, few to many viable seeds, offspring weak to vigorous
  Acacia decurrens
  Caragana arborescens
  Eucalyptus bicostata
  Picea abies, glauca, omorika
  Pinus monticola, taeda

Table 3.1 contains a partial list of species of trees in which selfing has been attempted. The results have been variable, partly because of the small scale of some of the experiments.

LACK OF ADAPTATION

Assume that a tree is heterozygous at 1000 loci. Selfing results in offspring which are homozygous at 500 or 50% of those loci. Even if the offspring live, many will be homozygous for deleterious genes. At any particular locus, the change from heterozygosity to homozygosity occurs so rapidly that selection has no chance to operate within a selfed line. Selection can cause some lines to disappear and others to survive, but random processes rather than selection control the outcome within a selfed line. In other words, the surviving selfed lines may not be well adapted to their environments.

As already mentioned, selfing is an extreme case. Inbreeding can occur with breeding populations of 5, 10, 100, etc., individuals. If so, some random gene fixation takes place that causes the end products to be unadapted to their environments. This seems to have been the case with some southern spruce species, several of which have very small ranges and very small breeding populations. The southern spruce species in general have no special characteristics indicative of adaptation to southern habitats, and a few are so poorly adapted to modern environments as to be on the verge of extinction (see section on genetic drift in this chapter).

Measurement of Inbreeding

Selfing is a special case of inbreeding in which $N = 1$ (N is the number of trees comprising a breeding population). With selfing, heterozygosity is lost at the rate of $\frac{1}{2}$ each generation. Thus, a selfed line retains $\frac{1}{2}$ of its original variability after 1 generation, $\frac{1}{4}$ after 2 generations, $\frac{1}{8}$ after 3 generations, etc.

The inbreeding coefficient, F, is used to measure the amount of inbreeding when N is 1 or larger. F is the loss of heterozygosity or its complement, the increase in homozygosity. Since the symbol Δ is often used to express a "change in" something, the per generation change in amount of inbreeding is expressed as ΔF. There are two general formulas for its calculation, the simplest being

$$\Delta F = 1/2N_{total}$$

This formula is used if N_{total} is the total number of trees in the breeding population and if each tree is used both as male and female parent. The other formula, used when different numbers of males (N_{male}) and females (N_{female}) are bred, is

$$\Delta F = \frac{N_{male} + N_{female}}{8(N_{male}N_{female})}$$

These formulas are identical if $N_{male} = N_{female}$ and if $N_{total} = 2N_{male} = 2N_{female}$.

To calculate inbreeding over a number of generations, it is most convenient to work with the heterozygosity coefficient, $H = 1 - F$. To calculate H_n ($= H$ at the nth generation), the general formula is

$$H_n = (H_{per generation})^n \qquad \text{or} \qquad \log H_n = n \log H_{per generation}$$

As a numerical example, assume that $N_{total} = 5$ and $\Delta F = \frac{1}{10}$. Then, $H_{per generation} = \frac{9}{10}$; $H_2 = (\frac{9}{10})^2 = 0.810$; $H_3 = (\frac{9}{10})^3 = 0.729$. After 3 generations, 72.9% of the original heterozygosity remains and the inbreeding rate is $F_3 = 1 - H_3 = 1 - 0.729 = 0.271$.

When working with evolutionary problems, it is often reasonable to assume that population size fluctuates from one generation to the next. This is probably true for eastern white pine, an intolerant species which reproduces in large numbers on open land but persists in much

TABLE 3.2

Inbreeding Coefficients of Populations
Maintained at Sizes of $N = 5$ to 250
for 10 or 100 Generations

Population size (N)	Inbreeding coefficient (F) after	
	10 generations	100 generations
5	0.6514	0.999973
10	0.4014	0.9941
25	0.1828	0.8673
50	0.0955	0.6335
100	0.0489	0.3946
250	0.0198	0.1815

smaller numbers as a component of the climax forest. Fire can destroy
a climax forest on a 40-acre tract, opening the way for abundant white
pine reproduction and a pure white pine forest which at maturity contains 40 trees per acre or 8000 trees on the 40 acres. But that stand
will be invaded by hardwoods, and the percentage of white pine will
be reduced in the next generation and reduced still further in the third
generation. In this way, the size of the breeding population of white
pine might be 8000, 400, and 3 trees in three successive generations.
If so, the inbreeding rate per generation is $\Delta F = 1/2N = 1/16,000$,
1/800, and 1/6 in the three successive generations. Total inbreeding
in the three generations is $F = 1/16,000 + 1/800 + 1/6 = 0.679791$ and
the average per generation rate is $\Delta F = 0.1679791/3 = 0.0559930 =$
1/19.7. Because $\Delta F_{\text{per generation}} = 1/2N = 1/19.7$, $2N = 19.7$, and $N =$
9.85. In other words, the effective average $N = 9.85$, which is much closer
to the smaller ($N = 3$) than to the larger ($N = 400$ or 8000) population
size.

Inbreeding coefficients of populations maintained at various sizes
for long periods of time are given in Table 3.2.

Use of the Inbreeding Coefficient

Breeders commonly select a few hundred plus trees from among
millions, graft or progeny test those plus trees, and thin their grafted
seed orchards or progeny tests to leave the offspring of only a few
of the best parents. The more rigorous the thinning, the greater the

possible genetic gain. However, inbreeding, with consequent loss of vigor, can also result if the thinning is too rigorous.

In annual plants it is relatively easy to maintain populations of 5 or 10 plants for several successive generations and then measure the results. With a tree species this is a much more difficult task and, in fact, has not been attempted. Instead, tree breeders make the working assumption that loss of vigor through inbreeding is proportional to the loss of heterozygosity as measured by the inbreeding coefficient F. It is relatively easy to obtain trees with an inbreeding coefficient of $F = 0.5$ by selfing for one generation and to measure the inbreeding depression of those selfed trees. It is also easy to calculate the inbreeding coefficient after n generations of inbreeding with population size N. Then a proportion is established such that

$$\frac{\text{Inbreeding coefficient, } n\text{th generation}}{\text{Inbreeding coefficient, 1st generation}} = \frac{\text{growth loss, } n\text{th generation}}{\text{growth loss, 1st generation}}$$

As a numerical example, assume that trees selfed for 1 generation have an inbreeding coefficient of $F = 0.5$ and suffer a growth loss of 40%. What is the growth loss of trees inbred for 3 generations if the population size is maintained at $N = 10$? First, calculate $\Delta F = \frac{1}{20} = 0.05$; then calculate $H_1 = 1 - 0.05 = 0.95$; then calculate $H_3 = 0.95^3 = 0.856$; and calculate $F_3 = 1 - 0.856 = 0.144$. The estimated growth loss is then calculated as

$$\frac{\text{Growth loss at generation 3 with } N \text{ of 10}}{40\%} = \frac{0.144}{0.500}$$

Growth loss at generation 3 with N of 10 $= 11.5\%$

Estimates of growth loss arrived at in this manner are probably high but are nevertheless valuable. If inbreeding proceeds slowly, as it might with $N = 10$, selection has some opportunity to operate and counteract the effects of the inbreeding.

In a paper titled "Isolation by Distance" Sewall Wright (1943) used inbreeding coefficients to estimate the amount of genetic diversity among subpopulations of continuous populations. In so doing he made use of one of the primary consequences of inbreeding, the fact that it engenders uniformity within and variability between groups of inbreds. He assumed various population densities and distances between subpopulations, then calculated expected amounts of inbreeding among the subpopulations. In this way he was able to calculate the expected amount of genetic diversity in parts of continuous populations.

Genetic Drift or Random Gene Fixation

Genetic drift is an evolutionary term meaning the random fixation of genes in small sized populations. If inbreeding is continued over long geological periods, it results not only in great uniformity but also in the development of nonadaptive traits within the inbred population.

Genetic drift can certainly occur in populations with $N = 250$ or less. That is apparent from Table 3.2, which shows that $F = 0.18$ if N is maintained at 250 for 100 generations. Presumably, it can also occur with population sizes up to $N = 5000$; at that size, however, loss of genes through inbreeding tends to be balanced by mutation.

In estimating the amount of genetic drift which has occurred in forest trees, it is necessary to remember that population sizes have fluctuated greatly in geological times. The eastern white pine example was quoted earlier. In that species and many others it is possible for N to vary from 1000 to 10,000 to 50 to 10,000 in successive generations. In such a case, genetic drift might occur during the periods when the species was uncommon, and its effects could persist as the species became more common.

Genetic drift is suspected to be the cause of the unusual behavior of several southern species of spruce. In that genus there are separate species in the mountains of southern Oregon and northern California, the mountains of Mexico, the mountains of Taiwan, the mountains of southern Japan, the eastern Himalayas, the western Himalayas, Turkey, and the Drina valley of Yugoslavia. Each of these species has a small natural range at present and has probably had a restricted distribution for many thousands of years. As a group, these southern spruces have no combination of characteristics indicative of adaptation to southern climates. For example, one has the shortest needles in the genus and a neighboring species has the longest needles in the genus. Some of these southern spruces are so obviously unadapted to their present environments as to be on the verge of extinction; they cannot even be planted successfully. One of them (Serbian spruce from Yugoslavia) can be selfed more successfully than most temperate zone trees; this indicates a long history of inbreeding and consequent past elimination of many deleterious recessive genes. Also, most of these southern spruce species are less variable than their northern counterparts.

Seed origin experiments have shown that eastern white pines from extreme southern Ontario and from an area in eastern Virginia behave peculiarly. When tested in comparison with trees from other places, the southern Ontario trees grew much faster than expected in view of the latitude and the climate in which they evolved; the Virginia trees grew much more slowly than did trees from farther north or farther south. The natural population in southern Ontario is isolated from the rest of the species and is small enough to have experienced genetic drift. Genetic drift is a possible but less likely explanation for the slow growth of the Virginia trees, since the Virginia population is relatively large.

Mechanisms for Inbreeding

Annual plants must produce seed each year, otherwise the species would become extinct. This can be ensured best by self-pollination. As compared with trees, annual plants are more apt to produce perfect flowers containing both stamens and pistils, so that self-pollination is possible. Furthermore, many annuals have evolved pollination and genetic mechanisms which make selfing more likely than crossing. That is the case with wheat, rye, and tomatoes.

Tropical rain forests normally contain a very large number of tree species, with very few trees of any one species per acre. Thus there is not so much opportunity for crossing as there is between trees of the same species in a temperate zone forest, which normally contains few species and many trees of one species per acre. Thus, to ensure good seed set, there is greater need for selfing in a tropical than in a temperate-zone species. Also, the climate in a tropical rain forest is relatively stable over long periods of time, so the uniformity engendered by inbreeding does not have serious consequences.

There are too few experimental data to indicate whether or not selfing is actually prevalent in the tropics, but studies of flower structure indicate that it may be. Trees of predominantly tropical families such as the Magnoliaceae, Meliaceae, Leguminosae, Bignoniaceae, Tiliaceae, Sapindaceae, Myrtaceae, Rutaceae, Araliaceae, and Cornaceae tend to have showy, perfect flowers which are insect pollinated.

Mechanisms against Inbreeding

In contrast, mechanisms to inhibit selfing are common in temperate and boreal tree species. These tend to grow in pure stands or in simple mixtures containing several trees of each species per acre, so adequate seed production is possible without selfing. Also, there is greater need than in the tropics for the variability engendered by outcrossing. Annual fluctuations in precipitation and temperature are great. In some areas there have been very great long-term fluctuations in habitat conditions because of fire, glaciation, etc.

Dioecism or the production of male and female flowers on different trees is the most effective mechanism to prevent selfing. It occurs in the poplars, willows, junipers, yews, hollies, ashes, boxelder, and many other groups.

Monoecism or the production of separate male and female flowers on the same tree, is also effective. Monoeciousness is characteristic of chestnuts, oaks, birches, alders, and sweet gums. In those genera male and female flowers tend to be scattered uniformly throughout the tree crowns. In Douglas-fir, pines, spruces, firs, and araucarias female flowers tend to occur mostly in the tops of trees whereas most pollen is produced on lower branches. Such separation of male and female flowers is probably especially effective, as the pollen must rise to effect selfing.

Wind pollination is characteristic of the pines, spruces, oaks, poplars, and many other temperate or boreal genera. This promotes crossing among trees several hundred feet apart more than does insect or bird pollination.

Yellow-poplar produces large, perfect, insect pollinated flowers. Selfing is inhibited by genetic self incompatibility and also by the fact that within a single flower the stigma is receptive before the anthers are ready to shed pollen.

Maples have several mechanisms to inhibit selfing. Boxelder is dioecious. Red and silver maples produce structurally perfect but functionally monoecious flowers, and the flowers of different sexes are usually borne on different branches. Sugar and Norway maples exhibit "dichogamy," or the production of male and female flowers at different times on the same tree.

In addition to the structural mechanisms there are genetic mechanisms to promote outcrossing. Consider Franklin's (1968) sample of 132 loblolly pines, of which only 4 responded well to selfing. Even when self-pollination occurred in the other 128 trees, seed sets were low or the seedlings grew slowly and probably died early under natural conditions.

Frequency of Natural Selfing

Most temperate zone forest trees contain deleterious recessive genes. Most such genes are rare. Therefore, if a heterozygous tree is crossed with another tree the offspring are apt to be normal. However, if a tree heterozygous for such a gene is selfed, $\frac{1}{4}$ of its offspring will be homozygous and exhibit a serious abnormality (white foliage, yellow-green foliage, etc.). This makes it possible to estimate the frequency of selfing under natural conditions. To do so, collect open pollinated

Table 3.3
Frequency of Natural Selfing in Five Tree Species

Species	Frequency of natural selfing (%)	
	Average	Range for single trees
Scotch pine	7	—
Slash pine	7	0–27
Loblolly pine	2	1–14
Japanese red pine	4	0–11
Douglas-fir	7	0–27

seed, grow the seedlings, count numbers of normal and abnormal seedlings, and multiply the frequency of abnormal seedlings by 4. (Some investigators feel that there is some pregermination mortality of selfed seedlings so that some potentially abnormal seedlings are not seen. For that reason it may be better to multiply by 5 rather than 4.)

Several tree breeders (Franklin, 1968; Ohba *et al.*, 1971; Sorenson, 1973) have used this method, with results as shown in Table 3.3.

Inbreeding as an Improvement Method

In those annual plants which are normally selfed, "line breeding" is a standard improvement method. First, hybrids are produced between genetically different parents which have a desirable combination of characteristics between them. Then the hybrids are allowed to self naturally, and several lines are established. Several generations of selection are practiced to obtain lines having the best combination of characteristics. This method does not seem applicable to most temperate zone trees.

In corn (maize), which is a normally outcrossing species, "selfing plus later crossing" is a common improvement method. First, several hundred corn plants are selfed. Second, the selfed offspring are again selfed. This process is repeated for several generations, during which time nearly all selfed lines decrease in vigor and many disappear. Third, the remaining selfed lines are crossed with each other in hundred of different combinations to produce hybrids which are carefully tested to determine which combinations are best. However, because the hybrid seed is produced on relatively nonvigorous selfed plants, commercial production of seed would be very expensive. Hence a fourth step is added. The hybrids are crossed with each other (AB × CD, AB × JM, CD × EF, AC × KM, etc.) in numerous combinations to produce double-cross hybrids. These are tested, and the ones which yield best are released for commercial production.

The "selfing and later crossing" method so useful with corn has been attempted in the shrubby genus *Rhododendron* by H. T. Skinner. He was able to carry the selfed lines only 2–3 generations because inbreeding depression was severe. Hybrids among some of the selfed lines were remarkably vigorous, however.

The "selfing and later crossing" method has been suggested for trees, but would be difficult to put into practice. It is most effective when plants carry a great many deleterious recessive genes which can be detected and eliminated through selfing. In other words, it is most effective in plants which suffer severe inbreeding depression. If a normal tree requires 5–20 years to flower, selfed trees may well require 15–60 years to flower and also require extra care during that time. Hence the method promises to be time-consuming and expensive.

Alternate Mild Inbreeding and Outcrossing as an Improvement Method

According to Sewall Wright, maximum evolutionary change can occur if there are alternating periods of mild inbreeding (i.e., with $N = 10$ to 100) and hybridization between the inbreds. The inbreeding results in some immediate loss of vigor, but results in some gene complexes which could not develop as a result of normal selection. These complexes, when combined with others, may produce an exceptionally well-adapted type of tree.

Climatic changes producing alternate periods of range contraction and expansion have probably made this mechanism effective in trees, especially in species with much disrupted ranges. This process may have been responsible for the development of the numerous pine and oak species found in Mexico.

The North Carolina State–Industry Tree Improvement Program involves many different companies, each with its own seed orchards. Most of these seed orchards contain a few score or a few hundred clones. After roguing, most will be reduced to so few clones that there could be serious inbreeding if each company were to work only with its own material for 5 to 10 generations. This is unlikely. It is more likely that each company will work with its own material for 2 to 3 generations, then exchange pollen or seeds with other companies. In that way, alternate inbreeding and outcrossing will be accomplished.

Inbreeding versus Improvement

Inbreeding can also be considered as a narrowing of the genetic base. Almost every improvement method results in a narrowing of the genetic base to some extent. This is true whether the improvement consists of selecting the best one of many species, the best one of many races within a species, the best one of many parental stands within a race, or the best one of many individual parents.

There is always a conflict. At any given moment, the genetic improvement will be greatest if the selection process is carried to the extreme of selecting the one best rather than the several best. This conflict has received major attention in the 1970's after corn blight caused such havoc in the numerous varieties possessing identical cytoplasm. There is a constant fear that the same thing could happen in other crops if genetic variability is not maintained.

As of now there is little reason to believe that the amount of genetic variability has been reduced to a dangerously low level in most tree species as a result of tree breeding activities. But it is necessary to take positive steps, such as the retention of large areas of wild stands or the establishment and retention of large provenance tests, to ensure that the variability in trees is maintained for future generations of people.

Hybrid Vigor

Definition

Hybrid vigor or heterosis is the exceptional performance of hybrid offspring of genetically different organisms. Strictly speaking, we can speak of hybrid vigor if the hybrids perform significantly better than the average of the two parents. We can also speak of "negative hybrid vigor" or "negative heterosis" if the hybrids are significantly poorer than the parental types. From the practical standpoint, however, the term is usually reserved for those cases in which the hybrids outperform the better of the two parents.

Four Explanations

Four explanations for hybrid vigor have been proposed. It is probable that all are true to some extent, since there is good evidence for each of the four in one organism or another. The four explanations are known as the dominance, overdominance, additive, and hybrid habitat hypotheses.

According to the "dominance" hypothesis, hybrid vigor is merely the absence of inbreeding depression. Most plants carry a large "genetic load" of deleterious recessive genes of which a few are homozygous and depress growth. After crossing two very different trees, most recessive genes from one are "covered" by dominant genes from the other so that there is no growth loss.

According to the "overdominance" hypothesis, the gene combination Aa is able to produce an effect which is impossible with AA or aa. In other words, hybrid vigor is due to the heterozygosity itself.

According to the "additive" hypothesis, hybrid vigor can result if a trait is composed of two components, each controlled by genes with additive effects. To illustrate, assume that seed yield per cone is composed of weight per seed and number of seeds per cone. Assume that tree 1 produces 30 20-mg seeds per cone and that tree 2 produces 60 10-mg seeds per cone. For both trees the yield is 600 mg per cone. With additive gene action, hybrid 1 $\times$ 2 trees will produce 45 15-mg seeds per cone, for a seed yield of 675 mg per cone. Thus the hybrid can produce more than either parent.

Note that the hybrid was intermediate between its parents in both weight per seed and number of seeds per cone, but nevertheless had an appreciably greater seed yield than either parent.

According to Edgar Anderson's "hybrid habitat" hypothesis, hybrid vigor is manifested principally in plants that grow on hybrid habitats. Muller (1952) furnished an example in trees. In parts of Texas, *Quercus mohriana* and *Q. havardii* occur, respectively, on limestone and sandy soils. There are occasional strips where the limestone and sandy soils are intermingled. Hybrid trees inhabit these "hybrid soils" but away from the zone of contact between soil types only the pure species are found.

Manifestation of Hybrid Vigor in the F_1, F_2, and Ensuing Generations

Assume that AA and aa trees are crossed with each other and that the hybrids are intercrossed with each to the fourth generation. The results will be as shown in the following tabulation.

Parental generation	AA $\times$ aa
F_1 generation	Aa
F_2 generation	1 AA : 2 Aa : 1 aa
F_3 generation	3 AA : 2 Aa : 3 aa
F_4 generation	7 AA : 2 Aa : 7 aa

In other words the proportion of heterozygous Aa trees will decrease from 100 to 50 to 25 to 12.5% in the first, second, third, and fourth generations, respectively.

If hybrid vigor is due to genes with overdominant effects, i.e. to heterozygosity itself, hybrid vigor will be most manifest in the F_1 generation and will decrease thereafter.

If hybrid vigor is due to genes with additive effects, it should be possible to practice selection for the best genes controlling each component. In the example given above it might be possible to select for trees having an average of 60 seeds per cone, each seed weighing 20 mg, for a total yield of 1200 mg per cone. That would be even more "hybrid vigor" than evidenced in the F_1 generation.

If hybrid vigor is due to dominance, it should be possible to select against recessive genes and thus fix the hybrid vigor. To the extent that the hybrid habitat hypothesis is true, hybrid vigor in hybrid habitats may be as large in the second and third as in the first generation.

Usefulness of Hybrid Vigor in Tree Improvement

Much of our theoretical and practical information about hybrid vigor has come from corn breeding programs. Especially in the United States, hybrid corn varieties almost replaced open pollinated varieties, starting in the 1930's.

Corn breeders placed their greatest emphasis on F_1 varieties, partly because overdominance was believed to be a major cause of hybrid vigor and partly because the F_2 and later generations had too much variability in a crop where uniformity was desired. Some breeders, however, pursued the development of open pollinated varieties and have had some success in fixing hybrid vigor. This indicates that genes with dominant and additive effects played a major role.

Tree breeders, influenced very much by the corn research, have concentrated on the production of heterotic F_1 hybrids. Many of these have proved difficult to mass produce. Dependence on the overdominance hypothesis discouraged work on the F_2 and later generations. However, if a major portion of hybrid vigor is due to other causes work on later generations holds promise.

Hyun et al. (1972), working in Korea with hybrids between pitch and loblolly pine, found F_1 hybrids to be very useful. In their latest work, they determined that F_2 hybrids slightly outgrew the F_1 hybrids. Both parental species are native to the United States, so the hybrids are being used outside their native range, presumably on "hybrid habitats." Thus there is good reason to believe that the hybrid vigor will persist in later generations.

Another factor must be considered. Most agricultural crops such as corn are machine harvested. Therefore uniformity is desirable. The F_1 generation is usually uniform, whereas the F_2 and later generations are more variable. Therefore, even if an F_2 population had as high an average yield as an F_1, it would not be as useful in agricultural practice. Most trees are harvested individually, and many tree plantations are thinned so that the final crop trees may be only a small percentage of those planted. Hence there is less need for uniformity, and an F_2 or F_3 population containing a reasonable percentage of heterotic individuals could be useful.

Recovery Ratios in the F_2 Generation

Assume that the A gene is more desirable than the a gene and that a cross is made between AA and aa trees. The F_1 generation will be entirely Aa and the F_2 generation will be variable, with the composition 1 AA : 2 Aa : 1 aa. The frequency of desirable AA trees in the F_2 will be $25\% = \frac{1}{4} = (\frac{1}{2})^2$.

Assume that genes A and B are most desirable and that a cross is made between AABB and aabb trees. The F_1 generation will be entirely AaBb. The F_2 generation will be variable, with a frequency of desirable AABB trees of $\frac{1}{16} = (\frac{1}{2})^4$.

These statements can be reduced to the general rule that in the F_2 generation the recovery ratio of trees homozygous for the most desirable allele at each of n variable loci is $(\frac{1}{2})^{2n}$. For differing numbers of variable loci the recovery ratios are as follows.

No. of variable loci	1	2	3	4	5	6	7
Recovery ratio	1/4	1/16	1/64	1/256	1/1024	1/4096	1/16,384

Tree breeders do not have much occasion to cross pure lines to produce F_1 and thence F_2 hybrids. However, they have much occasion to cross heterozygous wild types (analogous to F_1) to obtain trees homozygous for desirable genes. They also work largely with traits controlled by several pairs of genes.

The above figures illustrate why it is necessary to do much of the crossing and progeny testing work on a large scale. They also illustrate the necessity to pursue breeding work for several generations. Assume that growth rate and disease resistance are both controlled by 5 pairs of genes and that tree 1 (fast growth, low disease resistance) is crossed with tree 2 (slow growth, high disease resistance) to produce an F_1

and then an F_2 generation. There are 10 variable loci and a person would have to raise $2^{20} = 1,089,511,627,776$ trees to obtain one tree homozygous for all the fast growth and disease resistance genes. This is obviously impractical. The alternative solution is to raise a few thousand seedlings, obtain trees homozygous for 2 to 3 pairs of desirable genes, cross those, and obtain trees homozygous for 2 to 3 more pairs of desirable genes, etc.

Study Questions

1. Define selfing, inbreeding, and genetic drift.
2. Describe a mechanism by which selfing can reduce seed set and vigor.
3. What are the four primary consequences of inbreeding?
4. In a monoecious species $N_{total} = 4$. What is the inbreeding coefficient after 5 generations of inbreeding?
5. Each of 100 females is crossed with each of 4 males and the process is repeated for 2 generations. What is the inbreeding coefficient?
6. Selfing produces seedlings which grow only 70% as fast as normal seedlings. If a breeding population of $N = 5$ is maintained for 3 generations, what is the expected rate of growth of the third generation seedlings?
7. What evidence is there that genetic drift may have been an important factor in the evolution of some spruce species? Was it important in the evolution of white spruce or Norway spruce?
8. What evidence is there that inbreeding may be more common in tropical than temperate zone trees?
9. Describe the mechanisms by which inbreeding is minimized in the pines. In the maples.
10. Compare the usefulness of selfing as an improvement method in pine, wheat, and corn. Why the differences?
11. Why is alternate mild inbreeding and outcrossing so effective in evolution?
12. Name and explain four hypotheses of hybrid vigor. Which is the opposite of the explanation of inbreeding depression?
13. Under what circumstances is it possible to "fix" hybrid vigor in generations after the F_1?
14. Assume that growth rate and foliage color are each controlled by 2 pairs of genes and that a fast growing, poor-color tree is crossed with a slow growing, good-color tree. How many F_2 seedlings must be raised to obtain one which is homozygous for all fast growth, good-color genes?

4

Population Genetics—Selection

Definitions

Selection is a basic factor in evolution and tree breeding. An understanding of its effects can be obtained best by starting with the simplest situations—gene changes at single loci.

In bacteria, which are haploid, selection of a single mutant cell produces a strain which is true-breeding for that mutation. The effects are less simple in trees, which are diploid. In trees, selection operates on trees (zygotes) which have two genes at each locus. Trees may be AA, Aa, or aa and produce A or a gametes. Even if all aa trees are cut, some *a* genes survive in Aa trees.

Because of such complexities a special branch of genetics termed "population genetics" has developed. I shall introduce some of the simpler concepts, most of them from the work of Sewall Wright and R. A. Fisher, and adopted from C. C. Li's (1975) book "First Course in Population Genetics." Some definitions should be presented first.

A and *a* are the two genes (alleles) which may occupy one locus in a diploid tree. Commonly, A and *a* are considered to be dominant and recessive genes, respectively. However, that is not always true.

p and q are the relative frequencies of A and *a*, respectively, and are so defined that they total 1. Hence

$$p + q = 1, \quad q = 1 - p, \quad p = 1 - q,$$
$$p^2 + 2pq + q^2 = (p + q)^2 = 1$$

p' and q' are the frequencies of A and a in the first generation if p and q are the frequencies in the initial generation. Alternatively, p_0 and q_0, p_1 and q_1, p_2 and q_2 may be used to denote the frequencies of A and a in parental, first, and second generations, respectively.

$\hat{p}$ and $\hat{q}$ are the equilibrium frequencies when selection no longer produces a change.

s is the selection pressure, defined in terms of mortality of aa trees relative to mortality of AA and Aa trees. For example, $s = 0.1$ if survival rate is 50% for AA and Aa trees and only 45% for aa trees.

s_{AA}, s_{Aa}, and s_{aa} are selection pressures against (or for) AA, Aa and aa trees, respectively, also defined in terms of relative mortality.

$W = 1 - s$ is the fitness when defined in terms of a quantitative trait or survival. W_{AA}, W_{Aa}, and W_{aa} are the fitnesses of AA, Aa, and aa trees, respectively.

n is the number of generations.

Hardy–Weinberg Equilibrium

The Hardy–Weinberg "law" states: the gametic constitution of a population remains constant from one generation to the next in the absence of mutation, random changes or selection (Li, 1955). The law may seem intuitive and too simple to need explanation. The theorem used to prove it is interesting, providing a simple illustration of the mathematics of selection.

The parental generation produces A and a gametes in the ratio $p:q$. These produce an F_1 generation of AA, Aa, and aa trees which follows the binomial distribution $p^2:2pq:q^2$. The F_1 trees produce gametes as follows.

AA trees produce p^2 A gametes.
Aa trees produce pq A gametes and pq a gametes.
aa trees produce q^2 a gametes.
All trees produces $(p^2 + pq)$ A gametes and $(pq + q^2)$ a gametes.

Thus the ratio of A to a gametes produced by the F_1 generation is $(p^2 + pq):(q^2 + pq) = p(p + q):q(p + q) = p:q$, as in the initial generation.

Complete Elimination of Homozygous Recessive Trees

Selection against lethal recessives involves complete elimination of homozygous recessive aa trees each generation. The remaining heterozygous Aa trees produce some a genes, however. To calculate changes in gene frequency it is necessary to start with gametic frequencies, calculate tree frequencies, postulate elimination of tree types, recalculate gametic frequencies, etc. The exact methods are as follows.

1. Prepare table with headings as in Table 4.1.
2. Insert as p_0 and q_0 the initial frequencies of A and a.
3. Calculate $(p + q)^2 = p^2 + 2pq + q^2$ to obtain F_1 preselection frequencies of AA, Aa, and aa trees. Check to ensure that $p^2 + 2pq + q^2 = 1$.
4. Recopy the frequencies of AA and Aa trees and insert 0 (i.e., cut) as the frequency of aa trees.
5. Add frequencies of AA and Aa trees and insert as "Total."
6. Change "total" to "1" and adjust frequencies of AA and Aa trees accordingly.
7. To obtain p' and q' after selection in generation 1, p' = frequency of AA trees + half the frequency of Aa trees, q' = half the frequency of Aa trees, and $p' + q' = 1$.
8. Repeat steps 3–7 for generation 2.

Very simple problems can be worked best as fractions, as in the first numerical example in Table 4.1. For more complex problems, decimals are needed.

A shortcut is possible. Note in the upper numerical example in Table 4.1 that $q = 1/2$, $1/3$, and $1/4$ after selection in generations 0, 1, and 2, respectively. If calculated, q would be $1/5$, $1/6$, and $1/7$ in the third, fourth, and fifth generations, respectively. This can be made into a general statement for the calculation of q_n (q in the nth generation) in terms of q_0 (q in the zero generation).

$$q_n = q_0/(1 + nq_0)$$

In the absence of mutation, this type of selection leads to a complete loss of a genes so that the equilibrium frequency is $\hat{q} = 0$. However, mutations from A to a occur at the rate of u, which leads to an equilibrium frequency of $\hat{q}^2 = u$. In other words, at equilibrium the frequency of trees homozygous for a lethal recessive gene should equal the mutation rate. However, the frequency of such trees is often much higher than can be explained in terms of any probable mutation rate. When this happens, the high frequency of aa seedlings which die soon after germination is taken as an indication of either (1) selection favoring Aa trees or (2) inbreeding.

In Siberian pea tree, "drooping" branch habit is known to be a rare trait controlled by recessive genes; DD and Dd trees are upright,

TABLE 4.1

Methods of Calculating Changes in Gene and Tree Frequencies following Complete Elimination of Homozygous aa Trees

Generation before or after selection	Tree frequencies			Total tree frequency	Frequency of gene	
	AA	Aa	aa		A	a
General example						
0 before	p^2	$2pq$	q^2	1	p	q
0 after	p^2	$2pq$	0	$p^2 + 2pq$	$\dfrac{p^2 + pq}{p^2 + 2pq} =$	$\dfrac{pq}{p^2 + 2pq} =$
	$\dfrac{p^2}{p^2 + 2pq}$	$\dfrac{2pq}{p^2 + 2pq}$			$p/(1+q)$	$q/(1+q)$
Numerical example, $q_0 = 1/2$						
0 after					1/2	1/2
1 before	1/4	1/2	1/4	1	1/2	1/2
1 after	1/4	1/2	0	3/4		
1 after	1/3	1/3	0	1	2/3	1/3
2 before	4/9	4/9	1/9	1	2/3	1/3
2 after	4/9	4/9	0	8/9		
2 after	1/2	1/2	0	1	3/4	1/4
Numerical example, $q_0 = 0.6$						
0 after					0.400	0.600
1 before	0.160	0.480	0.360	1.000	0.400	0.600
1 after	0.160	0.480	0.000	0.640		
1 after	0.250	0.750	0.000	1.000	0.625	0.375
2 before	0.391	0.469	0.141	1.001	0.625	0.375
2 after	0.391	0.469	0.000	0.860		
2 after	0.454	0.546	0.000	1.000	0.727	0.273

whereas dd trees droop. Should one practice selection to eliminate the trait by cutting all the drooping trees? Assume that the initial frequency is $q_0 = 1/10$. The frequencies in successive generations are $q = 1/11$, 1/12, 1/13, etc., the frequencies of drooping trees in successive generations are 1/121, 1/144, 1/169, etc. The trees may be cut as a part of a silvicultural thinning, but even so some can be expected to occur in ensuing generations.

Partial Selection against Recessives

Most recessive genes have small effects so that aa trees cannot be recognized and eliminated with certainty. However, if they grow

TABLE 4.2

Method of Calculating Changes in Gene Frequency with Partial Selection at Rate s against Homozygous Recessive (aa) Trees

Generation before or after selection	Tree frequencies			Total tree frequency	Frequency of gene a
	AA	Aa	aa		
General example					
0 before	p^2	$2pq$	q^2	1	q
0 after	p^2	$2pq$	$(1-s)q^2$	$1-sq^2$	$q' = \dfrac{pq + q^2 - sq^2}{1 - sq^2}$
	$\dfrac{p^2}{1-sq^2}$	$\dfrac{2pq}{1-sq^2}$	$\dfrac{(1-s)q^2}{1-sq^2}$		$= \dfrac{q(1-sq)}{1-sq^2}$
Numerical example, $q_0 = 0.8$, $s = 0.25$ against aa trees					
0 after					0.800
1 before	0.040	0.320	0.640	1.000	0.800
1 after	0.040	0.320	0.480	0.840	
1 after	0.048	0.381	0.571	1.000	0.762
2 before	0.057	0.363	0.580	1.000	0.762
2 after	0.057	0.363	0.435	0.855	
2 after	0.066	0.425	0.509	1.000	0.722
3 before	0.077	0.402	0.521	1.000	0.722
3 after	0.077	0.402	0.391	0.870	
3 after	0.089	0.462	0.449	1.000	0.680

a little slower than average, elimination of all slow-growing trees will result in the elimination of proportionately more aa than AA or Aa trees. Mathematically, the relative survival rate of AA and Aa trees is considered as 1, and the relative survival rate of aa trees is considered as $1-s$, where s is called the selection coefficient. The general method for calculating generation-by-generation changes in gene frequency is illustrated in Table 4.2. The method is similar to that illustrated in Table 4.1 except that before selection frequency of aa trees is multiplied by $1-s$ rather than 0 to obtain the after selection frequency of aa trees.

The per generation change in q can also be calculated approximately as

$$\Delta q = q' - q = (-spq^2)/(1 - sq^2) = -spq^2$$

if s is small. One of these formulas can also be used to calculate average $\triangle q$ for a small range in q, as for a change from $q = 0.5$ to 0.45, $q = 0.45$

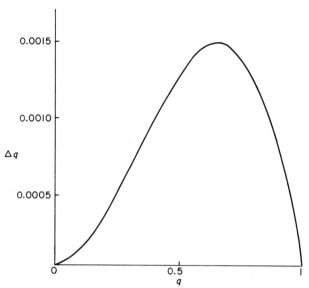

Fig. 4.1. Rate of change ($\triangle q$) in frequency of gene a after partial selection ($s = 0.01$) against aa trees. (After Sewall Wright, from Li, 1955).

to 0.40, etc. Or the following formula can be used to calculate the number of generations (n) needed to change q_0 to q_n.

$$n = \frac{(q_0 - q_n)/q_0 q_n + 2.303 \log[(p_n q_0)/(p_0 q_n)]}{s}$$

The rate of change per generation and the amount of change in a certain number of generations is shown in Fig. 4.1 and Table 4.3.

TABLE 4.3

Generations Required to Change Gene Frequency following Partial Selection against Homozygous Recessives

Change in q	Generations required if s is		
	0.5	0.1	0.01
0.99 to 0.90	4	25	250
0.90 to 0.70	3	17	168
0.70 to 0.50	3	13	132
0.50 to 0.30	4	21	208
0.30 to 0.10	15	80	802
0.10 to 0.01	185	924	9240

Note that the rate of change is greatest at intermediate gene frequencies. Progress is very slow when q is small.

As a corollary, consider that selection usually affects genes at several loci rather than at a single locus as is assumed to be the case in Table 4.3, and that changes in frequency at several loci may have resulted in a growth rate increase. Note that there is still room for considerable change in gene frequency and thus for improvement, even after several generations of selection.

Selection Favoring Recessives (against Dominants)

If gene A is dominant, AA and Aa trees are indistinguishable but either type can be distinguished from aa trees. Complete elimination of AA and Aa trees leaves only aa trees which thereafter breed true. In other words, $q = 1$, even after only a single generation of selection.

The effects of partial selection against dominants can be calculated in the same manner as already described except that $1 - s$ is the relative survival rate of AA and Aa trees if 1 is the relative survival rate of aa trees. Also, the equilibrium frequency is $\hat{q} = 1$. Per generation rate of change can be calculated as

$$\Delta q = (spq^2)/(1 - s + sq^2)$$

Partial elimination of AA and Aa trees is slightly (if s is small) to greatly (if s is large) more effective than partial elimination of aa trees. This is shown in Table 4.4.

Recessive genes are apt to be deleterious and therefore to be selected against in an improvement program but selected for in exploitive logging, $\longrightarrow$ which removes the best trees and leaves only the low-value ones to produce seed. Notice from the above tabulation that the genetic damage (i.e., change in gene frequency) caused by a few generations of dysgenic logging can be greater than the genetic good accomplished by the same amount of positive selection. (This is strictly true only with mass selection. As will be seen in Chapters 9–12, man can use other selection methods and can, therefore, often make large changes in gene frequency in a few generations.)

Selection for Genes with Additive Effects

Many economically important traits are under the control of many genes with small additive effects. In this chapter I discuss the theory of changes in gene frequency at single loci. [For additional information consult the books by Lerner (1958) and Falconer (1960).]

TABLE 4.4

Per Generation Changes in q with Partial Selection against Recessives and against Dominants

| q_0 | s | Per generation change in q with selection against | |
		Recessives	Dominants
0.5	0.2	-0.0264	0.0294
0.2	0.2	-0.0064	0.0079
0.5	0.5	-0.0717	0.1000
0.2	0.5	-0.0200	0.0308

For this section, genes A and a shall be defined as "good" and "bad" rather than as dominant and recessive. Thus, AA trees are better than Aa trees, which are better than aa trees. Also, the selection coefficient (s) shall be defined in such a way that the relative survival rates of AA, Aa, and aa trees are 1, $1 - s$, and $1 - 2s$, respectively. The methods used to calculate generation by generation changes in gene frequency are the same as those already outlined in Tables 4.1 and 4.2, except that the before selection frequencies of Aa and aa trees are multiplied by $(1 - s)$ and $(1 - s)$, respectively, to obtain the after selection frequencies of those genotypes.

Some formulas applicable to this type of selection are

$$\hat{q} = 0$$
$$\Delta q = q' - q = -spq/(1 - 2sq) = -spq \quad \text{(approximately if } s \text{ is small)}$$
$$n = (2.303/s) \log (q_0 p_n / q_n p_0)$$

n is the number of generations required to change from q_0 to q_n if s is small. Actually, the effects of this type of selection are almost the same as if the selection were directly against the gametes (pollen grains and egg cells).

Because $\Delta q = -spq$, it is at a maximum when the product pq is large, which happens when both p and q are about 0.5. In other words, selection progress is greatest at intermediate gene frequencies. Even so, changes in gene frequency take place relatively slowly, as is evident from the following tabulation based on an initial frequency of $q_0 = 0.5$ and selection coefficients of $s = 0.10$ and 0.25, which are probably to be considered quite large when applied to individual loci.

	Δq for generation		
s	1	2	3
0.10	−0.028	−0.028	−0.027
0.25	−0.084	−0.076	−0.068

Note that the per generation change in q was nearly the same for the third as for the first generation. As a corollary, if one generation of selection produces an appreciable amount of improvement, additional improvement of an almost equal amount can be expected from another generation of selection. Because a tree generation requires several years, there seems to be no plateau in the foreseeable future beyond which progress through selection is impossible.

"Fitness" and Fisher's "Fundamental Theorem"

Up to now, we have considered gene frequencies in terms of tree frequencies. To change gene frequencies, a portion (s) of the trees containing an unwanted gene have been eliminated. This is correct when considering a single locus. Actually, most traits are governed by genes at several loci. The presence of a particular allele at one of those loci determines not only whether the tree lives or dies, but how well it grows and reproduces. Thus, in talking of a population of trees it is possible to consider gene frequency in terms of growth rate, reproductive capacity, winter hardiness, etc.

The concept of "fitness" has been introduced to make this transition from gene frequency in terms of frequencies of trees or pollen grains to gene frequency as a measure of excellence. The word "fitness" is most applicable to natural populations, where fitness is a tree's ability to survive and produce offspring. However, the concept of "fitness" can apply equally well to artificial selection and can be used as a measure of growth rate, branch size, etc. Mathematically, the term can be defined as

$$\text{Fitness} = \text{survival rate} = 1 - s$$

where s is the selection coefficient.

In 1930, the late R. A. Fisher used this concept in his fundamental theorem of natural selection which states in essence that the per generation increase in fitness equals the genetic variance in fitness. Mathematically,

$$V_W = \Sigma f W^2$$

where V_W is the genetic variance in fitness, f_{AA}, f_{Aa}, and f_{aa} are the relative frequencies of AA, Aa, and aa genotypes, respectively, and W_{AA}, W_{Aa}, and W_{aa} are the relative fitnesses of AA, Aa, and aa genotypes respectively. As a further condition, Fisher defined f and W in such a way that $\Sigma f = 1$ and that $\bar{W}$ (average fitness) $= 1$. The theorem is strictly true only for genes with additive effects. Its mathematical proof is given in Li (1955).

Fisher was concerned with an hypothetical situation in which trees' fitnesses were governed entirely by their genotypes, i.e., no environmental control. In actual practice, traits, such as growth rate, branch size, are under environmental as well as genetic control. For that reason, his basic theorem has since been modified to become

Rate of improvement

$$= \text{selection differential} \times \frac{\text{additive genetic variance}}{\text{total variance}}$$

This formula will be mentioned many times in Chapters 9–12, dealing with the actual practice of selection.

Selection for and against Heterozygotes

Selection Favoring Heterozygotes

If genes exhibit positive overdominance, selection favors heterozygotes. Because the homozygotes differ in viability, they are unequally selected against. Thus, two selection coefficients, s_{AA} and s_{aa}, are used to denote selection against AA and aa trees, respectively. The general method for the calculation of generation by generation changes in gene frequency is similar to that already illustrated by Tables 4.1 and 4.2 except that the before selection frequencies of AA and aa trees are multiplied by $(1 - s_{AA})$ and $(1 - s_{aa})$, respectively, to obtain the after-selection frequencies of those genotypes.

This type of selection results in a stable equilibrium at an intermediate frequency of $\hat{q} = s_{AA}/(s_{AA} + s_{aa})$. The rate of change toward

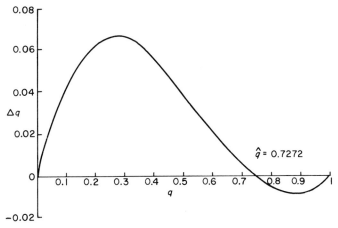

Fig. 4.2. Changes in $\triangle q$ in relation to q following selection for heterozygotes, with $s_{AA} = 0.4$ and $s_{aa} = 0.15$. (After Sewall Wright, from Li, 1955).

that equilibrium is slow if q is nearly 0, nearly 1, or nearly $\hat{q}$, but more rapid otherwise (Fig. 4.2).

At the equilibrium frequency of $\hat{q}$, a tree population is a hybrid swarm containing both parents, F_1 hybrids, F_2 hybrids, and various backcrosses of the hybrids to the parents. Some of these trees are less productive than others, but the mixture of AA, Aa, and aa trees is such that the population as a whole has a maximum average fitness when the equilibrium frequency is reached. Any attempts to manage the hybrid swarm by artificially favoring certain genotypes over others will reduce the average fitness. This can be shown in Table 4.5 which is based on an

TABLE 4.5

Average Fitness following Various Types of Selection in a Hybrid Population Exhibiting Overdominance with Selection Coefficients of 0.40 and 0.15 against AA and aa Trees, Respectively

Type of artificial selection	Average fitness
None, population at equilibrium of $\hat{q} = 8/11$	0.896
Partial selection favoring heterozygotes, $q = 7/12$	0.880
Population reduced to hybrids (Aa) the previous generation, $q = 1/2$	0.862
Selection favoring aa trees only, $q = 1$	0.850
Selection favoring AA trees only, $q = 0$	0.600
Clear cutting, with F_1 hybrids planted, $q = 1/2$	1.000

hypothetical population in which s_{AA}, s_{Aa}, and s_{aa} = 0.40, 0.00, and 0.15, respectively; alternatively W_{AA}, W_{Aa}, and W_{aa} = 0.60, 1, and 0.85 respectively.

Applicability of this model to actual practice depends on the extent to which natural survival ability (fitness) is associated with traits of economic value. The model is applicable if natural survival is primarily a function of growth rate, cold hardiness, resistance to insects, etc. The model is much less applicable if natural survival ability is governed by factors of little importance to man.

There are examples of large, persistent hybrid swarms, such as postulated above: white × Engelmann spruce and jack × lodgepole pine in western Canada, Norway × Siberian spruce in northern Scandinavia, and eastern cottonwood × European black poplar in Italy. In each case survival ability seems associated with economically important growth traits. According to Table 4.5, one of two silvicultural systems seems most desirable. Either (1) practice no artificial selection and let the hybrid swarms arrive at a stable equilibrium, or (2) clear cut and replace entirely with F_1 hybrids.

As a corollary, the long persistence of such hybrid swarms indicates that the hybrids possess superior growth characteristics (or at least superior ability to survive) and that their artificial production on a large scale should be contemplated.

Selection against Heterozygotes

Genes can also act in such a way that heterozygotes (Aa) are inferior to homozygotes (AA and aa). The following tabulation shows the effects of the simplest type of such selection, i.e., complete elimination of the heterozygotes.

Gener-ation	Tree frequency			Frequency of gene a
	AA	Aa	aa	
1 Before	0.810	0.180	0.010	0.100
1 After	0.988	0.000	0.012	0.012

Notice that the frequency of gene a was drastically reduced. The reduction was much greater than if aa trees had been eliminated, in which case the frequency of gene a would have changed from q = 0.100 to q' = 0.091.

From this example, it is possible to understand why many hybrid swarms do not persist but quickly revert to a relatively pure population of the more common parental type. In southeastern United States F_1 hybrid Sonderegger pine (longleaf $\times$ lobolly pine) is found many places where the two parental species grow together. The hybrids are fertile and can reproduce themselves but usually do not grow as well as the parents. Hybrid swarms do not persist (Namkoong, 1966). In Texas, *Quercus mohriana* and *Q. havardii* typically inhabit limestone and sandy soils, respectively (Muller, 1952). Hybrid swarms persist in narrow zones where the two soil types overlap and selection favors the heterozygotes; away from those narrow zones selection favors the homozygotes and hybrids quickly disappear. Tucker (1952) reported similar cases involving oak hybrids in California.

Selection in Small Populations

All examples quoted so far in this chapter apply to populations of infinite size. Actually, tree populations are of finite size, sometimes very small. In small populations, gene frequency may be determined more by chance than by selection. Sewall Wright showed that the fate of a gene depends on a combination of population size (N) and selection pressure (s) such that

If $s = 2.5/N$, selection is effective and random fluctuations are small
If $s = 0.25/N$, selection is moderately effective
If $s = 0.025/N$, selection is less effective than chance

In actual numbers, $N = 400$ is on the borderline between "small" and "large" if $s = 0.01$, and $N = 100$ is on the borderline between "small" and "large" if $s = 0.04$.

In an actual improvement program, the number of clones or families included in a seed orchard is likely to be between $N = 40$ and $N = 200$, so there will be many cases in which changes in gene frequency are due to chance rather than to the artificial selection. C. C. Li noted that strong selection for any one trait usually resulted in such small population sizes as to make chance variations in other traits almost inevitable.

Have population sizes in natural forests also been so small as to make chance a large factor in evolution? Here it should be noted (see Chapter 5) that most seeds and pollen grains travel relatively short distances, so that the trees which mate with each other in any one generation are apt to be confined to a relatively small area. Sakai and

Miyazaki (1972) showed this to be true with their peroxidase studies of Japanese aborvitae. The peroxidase isozymes vary little with environment but greatly among trees; thus, the isozymes are an ideal way to study inheritance patterns in a natural forest. They found great similarities among trees growing within 50- to 75-ft circles, indicating that local breeding populations may be very small. Their results indicate that chance may have a great influence on the frequency of many genes.

How to Increase Selection Pressure

Measurement of the selection coefficient s is difficult and possible only with a few genes having very large effects. However, there are ways in which it can be increased, as follows.

1. Grow trees in a uniform environment so that nongenetic effects are minimal.
2. Make selection rigorous by selecting the best trees of many.
3. Grow offspring of selected trees and select on the basis of those offspring.

The first two methods are so obvious as to need no comment. The third method needs some explanation. If a trait is under environmental and genetic control and if selection is based only on phenotype, there is a high degree of uncertainty as to whether any given tree is AA, Aa, or aa. But if many offspring of each parent are grown, the AA, Aa, and aa parents can be identified with greater certainty. Having been identified, the desirable genotypes can be selected more rigorously. This method of selection called "family selection" is covered in more detail in Chapters 9–12.

A Word Picture of Selection

Imagine a forest of eastern white pine, situated in southern Canada or northern United States, many eons ago. The pine cones ripened in mid-September and promptly shed their seed. The seed probably needed to be "after-ripened," that is to be stored moist at temperatures slightly above freezing for a 30- to 40-day period. There were too few such days in the fall for the seed to germinate before winter, but there were enough in the fall and early spring to ensure germination the following May. Winter set in by December or sometimes early November, and from then until early April the temperatures were generally much below freezing, often −20° or −30°C. Only trees with great winter hardiness could survive unscathed and thrive the next year.

Some seeds from that forest blew south and grew into mature trees. Seeds from those trees blew still farther south. No two trees were exactly alike genetically, and during the southward migration a process of sorting took place. There seems to be a physiological association such that trees that can withstand very cold temperatures are not the fastest growing. Conversely, fast-growing trees tend to be less winter hardy. Since there was less need for winter hardiness, trees with many rapid-growth genes grew a little faster and produced a little more seed than their neighbors. Thus, those rapid-growth genes gradually increased in frequency. The trees varied in other respects. Seeds with a 30- to 40-day after-ripening requirement did not normally germinate until spring. However, there were occasional winters when November and December were mild and there was a very warm spell in January or February. During such winters, seeds with a 30- to 40-day after-ripening requirement could germinate more easily in midwinter than those with a 40- to 50-day requirement. If they did, the tender young seedlings were almost sure to die during the remainder of the winter. Hence there was a gradual increase in the proportion of trees having a longer after-ripening requirement.

Today we see the evolutionary end products. There have been several provenance tests of eastern white pine, in which trees grown from seed collected in many parts of the natural range were tested under similar conditions. Laboratory tests conducted in Indiana showed that southern Appalachian trees could withstand midwinter temperatures of −30° to −40°C without serious damage, whereas trees grown from Canadian or northern United States seed could withstand much lower temperatures, i.e., −50° to −60°C. If trees of various origins were grown in the northern Great Lake states, northern trees withstood the severe winters best and grew tallest. However, if trees of various origins were grown in the central or southern states, southern Appalachian trees grew 40–60% faster than northern ones. Also, seeds from the southern Appalachians need a much longer period of moist storage at temperatures of 3° to 4°C in order to germinate than do seeds from southern Canada (Fowler and Dwight, 1964; Wright, 1970a; Maronek and Flint, 1974; Garrett et al., 1973).

There are anomalies that indicate that this picture of selection is oversimplified. Eastern white pines from North Carolina and Tennessee have escaped winter injury and grown rapidly for 16 years when planted in relatively cold areas, such as central Pennsylvania and southern Michigan. Virginia is a southern state with relatively mild winters and favorable growing conditions, yet, wherever planted, trees from certain parts of Virginia have grown relatively slowly.

Recapitulation

The effects of selection vary considerably according to whether genes have additive effects, are dominant–recessive, or show overdominance. Therefore it is desirable when doing selection work to determine the mode of inheritance.

One of the most important applications of the theoretical calculations of per generation changes in gene frequency is the demonstration that selection for successive generations can often produce an almost equal amount of change in each generation. In practical terms, the attainment of a certain amount of improvement in the first generation should be considered as an impetus to continue the work, not as a sign that all possible improvement has been achieved.

Sewall Wright's work on selection in small populations and R. A. Fisher's fundamental theorem are theoretical. However, in both cases the theoretical work has very strong practical implications, which are considered in the day-to-day conduct of most breeding programs. The same can be said of many of the other theoretical relations covered in this chapter. While it is difficult to obtain realistic estimates of p, q, s, etc., their consideration can give an insight into the inner working of selection and an ability to forecast what is apt to happen under particular circumstances.

Study Questions

1. Define p, q, $\hat{p}$, $\hat{q}$, s.
2. Show why $p^2 = 1 - 2pq - q^2$.
3. Using the methods illustrated in Table 4.1, calculate q_2 if $q_0 = 2/5$ and $s_{aa} = 1$.
4. Using the methods illustrated in Table 4.2, calculate q_2 if $q_0 = 2/5 = 0.4$ and $s_{aa} = 0.1$. Make all calculations accurate to three decimal places.
5. If $q_0 = 0.4$, $s_{aa} = 1$, what is q_{1000}?
6. If dominants are completely eliminated, how many generations are needed to reach equilibrium frequency of $\hat{q} = 1$?
7. Assume that a trait is under the control of a genes with additive effects and that $s_{Aa} = 0.1$. What is s_{aa}? Using the approximate formula, calculate $\triangle q$ from generation 0 to generation 1 if $q_0 = 0.4$. Calculate $\triangle q$ from generation 3 to generation 4.

8. If growth rate is controlled by genes with additive effects and a breeder achieved a 3% gain the first generation, should he continue the work for five more generations? Assume that the 3% gain was worth $5X$ dollars and cost $1X$ dollars to achieve.

9. Explain why it is necessary to know the method of gene action.

10. How can selection pressure on a gene be increased?

11. Japanese red and Japanese black pines cross easily and the hybrids are fertile, yet the two species remain distinct in Japan. Why?

12. Why is mass selection not advisable in a hybrid swarm?

13. Describe the process of evolutionary change in growth rate, winter hardiness, and after-ripening requirement in eastern white pine.

5

Population Genetics—Mutation, Migration, Isolation

Mutation

Ultimate Fate of Single Mutations

Assume a population of AA trees, in one of which the mutation A to a occurs. That tree may die, leaving no offspring. It may mate with its AA neighbors in which case half its offspring will be homozygous AA. It may self, in which case 25% of its offspring will be homozygous AA. There is a probability of 0.368 that one of these possibilities will occur and that the mutant gene a will become extinct within a generation of its origin if the population is stable and each tree leaves an average of one offspring. If the mutant gene survives one generation, there is a probability that it will become extinct the second, or the third, etc., generation.

These probabilities of extinction are increased if the mutation is deleterious, decreased if it is beneficial. R. A. Fisher (from Li, 1955) calculated them on the assumptions that (1) the new mutation was neutral, conferring no selective advantage or disadvantage to the tree or (2) the new mutation conferred a 1% selective advantage on the tree. His results are shown in the following tabulation.

Generation:	1	3	5	15	63	127	∞
Probability of extinction							
No selective advantage:	0.37	0.63	0.73	0.89	0.97	0.98	1.00
1 % selective advantage:	0.36	0.62	0.72	0.88	0.96	0.97	0.98

TABLE 5.1

**Generations Required to Change q from 0.1
to 0.2 as a Result of Mutation at Varying Rates**

Generations needed to change q from 0.1 to 0.2	Mutation rates	
	u	v
4463	0.00003	0.00002
44.6	0.003	0.002
59	0.002	0
12	0.01	0

If one remembers that most new mutations are deleterious when they arise, it can be seen that ultimate extinction is the fate of the vast majority of new mutants. Those that survive are the ones which manage through combination with other genes to become advantageous.

Rate of Change with Recurrent Mutation

Thus it can be seen that a tree breeder is concerned not so much with single mutations as with the constant recurrence of mutation. The rate of change can be calculated by use of the formula $\Delta q = up - vq$, where p and q are the frequencies of genes A and a, respectively, u is the A to a and v is the a to A mutation rate.

With other formulas developed by Sewall Wright (from Li, 1955) it is possible to calculate the number of generations required to cause a given change in gene frequency as the result of differences in mutation rate. Some examples are given in Table 5.1.

The last line of Table 5.1 shows the relatively high mutation rate which might be expected as the result of irradiation. Even in that case, 12 generations are needed to produce the same change which might result from moderate selection ($s = 0.1$) in 8 generations. Thus, mutation rate is usually disregarded when calculating rate of evolutionary change or practical improvement.

Migration

Change with One-Way Migration

Migration is the movement of genes by pollen, seeds, or transplanting of trees. Although migration and mutation are different processes, they resemble each other in one respect. Both are ways in which new genes can be introduced into a population. But, whereas mutation rates are always very low, migration rates can be considerable. In migration we consider only the movement of genes into a population; pollen and seeds are nearly always produced in such excess that their removal from a stand does not affect the genetic constitution of the stand.

The simplest situation is one in which pollen or seeds are transported from a large donor population to a smaller recipient population. Assume that q_D and q_R are the frequency of a in the donor and recipient populations, respectively, and that m is the ratio of donor/(donor + recipient) seeds [m = donor/2(donor + recipient) pollen because the pollen supplies only half the genes for a tree]. Then

$$\Delta q_R = m(q_D - q_R)$$

This formula can be used to assess the importance of contamination of a seed orchard by pollen from nearby wild forests. It can also be used to estimate the effects of migration from a large continental population to a smaller island one.

Migration and Selection—Conflicting Processes

Two populations of trees may be situated on very diverse habitats and tend to become genetically different because of differences in selection pressure. However, if there is continuous intermigration, they tend to remain similar genetically. For example, some red maples might grow in a swamp and others in an upland a few hundred feet away. Or in a mountainous region, some ponderosa pines may grow at 5000 ft elevation whereas others grow at 7000 ft elevation 2 to 3 miles away. The site conditions might well be such as to generate considerable genetic differences between the swamp and upland red maples or between the 5000- and 7000-ft ponderosa pines. However, whether those genetic differences will develop will depend on the amount of gene exchange.

Sewall Wright developed formulas by which the joint effects of selection and migration could be considered. His simplest formulas pertain to a gene with additive effects so that $s_{AA} = 0$, $s_{aa} = 2s_{Aa}$. Assume that s is positive in one local population and negative in its neghbor, that $q = 0.40$ in the species as a whole, and that migration rate (m) varies. The equilibrium frequencies in the two populations will be as shown in Table 5.2.

In the first two examples in Tables 5.2, where m is approximately equal to or much less than s, the local populations will tend to become

TABLE 5.2

Equilibrium Frequencies in Two Partially Isolated Populations with Varying Selection Pressures for and against Gene a and with Varying Amounts of Migration from a Larger Homogeneous Population

	Equilibrium frequencies if	
Relative sizes of m and s	a is favorable	a is un- favorable
$m = s$	0.63	0.22
$m = 0.01$, $s = -0.15$ or $+0.15$	0.96	0.03
$m = 0.15$, $s = -0.01$ or $+0.01$	0.42	0.38

different genetically. In the third example, where m is much larger than s, the final equilibrium frequencies will be nearly the same.

Seed Dispersion Mechanisms

Most tree seeds are produced high in tree crowns. In many species the seeds are winged, to facilitate dispersal by wind. The wings are not usually large enough, however, to slow the rate of fall to less than 3 ft/sec. At that rate, a seed falling vertically 100 ft could travel sideways 2900 ft if exposed to a hurricane velocity wind of 88 ft/sec (60 mph). Wind velocity is always much reduced within a forest, so horizontal velocities of 10 ft/sec or less are more likely. With such velocities, small-winged seeds of birch or arborvitae could travel horizontally as much as 300 ft while dropping 100 ft.

Seeds of pines, spruce, firs, Douglas-fir, maples, ashes, and basswoods are also winged but are heavy enough to fall faster. Unless there are consistently strong winds from the same direction, most seeds of those trees come to rest within a few hundred feet of the parents.

Also in the pines it is necessary to remember that squirrels commonly chew off and drop entire cones. The seeds, although winged, do not become windborne. In eastern white pine during a normal seed year, almost no cones will be permitted to open normally on the trees; only during exceptionally heavy seed years is there much opportunity for wind dispersal.

Dispersion after reaching the ground must be considered in arborvitae and birch. Their small, winged seeds are often shed during the winter when there is snow on the ground. If, as sometimes happens, the snow is crusted, the light seeds can blow a few hundred feet on the surface.

Poplars and willows produce extremely small seeds with attached tufts of hairs. They have a very slow rate of fall and may blow a quarter mile. Also, they grow in river bottoms subject to flooding. If a flood occurs at the time of seed ripening, the seeds may float a few hundred miles downstream. They normally live only 2 to 3 days unless stored properly, and most of these waterborne seeds probably die.

Oaks, walnuts, and hickories produce very heavy seeds which drop nearly vertically and normally disperse only a few feet from their parents. However they are horded by squirrels, who may move them a few hundred feet. Nevertheless, average dispersion distance is probably small because the squirrels eat most of the seeds they carry away.

Some trees, such as junipers, cherries, and hawthorns, produce fruits that are eaten by birds. The birds usually digest the pulp rather than the seeds, which pass through the digestive tracts and may fall to the ground a considerable distance from the parent trees. Moreover, the bird-disseminated seeds often germinate best. Thus average dispersion distance may be high.

Coconut palms live naturally in the slightly saline soils next to tropical beaches. The seeds are large and can float in sea water. Furthermore they are durable and germinate after floating across an ocean.

Pollen Dispersion Mechanisms

WIND

The majority of commercially important timber trees of the temperate and subtropical zones produce wind-disseminated pollen. That applies to nearly all conifers and to the poplars, oaks, hickories, walnuts, birches, alders, and most ashes. In such species the male flowers open and shed their pollen during warm, dry periods—that is during the day. Little pollen is shed at night. Most pollen comes to rest on the day it is shed. The amount of airborne pollen at night is a small fraction of that present during daytime hours. Of the little pollen that manages to remain airborne overnight, even less retains its viability. All laboratory trials agree in showing that pollen viability is affected adversely by exposure to high relative humidities, such as occur at night.

In contrast to seed, pollen is microscopic. Douglas-fir pollen, among the largest, is about 100 μm in diameter. Grains of many trees are as small as 8–10 μm in diameter. The surface area of a pollen grain is so large in relation to its weight that its rate of fall is very much less than that of a seed. The principles governing pollen dispersion are more similar to those for smoke particles, dust, disease spores, etc., than to those which govern the movement of seed. They have been explained very well by Gregory (1945), who studied fungus spores.

Pollen is dispersed into the air stream, which as vertical movement (turbulence) as well as horizontal movement (wind). Some air movement is necessary for much pollen to be shed, since the anthers are rigid enough to retain their pollen if not shaken slightly. If the air is moving upward, the pollen can be carried aloft and travel great distances in the upper air stream. If the air is moving downward, the pollen will be carried down much more rapidly than if allowed to fall freely. As it approaches the ground, it will encounter a "capture zone" occupied by vegetation. In this capture zone the air is nearly still and the pollen can fall freely with virtually no horizontal movement. In a forest the capture zone is approximately as thick as the forest is tall.

According to Gregory's model, pollen is constantly moving up and down, and some is deposited with each downward movement. In deriving his theoretical dispersion curves, air turbulence was the principal factor considered—the greater the turbulence the less the dispersion distance. Entirely disregarded were such factors as wind velocity and direction, diameter or shape of the pollen grains, or rate of pollen fall in still air.

Both Gregory's empirical data about fungus spores and the empirical pollen dispersion data cited by Wright (1952) fit Gregory's theoretical model. Wright found no essential differences in pollen travel distances among species with large and small pollen grains, or between species with winged and unwinged pollen. He found that the average distance was slightly less (probably not significantly so) on some windy days than on other days with lower average wind velocity. One day, when the wind was of a constant direction throughout the study period, considerable amounts of pollen traveled to the leeward of the source tree.

INSECTS AND BIRDS

The basswoods, magnolias, yellow-poplar, most maples, and eucalyptus are among the insect-pollinated trees. Except for the maples, they produce perfect flowers. Also they produce slightly sticky pollen, which is shed in clumps rather than as dust. In the Amur corktree, the clumps are large enough that on a quiet day a person standing under a tree can actually hear the clumps as they fall onto leaves. The stickiness causes the pollen to adhere to the legs and bodies of insects which visit the flowers.

The above-named species have no special adaptations to promote crossing. An insect may crawl from anther to stigma of the same flower. After visiting the flower, the insect is most apt to visit a nearby flower on the same tree or on a neighboring tree. Thus the frequency of self-pollination may be high although genetic self-sterility mechanisms may inhibit self-fertilization.

Many trees in the pea family have butterfly-shaped flowers so constructed as to promote crossing, at least between different flowers on the same tree. In these trees, the anthers and stigma are enclosed within two petals called the "keel" and so positioned that an insect is apt to brush first against the stigma, then against the anthers, and then to pollinate another flower.

Beetles and flies usually visit flowers incidentally. This is not so with honeybees, which obtain all their food from pollen and nectar. Bateman (1949) described the flight and pollination habits of honeybees. A worker emerges from the hive in the morning, visits one flower, then another flower of the same species, then another flower of the same species, etc. To do this she may travel several hundred feet from hive to flower and back. She is most likely to switch species only when one is through flowering. Upon returning to the hive, she communicates her choice to other workers, which will usually make the same choice. Thus all bees from one hive are apt to work the same species on any one day—other species will be visited by workers from different hives. That is why beekeepers can sell basswood honey, clover honey, etc.

Honeybees carry this tendency even farther. They can recognize minor variations in flower structure and color. If a worker first visits a white-flowered plant of an otherwise pink-flowered species, she is most apt to visit other white-flowered plants, even though they are rare. To test this, Bateman planted various types of snapdragons in the same garden and collected seed later. He knew from artificial crossing experiments that many intertype crosses were as easily made as intratype crosses. Yet the actual percentage of intertype crossing was much lower than expected. In other words, the worker bees had recognized fine differences in flower appearance, and each worker had tended to concentrate on a single type as long as possible.

Measurement of Short Pollen Dispersion Distances

Seeds, being relatively large, can be observed and trapped easily. Specialized techniques are needed to study pollen dispersion.

The "marker gene" method is easiest and most satisfactory but is seldom used with trees because of the difficulty of finding mature stands containing a tree carrying such a gene. A marker gene is a dominant gene with a relatively large effect. It is especially useful if the effect can be seen in the seed before germination. That is the case with the "purple aleurone" gene of corn. No matter what the genotype of the female corn plant, a kernel resulting from pollination by a pollen grain carrying that gene is purple.

To use the marker gene technique, plant one or a few plants carrying the gene in the center of a field of normal plants. Then harvest seed from plants at varying distances from the marker plants, and count the proportion of "marker" to normal seeds (corn) or seedlings (most other plants). If the marker plant is heterozygous for the marker gene, half its offspring will carry the marker gene. Therefore, in the offspring of any one normal plant, the frequency of pollination by the marker plant is double the frequency of seeds or seedlings with the marker gene.

The marker gene technique has been used with corn, radishes, and cotton. Langner (1953) also used it with trees. He located a mature stand of Norway spruce which contained a few yellow-leaved trees in the center. He found these to be heterozygous for the dominant gene "aurea" which caused yellowing of the foliage of young seedlings. He collected and grew seeds from many normal trees scattered throughout the stand, then counted the number of yellow and normal seedlings in each family.

Another technique is trapping pollen shed by isolated trees. It has serious limitations in that it can be used only with wind-pollinated species. Also, since it is practically impossible to distinguish pollen shed by one tree from that of other trees in the same species (of often genus or family), the source tree must be isolated and growing under atypical conditions. For example, Wright (1952) used as source trees Lebanon cedar in Philadelphia and pinyon pine and Douglas-fir growing in Indiana, each chosen because of its rarity. Nevertheless this is the method most commonly used with trees (Wright, 1952; Stand, 1957; Wang et al., 1960; Sarvas, 1962).

To use this second method, the first step is to locate a suitable isolated source tree. The second step is to establish a series of pollen collecting stations on four (or more) transects running different directions (usually north–south and east–west) from the source tree. It is best if the stations are at geometrically increasing intervals from the source, e.g., 50, 100, 200, 400, 800, 1600 ft. Because all pollen-containing slides must be collected daily and within a relatively short time, the network of stations can not be too extensive nor can the stations be situated far above the ground. The third step is to place a sticky microscope slide horizontally at each station. The sticky substance is usually vaseline or gelatin, and is not very weatherproof. Therefore the slides should be collected and replaced daily, preferably within a 1–1½ hr time span. The fourth and final step is to count the number of source tree pollen grains on a given area on each slide. The area chosen should be such that it results in some frequencies of 500–1000 source tree pollen grains per slide. As an aid in identification, it is well to have a specially made reference collection prepared with the same type of sticky substance used on the field slides; the appearance of pollen varies dramatically with the mounting medium.

Wright (1952) was unsuccessful with a variation of the second method. Hoping to use a nonisolated source tree growing under typical conditions, he spent several hours collecting pollen and storing it; he later climbed the source tree to release the stored pollen after the species had flowered. The amount of pollen, although seemingly large, was too limited to give high enough recovery rates for statistical purposes.

The third or "tracer element" method has been proposed but has not yet proved practical. In this method a radioactive tracer element would be injected into the soil or sapstream in such a way as to be absorbed by the pollen grains. Then, source tree pollen could be distinguished from all other pollens. There are major difficulties, one of which is our lack of knowledge about the mineral nutrition of pollen. In working with radioactive tracers there is always a certain amount of background "noise" due to cosmic radiation, etc. So, to be able to positively identify source tree pollen, it is necessary that each grain be labeled with several atoms of the tracer element. Also, dispersion data are not reliable unless based on counts of hundreds or thousands of grains from a source tree. It has proved difficult to put the necessary amount of tracer elements into live pollen.

Measurement of Long-Distance Pollen Transport

Polunin (1951) carried sticky slides aloft in an airplane and exposed them far north of the Arctic Circle, several thousand feet above the ground. He demonstrated that small quantities of pollen exist high in the air and many hundreds of miles from the nearest possible source.

Erdtman (1954) ran a vacuum cleaner daily on a transatlantic voyage. He aimed the cleaner over the clean air coming over the prow of the ship and after each exposure cleaned the dirt bag so that its contents could be studied later. On every day, even though more than 1000 miles from shore, he could recover small amounts of pollen.

Mathematics of Pollen Dispersion

From studies such as those of Polunin and Erdtman it is not possible to determine the source frequencies or to determine the amounts dispersed to various distances. That has been possible only with short-distance studies, where transects are 1 to 2 miles long each way from a source tree.

In an actual pollen dispersion curve, as derived from short-distance studies, the frequencies are very high near the source, decrease rapidly at distances of a few hundred feet, then decrease more slowly to frequency 0 at infinite distance. With such a curve it is meaningless to calculate average dispersion distance in any one direction because the average would depend too much on location of the sampling stations.

Instead, it is best to consider that the pollen travels plus distances in one direction and minus distances in the opposite direction, so that the average distance is 0. Since the pollen travels in all directions, the curve should be a three-dimensional one, with the high point in the center and curved lines radiating in all directions, somewhat as streamers from a maypole.

Bateman (1947a,b) suggested that pollen dispersion data be fitted to the curve described by the equation

$$F_D = F_0 \, e^{-kD}$$

where F_D is the frequency at distance D, F_0 is the source frequency which equals the frequency at distance 0, e is the base of the natural logarithms, and k is a constant. The logarithmic statement of that formula is (if all logs are to base 10):

$$\log F_D = \log F_0 - 0.4343kD$$

That formula is similar to the general equation for a straight line, $Y = a + bX$. Following suggestions furnished by Sewall Wright, J. Wright (1952) calculated the slope (b) and intercept (a) as follows.

$$b = 0.4343k = \Sigma \, (wXY) - \frac{\Sigma \, (wY) \, \Sigma \, (wX)}{\Sigma \, w} \left(\Sigma \, (wX^2) - \frac{[\Sigma \, (wX)]^2}{\Sigma \, w} \right)^{-1}$$

$$a = \log F_0 = \Sigma \, \frac{(wY) - b \, \Sigma \, (wX)}{\Sigma \, w}$$

where X is distance, Y is log frequency, and w is weight which equals frequency.

Except for the addition of the weight, w (frequency), the calculations are identical with those used in any simple regression problem.

Sewall Wright thought it desirable to weight each observation directly as the pollen frequency to compensate for the fact that the low-frequency counts at some distance from the source were less reliable than the high-frequency counts close to the source. Other weighting schemes are possible. For example, each observation could have been weighted inversely as the distance or as the square root of the frequency.

As already mentioned, $b = 0.4343k$, which is the slope of the straight line if the data are transformed logarithmically and plotted. But for the untransformed data, the constant k is of greater interest because of its relation to the standard deviation (σ), as follows.

$$\sigma_D{}^2 = 2/k^2$$

This σ applies to the curve

$$F_D = F_0\, e^{-kD}$$

rather than to the normal curve

$$F_D = F_0\, e^{-kD^2}$$

For that reason, the areas and ordinates as given in ordinary statistical tables do not apply here. They may be calculated, however, by means of calculus.

If the curve described by the formula $F_D = F_0\, e^{-kD}$ is rotated around a vertical axis to form a solid, 91% of the volume of that solid is contained within $\pm 1\sigma$ of the axis. Or speaking in terms of pollen, 91% of the pollen released by a source tree will come to rest within a distance of 1σ of the source tree. Thus, σ makes a convenient statistic when describing pollen dispersion distances.

Measured Pollen Dispersion Distances in Some Trees

Among the earlier workers on pollen flight of trees were Dengler and Scamoni (1944). They did not work with isolated trees, but assumed that pollen would fly only to the windward. Therefore they substracted the average frequency to the leeward from the average frequencies to the windward to obtain estimates of the amount contributed by their source trees at the wind stations. Some of their assumptions now seem incorrect, so their data are discounted.

Among those who have worked with isolated trees are J. Wright (1952), Langner (1953), Strand (1957), Wang *et al.* (1960), and Silen (1962). Their results are summarized in Table 5.3. Except for American elm, the results are consistent in showing that most pollen comes to rest within a few hundred feet of its source. The American elm data

TABLE 5.3

Pollen Dispersion Distances in Forest Trees

Species	Distance within which 91% of the pollen comes to rest (ft)	Reference
Atlas cedar	238	Wright (1952)
Lebanon cedar	145	Wright (1952)
Green ash	55	Wright (1952)
Norway spruce	126	Wright (1952)
Norway spruce	>300	Langner (1953)
Norway spruce	>300	Strand (1957)
Pinyon pine	55	Wright (1952)
Slash pine	225	Wang et al. (1960)
Scotch pine	>200	Strand (1957)
Douglas-fir	59	Wright (1952)
Douglas-fir	>500	Silen (1962)
American elm	2200	Wright (1952)

are somewhat suspect in that there were many native American elms within 1 to 2 miles of the source trees; they could have accounted for most of the pollen trapped at 1000 to 3000 ft from the source.

Where these workers considered wind velocity, there was no relation between wind velocity and dispersion distance. Nor were there obvious relations between dispersion distance and tree height or pollen grain size.

The pollen dispersion distances measured for trees are of about the same magnitude as those determined for herbs, such as radish, cotton, and corn.

Implications of Dispersion Data

Occasional long-distance transport of seed is important from the ecological standpoint because it explains why species can invade bare areas rapidly. A few seeds may travel far to an unvegetated area and form the nucleus of a forest. However, once the forest is established, it supplies most of its own germ plasm, and very low migration rates from outside are of little consequence. The data from short-distance experiments are of most value genetically.

The fact that most pollen comes to rest within a few hundred feet of its source means that those ecologists who have studied plant migration by analysis of the pollen deposited in peat bogs have probably sampled ancient forests within short distances of the bogs.

The relatively short pollen dispersion distances explain why investigators have found that up to 25% of the open-pollinated seed produced by some trees resulted from self-pollination. Also, in a natural stand, open-pollinated seed does not result from pollination by a random sample of the entire species, but instead by close neighbors. Therefore it can be assumed that most trees growing within a few hundred feet of each other are related and should tend to be more similar genetically than trees growing several miles apart. That expectation is not always realized, however.

One can assume short average dispersion distances when planning a seed orchard. However, it is also necessary to consider the size of the orchard in relation to the surrounding natural forest and the age at which seed production is expected. A 10-acre orchard, totally isolated except for a few large trees several hundred feet away, should experience little contamination from outside pollen once the seed-orchard trees are large enough to produce reasonable amounts of pollen. However, Squillace (1967) found that a 400-ft isolation strip around a 5-acre slash pine seed orchard was ineffective during the orchard's early years. The orchard was located in a region of extensive mature slash pine forests which produced very large amounts of pollen. Even though the percentage of pollen which traveled more than 400 ft may have been small, the actual amount of pollen arriving in the orchard from outside was large enough to ensure normal seed set after removal of all male flowers from the orchard trees.

Isolation

Isolation is the reverse of migration. For populations to become genetically different, some isolation is necessary. Otherwise continued interchange of genes will counteract the effects of selection. There are several types of isolating mechanisms, discussed below.

Polyploidy

A polyploid is a plant with more than two complete sets of chromosomes. Tetraploid ($4n$) plants are the most common. A tetraploid can cross with a diploid ($2n$) to produce triploids ($3n$). However, triploids are usually sterile and do not form a pathway by which $4n$ and $2n$ plants can exchange genes. Therefore, doubling of the chromosome number is one way in which different parts of a population can become reproductively isolated in a single generation.

Differences in Flowering Time

Assume that a tree of genotype AA normally flowers during a given 4-day period in the spring. Assume also that mutation A to a occurs, resulting in tree Aa which flowers 1 day earlier. There will be only 75% overlap between the AA and Aa trees in blooming dates, but that is sufficient for the two types to intercross. Even if the Aa types self, producing aa trees which bloom 2 days earlier, the aa and AA trees can still intercross.

This example shows that it is difficult to envision a genetic mechanism by which mutations for earlier or later flowering can by themselves result in reproductively isolated populations. However, if such mutations occur and parts of a population start to migrate in opposite directions, selection can fix early flowering genes in one part and late flowering genes in another part. After several generations the differences may be large enough to prevent interbreeding if the subpopulations again merge.

This is presumably the way in which the differences in flowering time between silver (early) and red (later) maples arose. By storing pollen, it is easy to cross these species artificially. Yet in nature they occupy the same ranges without interbreeding.

Pollinating Mechanisms

Honeybees differentiate among flowers that differ very slightly. Once a worker visits a particular type of flower, she is apt to visit next the same type of flower. This makes possible an isolating mechanism not found in wind-pollinated plants.

Assume a population of many thousand AA plants, in a few of which the mutation A to a occurs, affecting flower color. A worker who visits one Aa plant will search for another Aa plant, thus producing many more aa plants than would be expected due to chance alone. The next generation some workers will visit mostly AA plants, some will visit mostly Aa plants, and a few will visit mostly aa plants. Thus the aa type of plant will tend to be self-perpetuating within the basic AA population. By a similar process, the aa subpopulation may give rise to an even more different a'a' subpopulation. Thus, races that differ in flower color or structure may arise without geographic isolation. Once the races are reproductively isolated, they may continue to diverge in vegetative characteristics. The same may happen in plants that are pollinated by birds that can recognize fine differences.

TABLE 5.4

Relation between Pollinating Agent and Percentage
of Diagnostic Traits Relating to Flowers

Pollinating agent	Percentage of diagnostic traits related to floral structure or color
Bees, long-tongued flies	40
Birds	37
Promiscuous insects	15
Wind	4

Grant (1949) pointed out the importance of pollinating mechanisms in evolution. He studied a very large number of species and in each counted the number of diagnostic traits related to flower structure or color. Diagnostic traits are those that are clear-cut enough to be used by taxonomists in identification. He found a clear relation between pollinating agent and the frequency of floral diagnostic traits. This is shown in Table 5.4.

Grant's generalization can be demonstrated easily in trees. Eucalypts are primarily bee-pollinated, and the primary botanical subdivisions of the genus are based on a floral character—structure of the anthers. Willows are insect-pollinated, whereas poplars in the same family are wind-pollinated. It is difficult to identify many willows without flowers, yet flowers are rarely mentioned in poplar identification keys. The maples, elms, and magnolias are insect-pollinated, and in each of these genera species can be identified as easily or more easily with floral characters as with leaf or growth characters. However, flowers are not needed for the identification of wind-pollinated pines, spruces, or oaks.

Range Gaps

Species inhabiting arid mountainous regions normally have much disrupted ranges with frequent range gaps. The populations inhabiting separate mountains are reproductively isolated and thus can become genetically differentiated as the result of selection or random processes.

The state of Michigan is composed of two peninsulas, each nearly surrounded by one or more of the Great Lakes. The peninsulas are separated from each other by the Straits of Mackinac, which are about 4 miles wide from tree line to tree line. That separation has been sufficient to result in large genetic differences between Upper and Lower

Peninsula races in four different conifer species (Wright, 1972). Data on the effectiveness of smaller range gaps are scanty.

Isolation by Distance in Continuous Populations

Most commercially important species have continuous distributions, at least in large portions of their ranges. Via pollen and/or seed, genes can migrate from one end to the other end of a continuous distribution, and thus prevent genetic differentiation. However if, as seems to be the case, average pollen or seed dispersion distances are small, per generation migration rates are slow, and distance itself can be an isolating agent.

Sewall Wright (1943, 1946) has furnished a theoretical basis for calculating the effects of isolation by distance. To do this, he introduced the concept of the "neighborhood," as the largest subpopulation within which mating takes place at random. Mathematically a neighborhood can be defined as a circle of radius σ if pollen and seed are dispersed equally and if the population is already continuous. (See Table 5.3 for sample values of σ.) If only pollen is dispersed, the circle has a radius of only $\sigma/(2)^{1/2}$. For species with linearly continuous ranges along streams, seashore, etc., the neighborhood is a strip $2\sigma(\pi)^{1/2}$ long (pollen and seed dispersed equally) or $\sigma(2\pi)^{1/2}$ long (only pollen dispersed).

According to Sewall Wright's calculations, subpopulations of a continuous population tend to become most different genetically if one of the following conditions is met.

1. Neighborhood size is small due to short pollen or seed dispersion distances. Short dispersion distances mean slow gene migration rates, and therefore more opportunity for genetic differentiation.

2. Neighborhood size is small due to low population density. With low population density gene migration is slowed down because there are fewer pathways between parts of the population.

3. A species has a linearly rather than an areally continuous range. With a linearly continuous range there are only a limited number of pathways by which a gene can migrate from place to place. Those pathways are direct. With an areally continuous range there are the same number of direct pathways and many times that number of indirect pathways, so that a gene can migrate faster from one place to another.

J. Wright (1952) used empirical pollen dispersion data in combination with estimates of population density to estimate neighborhood sizes in some tree species. He estimated neighborhood sizes of 1 to 25 trees in white ash, pinyon pine, and some other uncommon species with areally continuous ranges; also in several species with linearly continuous ranges. With such neighborhood sizes, genetically different local races might be expected to develop within a few miles of each other, even without strong differences in selection pressure. He estimated that most commercially important temperate-zone timber species had intermediate neighborhood sizes of 25 to 200 trees, permitting slight to moderate local genetic differentiation within distances of 50 to 100 miles. In such species as jack pine and slash pine, which commonly occur in dense pure stands, neighborhood sizes of 200 to 1000 trees may be frequent; with such large neighborhood sizes local genetic differentiation should be slight unless there are large differences in selection pressure.

Estimations of δ, population density, and selection pressure are subject to such great experimental error that it is impossible to obtain exact confirmations of theoretical expectations from empirical studies. However, there have been a few general confirmations. Pinyon pine, with its short pollen dispersion distance and its very low population densities, has become so diversified genetically as often to be considered as four separate species inhabiting the same general region. Racial differentiation seems more pronounced in species with linearly continuous ranges, such as green ash and eastern cottonwood, than in several other eastern American hardwoods which have areally continuous range. The amount of local genetic diversity is limited in species, such as jack pine and slash pine, which have large neighborhood sizes and areally continuous ranges.

Geographic versus Altitudinal Isolation

Two points on the same mountain differing in elevation by 2000 ft probably differ climatically from each other by as much as two points several hundred miles apart. Differences in selection pressure may be similar in both cases. However, points on the same mountain may be only a few miles apart and subject to constant gene interchange, whereas the rate of gene interchange between points several hundred miles apart must be smaller.

Therefore it is to be expected that differences in elevation per se will result in much less genetic differentiation (perhaps in none) than differences in geographic location, even when climatic differences due to elevation are as great as those due to geographic separation.

Among the most critical experiments on altitudinal races are the following. Callaham and Hasel (1961) grew ponderosa pines from seed collected along a transect up the Sierra Nevada Mountains of California. The slope of the transect was gentle, and the high- and low-elevation ends of the transect were separated by more than 50 miles. At low-

and medium-elevation test sites the high-elevation trees grew more slowly than the low-elevation trees, but the differences were much less than reported between Washington and California trees in other studies. Langlet (1936) grew Scotch pines from many parts of Sweden, where there is a gradual increase in elevation from sea level along the coast to 4000 ft along the Swedish–Norwegian border. Thus at any given latitude high- and low-elevation sources were separated by 100 to 150 miles. At a given latitude, percent dry weight of the foliage increased with elevation of the source, but the differences were much smaller than found for seedlings grown from south Swedish and north Swedish seed. McGee (1974) grew northern red oak from seed collected at four elevations within 60 miles of Asheville, North Carolina; he also tested the seedlings at four different elevations. Average dates of bud burst were April 18, 20, 25, and 29 for trees originating at 40, 810, 1140, and 1380 m, respectively. Those differences are as great as found in other studies in which northern red oaks were grown from acorns collected in many different states.

In contrast to those authors, Kung and Wright (1972) failed to find consistent intraregional elevational trends in four western American species (ponderosa pine, the limber–southwestern white pine complex, white fir, and Douglas-fir). In each case they grew trees from seed collected in many parts of the species natural ranges. Their experiments included several pairs of seedlots collected at elevations differing by 2000 to 3000 ft within the same general area. Most of the seeds were collected in very mountainous country with great elevational differences within a few miles. Hence the elevational differences were not accompanied by geographic isolation. They found, for example, that trees grown from seed collected at high elevations in southern Colorado differed from trees grown from seed collected at low elevations in northern Idaho, but found no significant differences among seedlots originating at different elevations within southern Colorado or northern Idaho.

There are provenance tests for several species inhabiting the non-mountainous portions of eastern United States, and including seedlots collected at elevations differing by a few hundred feet within the same state. No consistent elevational trends have been reported in such studies.

Study Questions

1. Why is it likely that any single mutation will ultimately become extinct?
2. Why are differences in mutation rates usually disregarded when considering rate of evolutionary change?
3. In what respects are mutation and migration similar? Different?

4. What are the joint effects of migration and selection on genetic differentiation? Why are altitudinal races less well-developed than geographic races?

5. What factors affect the distance viable seed is dispersed in pine? Oak? Eucalyptus? Cottonwood? Willow? Birch?

6. How does pollen release vary with time of day? Weather conditions? Pollinating agent?

7. Describe Gregory's hypothesis of pollen dispersion. What factors are considered most important? Not important?

8. The eucalypt genus is largely bee-pollinated, is nearly confined to $\frac{1}{3}$ of the continent of Australia, and contains somewhere between 300 and 500 species. How is it possible that so many species developed within such a limited area?

9. The cottonwoods and balsam poplars form one major subdivision of the poplar genus. In general, there is just one species of cottonwood or balsam poplar in any given region. The willow genus belongs to the same family and occupies roughly similar sites. Yet there are usually several willow species in any given region. Why?

10. Why do taxonomists use more flower characteristics when identifying maples, elms, and eucalypts than when identifying larches or oaks?

11. Describe two different techniques for measuring pollen dispersion distance. What are the advantages and limitations of each?

12. How does the formula used by Bateman to describe pollen dispersion curves differ from the formula for the normal equation? What is σ? Name the two reasons why the percentage of observations included within $\pm 1\sigma$ of the source is 91% with the Bateman formula rather than 67%, as for the normal curve.

13. What is the general range of σ in forest trees? How does that compare with some herbaceous plants? Give implications of such a range of σ values.

14. If σ is so small, why did Squillace find a great deal of contamination by outside pollen in a seed orchard with a 400-ft isolation strip?

15. List and describe three isolating mechanisms in forest trees. Consult the chapter on polyploidy and tell which isolating mechanism is probably most operative in maple, eucalypts, pines, oaks, basswoods, spruces, willows, and poplars.

16. Why are differences in flowering time a secondary rather than a primary isolating mechanism?

17. How small a range gap has been effective in promoting geographic differentiation in forest trees?

6

Useful Techniques—
Controlled Pollination

Controlled pollination is the artificial transfer of pollen from one flower to another under conditions such that both male and female parents are known. It is primarily an experimental technique used in full-sib family selection and in hybridization between races or species. However, S. K. Hyun, with low labor costs, has used artificial pollination to produce commercial quantities of hybrid pine seed in Korea. Artificial production of willow, poplar, and some birch hybrids on a large scale is also feasible.

Briefly the steps involved are selection of parent trees, climbing to reach the flowers (sometimes unnecessary when working with small grafted trees), placement of pollination bags, pollination, removal of pollination bags, placement of fruit-protection bags, harvesting of seed, and cleaning of seed.

Detailed procedures vary considerably among species. They vary also with personal preference and available equipment. Often the procedures are described briefly in the "Methodology" sections of larger papers. Several pre-1960 papers devoted solely to techniques are cited in Wright (1962). Among the recent papers giving detailed information about specific species are those by Livingston and Ching (1967), Gerhold (1968), and Forbes (1974). The book "Seeds of Woody Plants in the United States" (USDA, Forest Service, 1974) is very good on seed problems.

Selection of Parent Trees

Parents should be selected primarily on the basis of form, growth rate, etc. But if seed is the goal they must also be selected for fruitfulness. It is desirable that female parents produce enough flowers to support 15 to 20 pollination bags, each including several flowers or flower clusters. In some species (e.g., white ash), many trees habitually produce flowers but no seed. In such cases, individual tree flowering and fruiting records maintained for 3 to 5 years are helpful.

Generally, selection for flowering ability can be done from the ground 1 to 2 weeks prior to bagging, as female flower buds often differ enough from vegetative buds as to be recognized from a distance with the aid of field glasses. However, climbing is necessary in some spruces and maples in which female flower buds can be recognized only by close inspection.

Climbing

Most tree flowers are borne some distance from the ground and can be reached only by climbing. Because of the safety hazards, most breeders try to keep their climbing to a minimum and work on trees less than 40 ft tall (see Figs. 6.1 and 6.2).

Avoidance of Climbing by Grafting

As discussed in Chapter 7, grafted trees of some species fruit earlier than do seedlings. This makes it possible to do controlled pollination work on trees 10 to 20 ft tall rather than 20 to 40 ft tall. Also, the grafted trees can be planted in one convenient location instead of being scattered over a large area.

Whether or not this is a desirable procedure depends on the species. The ease with which controlled pollinations can be performed on small trees may more than make up for the time spent in grafting and the delay in starting the pollination work. On the other hand, seedlings of some northern species flower early enough to be climbed with ease.

Apparatus

Three-legged step ladders, 6 to 12 ft tall, can be used with many hardwoods which bear flowers over the entire crown. They can also be used with many young pines or spruces, or with grafted trees.

Fɪɢ. 6.1. Heavy flowering on a grafted Norway spruce tree in Finland. The erect conelike structures are female flowers (strobili) at a stage of maximum receptivity. At that stage they are about 2 in. (5 cm) long and brilliant red. After pollination the scales will close and the cones will become pendant. (Photograph courtesy of Lauri Kärki of the Finnish Forest Tree Breeding Foundation.)

Truck-mounted ladders or hoists of the types used by firemen and telephone linemen are commonly used in seed orchards located on level ground with large spaces between trees. They are expensive and difficult to use under forest conditions (see Fig. 6.3).

Wood or aluminum sectional ladders are commonly used. The sec-

Fig. 6.2. Grafted Norway spruce in Finland with paper pollination bags in place. The grafted trees of this species fruit much earlier than seedlings, making controlled pollination a relatively easy task. The Kraft paper bags are weatherproof enough to protect the flowers for 3 to 4 weeks. (Photograph courtesy of Lauri Kärki of the Finnish Forest Tree Foundation.)

tions are usually 5 to 10 ft long and so designed as to fit on top of each other. Each section has two straps by which it can be fixed to the bole of the tree. The climber carries sections up and attaches them to the tree one at a time until he reaches large live branches.

A device designed in Sweden consists of two stainless steel hoops, each with an attachment by which it can be strapped to the climber's foot. The inner surface of the hoop has rubber grips or steel cleats which hold tightly to the bark when pressure is applied to the foot side of the hoop. These are used in conjunction with a safety belt. By advancing the belt, one hoop, and then the other hoop in succession a climber "walks" up a tree (see Fig. 6.4).

Rope climbing was practiced for many years at Placerville, California, but requires a high degree of skill and is not commonly used elsewhere. The climber throws a stout rope over a large lower branch,

Fig. 6.3. Picking ponderosa pine cones from a truck-mounted ladder. This particular model is hand operated, weighs about 800 lb and can be extended 35 ft. For travel, the unextended ladder projects over the cab of the $\frac{1}{2}$-ton pickup truck. Truck-mounted ladders of various types are often used for pollination and fruit picking in seed orchards. (Photograph courtesy of Ralph Read of the United States Forest Service, Lincoln, Nebraska.)

ties one end to his safety belt, and climbs from the other end. He does that by looping the rope around his instep in such a way that the loop becomes tight as the foot is lowered and loose as the foot is raised.

Ropes are also used to facilitate movement while in the tree crown.

FIG. 6.4. The Swedish climbing device consisting of two hoops attached to a climber's feet. This climber is "walking" up a Scotch pine tree in Finland with his right foot constantly above his left foot. (Photograph courtesy of Finnish Forest Tree Breeding Foundation.)

In this case, a climber goes up the bole, attaches the rope near the top, and keeps the other end tied around his waist. Then, if necessary to work at some distance from the bole, the rope prevents a fall to the ground.

Spurs, of the type used by telephone linemen, are occasionally used in conjunction with safety belts on trees with thick barks. However, they are not favored because they cause damage, especially to trees which must be climbed repeatedly over a period of years.

Safety belts are always used when climbing in trees with single central stems. When a climber reaches the proper height, the belt is attached around the bole and climber. This is not only a safety precaution but leaves the climber with both hands free.

Safety Precautions

Experienced tree breeders have suffered broken backs, broken ribs, and near-electrocution (from working with a metal pole near a power line). An inexperienced tree breeder was killed in a fall from a very large tree. Hence, stringent safety precautions are necessary. Among them are the following.

Climb small trees if possible.
Use a safety belt.
Work in pairs so that help is available if needed.
Know the holding power of dead and live branches and of the central stem of each species.
Use extra care if near power lines, or above rocks or concrete.
Avoid climbing on rainy or windy days.
Use good ladders and set them firmly.

On most commercial tree species, climbing can be enjoyable and safe if suitable precautions are observed. There are a few trees, however, in which controlled pollination work is inherently dangerous and has been limited by the safety factor. Most true firs bear female flowers only near the tip of the crown so that a climber is literally at the tree top when pollinating them. Most larches have brittle branches and produce relatively few flowers on lower branches. Most eucalypts grow so rapidly (10+ ft per year) that flowers are apt to be far from the ground.

Bagging to Prevent Outside Pollination

Time to Apply and Remove Pollination Bags

Bags are usually put in place 1 to 2 weeks before time for pollination. A good practice is to put the bags in place as soon as the flower buds are recognizable as such and before the female flowers are actually exposed to the air. This does not pose a problem in most pines, poplars, ashes, etc., in which the female flower buds are easily recognizable before they open. There is a problem in the spruces, however. In them female flower buds are borne laterally on upper branches, as are vegetative buds, and the two types are very similar until nearly ready to open. In such a case it may be necessary to dissect a few buds per tree and then apply pollination bags on upper branches of flowering trees in the hope that each bag contains female flower buds as well as vegetative buds.

In any given pollination bag, the flowers usually remain receptive for a few days only. If the conelet scales have grown shut or the stigmas have withered, receptivity is probably past and it is safe to remove the bags. Stage of maximum receptivity is shown in Figs. 6.1 and 6.5. In many species which are sensitive to bag damage, the sooner that is done, the better. However, in a few species, such as red maple and eastern white pine, pollination bags can remain in place until the fruits mature.

FIG. 6.5. An unusual hermaphroditic ament of hybrid willow (*Salix matsudana* × *S. alba*). The male flowers with unopened anthers are at the top, and the female flowers with broad stigmas are at the base. These flowers are at a stage of maximum receptivity. (Photograph courtesy of Mirko Vidaković of the University of Zagreb, Yugoslavia.)

Flower Bud Recognition

As compared with vegetative buds, female flower buds on the same tree usually differ in size, shape and location. In pines they are borne at the ends of the expanding shoots but slightly to the side of the terminal bud or (multinodal species) laterally at the nodes of the newly expanding growth; they are usually smaller and blunter than vegetative buds. Larch female flower buds are borne laterally on "short shoots," are ovoid and larger than vegetative buds. Red maple flower buds are borne laterally in groups of 2–3, are globose and slightly larger than vegetative buds. Flower buds of Asiatic magnolias are borne terminally, are larger than and expand sooner than vegetative buds. In other words, species vary and a breeder must become acquainted with each new species.

Emasculation

Emasculation means depriving masculinity, in other words removal of male flower parts. It is usually not needed in dioecious or monoecious genera in which the sexes are contained in different flowers and usually not included in the same pollination bag.

Emasculation may be needed in perfect-flowered species, however. The methods vary with flower size and structure. Magnolias and yellow-poplar have large, perfect flowers and tend to be protogynous (i.e., stigmas receptive before pollen is shed). The sepals, petals, and anthers can be cut (with fingernail or knife) from flower buds that are nearly ready to open, and the exposed pistils can be pollinated immediately. Eucalypts have small perfect flowers, and each flower consists of an ovary topped by a staminal ring with stamens on its inner surface. Atop the staminal ring is an operculum (lid) which seals the flower bud prior to the time of pollen shedding. Emasculation is accomplished by cutting through and removing the staminal ring just before the operculum falls.

Elms and basswoods have small, perfect flowers. Emasculation is tedious because it involves removal of several stamens per flower by means of tweezers, and therefore it is to avoided if at all possible. That can be done by determining the self-compatibility of many trees and using as female parents only those that are self-sterile.

Bagging Materials

Pollination bags are commonly 8 to 12 in. long and large enough in diameter to enclose a branch containing several female flower buds.

Several materials are used. Whatever the material, it should be weather-proof and permeable to water vapor. Polyethylene and rubber, which are impermeable, are unsatisfactory because water droplets condense on the inner surface and cause the pollinating equipment to become clogged. Bags should also be light in color to prevent overheating (Fig. 6.6).

Synthetic sausage casings are most common. The material is purchased as a seamless tube that can be cut to any length and stapled at one end. They are satisfactory with most pines but tend to overheat and should be covered with paper bags when used with spruce. They are transparent, which makes it easy to observe flower development.

Terylene is a trade name for a nonwoven cloth-like material that

FIG. 6.6. Control pollinating loblolly pine. Note that the needles are clipped (upper branch) before a branch bearing female strobili is enclosed in a sausage casing pollination bag. Note also that cotton surrounds the stem where the bag is tied to the branch. (Photograph courtesy of Leroy C. Saylor of North Carolina State University at Raleigh.)

is white and very permeable to water vapor. Terylene bags do not become as hot inside as do sausage casings, but some workers have found that they are not absolutely pollen-proof. Some paper bags are satisfactory, if made with waterproof glue. Weatherproof and pollen-proof bags can also be made from vegetable parchment or canvas. Transparent windows are sometimes inserted in such bags to permit observation of flower stage development.

Gerhold (1968) uses small 1×3-in., nonwaterproof, clear plastic bags (termed mini-bags by Gerhold) for pine. These are just large enough to slip over the tip of a branch and enclose 2 to 3 flowers each. They are fitted over slit plugs made of Styrofoam, which make a tight seal with the branch (Fig. 6.7).

Application of Pollination Bags

In pines, female flowers are borne terminally on the succulent new growth. Pollination bags may be placed to entirely cover the new growth and tied around the sturdier growth of the previous year, but if so the new growth becomes crowded and misshapen. Thus many breeders fasten the lower ends of the bags halfway up the new growth, making sure to enclose several shoots per bag (one shoot per Gerhold mini-bag). In species with laterally borne female flowers, bags are placed so that their bases are fastened around sturdy old growth.

Bags must be tied securely in place by means of string, soft wire, or patented paper-covered wire (Twist'Ems). Otherwise they tend to whirl and cause branch breakage on windy days or to blow away. The seal between bag and branch should be made pollen-proof with cotton or Styrofoam.

Reaching Branches

A breeder working near the bole of a tree may find that the twigs bearing flowers are beyond his reach. Thus it is convenient to carry a 3 to 6 ft wire, with hooks at both ends, to reach the outer branches (Fig. 6.8).

Pollen Handling

Pollen Forcing

If two trees differ in flowering time, it is easiest to use the earlier-blooming one as the male parent or to store the pollen of the late-bloom-

FIG. 6.7. Control-pollinating Scotch pine using the small mini-bags recommended by Gerhold. Each bag encloses a single flowering shoot. By slight pressure on the rubber bulb a cloud of pollen is injected into the bag around the enclosed female strobilus. (Photograph courtesy of Ralph Read of the United States Forest Service, Lincoln, Nebraska.)

ing one for use the next year. However, there are occasions when it is desirable to force (i.e., hasten) flowering of the late blooming tree.

Several hardwoods including elms, poplars, and willows, may be forced easily by bringing large branches into a greenhouse 4 to 6 weeks prior to normal flowering time and placing the branches with their cut

FIG. 6.8. Control pollinating 20-year-old eastern white pine in Ohio. This is as close to the top as Dr. Kriebel dares to climb in this species. He uses a wire hook to bring branch tips within reach. Each sausage casing bag encloses several expanding shoots and up to 20 female strobili at the tip of a branch. (Photograph courtesy of H. B. Kriebel of the Ohio Agricultural Research and Development Center.)

ends in water or nutrient solution. Forcing by 2 to 3 weeks is possible if the ends are recut and the water changed occasionally. Some spruces can also be forced in this way.

Pines and some other trees are best forced by placing sausage casing bags over male-flowering branches 2 to 3 weeks prior to anthesis. These cause the temperature around the branch to become higher and to thus hasten flower development. Forcing by a few days is possible in this way.

Pollen Collection

Most controlled pollination work is done as part of a family selection program (see Chapter 9) in which the breeder wishes to test the offspring of many different females and males. Usually it is better to obtain 100 to 200 good seeds from each of many crosses than to obtain larger numbers of seeds from fewer crosses. At least in cool temperate regions the flowering period for any one species is short, and the crossing work must be done in a few days. Hence pollen collection techniques should be simple.

If one is using many different male parents, it is desirable to limit the amount of pollen collected from any one to a reasonable amount. Usually 10 to 20 cm³ per tree will suffice. Collection of greater amounts of pollen not only requires extra time but may complicate the problem of keeping the pollen dry.

A very simple procedure useful for species with large male inflorescences (most pines, spruce, and poplars) was suggested by John W. Duffield. He collected nearly ripe catkins, packed them loosely in a sausage casing pollination bag, and hung the bag in the open to permit drying. It could be hung in a room, in the back of a truck, or even from the branch of a tree. When pollen shedding started, sometimes in a few hours, he inverted the bag and poured the pollen through cheesecloth into a glass vial. With this technique he could collect pollen and use it on the same day, even though far from a laboratory. Also, contamination was minimal.

Several other conifers produce small male catkins. With them it is necessary to cut entire branches, spread them on paper to dry, remove the branches and screen the pollen. A somewhat similar technique is used with many hardwoods that produce small amounts of pollen. Branches are cut, placed with their bases in water, and the pollen is allowed to ripen and fall onto sheets of paper.

More sophisticated procedures are used by many breeders. These include the canvas extractors developed by the Institute of Forest Genetics at Placerville, California. The extractor consists of a canvas bag fitted with a window and a funnel at the lower end. Inside the funnel is a wire screen and a rubber hose fitted with a pinch clamp is attached to the bottom of the funnel. The extractors are hung in a warm dry room and shaken whenever pollen is needed.

Some special cases should be mentioned. Pollen of yellow-poplar loses viability so rapidly (within a few hours) that it is best to carry flowers, slit anthers which are nearly ready to open, and use newly opened anthers as pollinating brushes. Norway maple produces such small amounts of pollen as to make its collection impractical. In that species, too, it is necessary to carry fresh flowers and use opened anthers as pollinating brushes.

Pollen to be applied through a blower is usually screened to remove debris. Fine cheesecloth, which can be discarded after use, is convenient for this purpose.

Pollen is dustlike and therefore subject to contamination, especially when dealing with several different lots. A few simple precautions usually suffice. Wash hand, table surfaces, vials, applicators, etc., between different lots of pollen. If extracting in the open, work in a closed room and keep individual lots of pollen well separated. If a hood is available, work in it with a blower on. It is well to check the amount of contamination by exposing samples of talcum powder in the extraction room and later applying this sample to otherwise unpollinated female flowers. If sound seeds result, precautions against contamination must be tightened.

Pollen Storage and Shipment

A pollen grain is living although dormant. It stores best if kept cool and dry. For day to day storage it can be kept in cotton-stoppered vials at a relative humidity of 25% or less and at temperatures of 3° to 4°C, or at temperatures of −10° to −20°C if a deep-freeze is available. It is best to keep the vials in several small desiccators which can be brought to room temperature quickly before being opened. This prevents condensation on the individual vials.

When planning to use the same lot of pollen on successive days, it is best to store it in several small vials, each of which can be discarded after a day's use.

It is always desirable to use fresh pollen, but it is sometimes necessary to store pollen for a year or more before use. Grape, spruce, and pine pollen stored a year or more at temperatures around −18°C and at very low relative humidities have resulted in sound seed, although the seed sets were not always as high as with fresh pollen.

Livingston and Ching (1967) found that the type of pretreatment given freshly collected pollen affected its longevity during prolonged storage. According to their results, Douglas-fir pollen could be stored

2 years at −18°C if (1) fresh pollen was air-dried to 8% moisture content for 4 hr, (2) pollen was then stored 30 days at 0°C, (3) pollen was then placed in vacuum for 1 hr at −77°C, and (4) pollen was then sealed in airtight vials.

Pollen of many conifers, poplars, and perhaps other species can be shipped successfully over long distances if dried, placed in a pollen-proof envelope, and shipped via air.

Pollen Testing

Laboratory tests are time consuming and frequently overestimate the ability of pollen to effect fertilization when applied to actual tree flowers. Therefore, laboratory testing is most useful when determining the viability of pollen which has been stored for 1 year or more and as a part of special studies aimed at better storage procedures. A common procedure, which has given good results with several kinds of trees, is as follows. (1) Prepare a 2% agar medium containing 5 to 10% sucrose, sterilize, solidify in a thin layer, cut into small blocks, sprinkle them with pollen and place blocks in Petri dishes. (2) Place a few extra drops of water in the dishes and cover them. (3) Incubate 48 to 72 hr at 24°C. (4) Count germinating pollen gains at 100 × magnification.

Pollination Procedures

Judging Receptivity

Lacking exact information about a particular species, the time when female flowers are receptive and should be pollinated is:

Similar to the time of pollen shedding on the same tree (most species but not some maples).

The period when the stigmas are fresh and not withered (most hardwoods) (Fig. 6.5).

The period, usually 2 to 3 days after the strobili emerge from the flower buds, when bracts and scales are equally long, and there are plainly visible spaces between scales (most pines).

Slightly before flower bud opening (yellow-poplar, magnolia).

The period between flower bud opening and bud-scale closure (most spruces) (Fig. 6.1).

In most trees the receptive period lasts 1 to 3 days for a single flower or for the flowers on one twig. In the majority of cool–temperate spring-blooming species, the receptive period for one tree or for the trees of one species in a locality lasts 3 to 5 days. The flowering period on a single tree lasts 1 to 2 weeks on some summer-blooming northern hardwoods, and 4 weeks or longer on warm–temperate and subtropical pines.

Pollen Applicators

With wind-pollinated species producing abundant pollen, a blower is used to apply pollen. The simplest is a small medicine dropper, flamed at one end to result in a hole about 0.2 mm in diameter. Such droppers are inexpensive so that it is possible to have a separate dropper for each lot of pollen, to be kept in the same vial as the pollen for the duration of the season.

Small plastic wash bottle with a fine-bore nozzle are also inexpensive and useful. The nozzle is part of a tube bent at a 135° angle and reaching nearly to the bottom of the wash bottle. To use, suddenly invert the bottle to fill the tube and squeeze the bottle.

Many pine breeders use an hypodermic syringe fitted with a rubber bulb to which a U-shaped tube is attached. This causes the pollen to be mixed with air inside the syringe and results in the extrusion of a pollen cloud.

In most insect-pollinated species pollen is sticky and difficult to collect in quantity. It is best applied with a small brush or swab.

Pollen Extenders

Small amounts of pollen may be diluted or extended by mixing 1 part active pollen with 3 parts talcum powder, lycopodium spores, pollen of a different genus, heat- or age-killed pollen.

Actual Pollination

When using a brush, remove the pollination bag, apply pollen individually to each receptive pistil, and replace the pollination bag. Brush pollination of small flowers requires a steady hand and is apt to be most successful if the weather is warm and calm.

When using a blower, punch a small hole in the pollination bag, blow a small amount of pollen into the bag, and seal the hole with tape or cotton. Apply enough pollen that some is visible on each female flower and shake the bag to distribute the pollen evenly. With Gerhold's mini-bags, pollen can be blown in through the throat of the bag (Figs. 6.6–6.8).

Most breeders pollinate each bag only once, preferring to spend the extra time pollinating more bags. However, if flowers within a bag vary considerably in time of receptivity, repollination after 1 to 2 days may be desirable. Pollination work should be done when the weather is dry, as wet pollen loses viability, and wet pollinators become clogged.

Tree and Branch Labeling

Each female parent should be numbered and mapped so as to be relocatable. This may be done in various ways.

On each female parent, every pollinated branch must be labeled in such a way as to denote the male parent. The labels should be applied immediately below the pollination bags so that each label applies to all flowers apical to it. As 6 weeks to 18 months elapse between pollination and seed ripening, the labels should be weatherproof. Ordinarily, only 2 to 10 different males are crossed with a given female in any one season. Therefore, no matter how many labels are applied, the labels on any one female parent need carry only 2 to 10 different numbers or colors.

Small plastic spiral rings (often called "chicken rings" and sold by poultry supply houses) are among the most satisfactory labels. They are durable, inexpensive, easily applied, and are sold in a variety of colors. Thus, orange, red, blue, etc., rings can be denoted in the records as O, R, B, etc., meaning male parents numbers 1, 2, 3, etc. Wired sheet aluminum labels are also used. Some wired cardboard labels are durable enough for species maturing their seed in 3 to 4 months.

In a dense tree crown the small branch labels are often difficult to relocate from a distance, so brightly colored plastic streamers are often tied to the pollinated branches at the time of pollination.

Pollination Records

It is well to design a special "pollination and seed set" record form to be used in a looseleaf notebook small enough to be carried while climbing. Most breeders prefer to use a separate page for each female parent, with appropriate space at the top of the page to enter the female parent's number, identity, location, etc. The remainder of the page consists of ruled columns and lines, with one line devoted to each pollen parent used on the tree. The columns should have heading as follows.

1. Label number or color.
2,3,4. Dates of pollination, fruit harvest and seed extraction.
5. Number of male parent.

6,7. Numbers of bags and of flowers pollinated.

8,9. Numbers of bagged branches maturing fruit and of fruit harvested.

10,11. Numbers of full and empty seeds harvested.

12. Seedlot number (to be assigned after seeds are harvested and cleaned).

13,14,15. Initials of the people doing the pollinating, harvesting and extraction work.

Breeders experienced with a single species find it possible to eliminate some of these columns. Conversely, if working with a new species, it may be desirable to record additional data about stage of flower development, weather conditions, and other things which could affect the success of the controlled pollination work.

Indoor Controlled Pollination

Poplars, willows, and elms flower in early spring and ripen seeds 5 to 6 weeks later. The seeds are small and require relatively little nutriment from the seed parent. In these species seeds can ripen on branches cut prior to flowering and kept indoors. This makes it possible to do pollination work under convenient conditions.

The branches should be 3 to 6 ft long and kept with their bases in water, nutrient solution, or moist sand. Results are best if the work is done in a moist, cool greenhouse. Too much warmth may hasten fruit ripening more than seed ripening, so that the ripe fruits drop with immature embryos. Clogging of the water conducting tissue can be a serious problem, so application of a weak fungicide to the freshly cut branch bases is advisable. During the process of seed maturation some rooting usually takes place on willow and cottonwood branches.

With small-seeded species, which do not root so readily and which mature seed in the fall (e.g., alders, and most birches), a variation of this technique is useful. The branches are brought indoors and grafted onto seedling understocks. Bottle grafts are frequently used (see Chapter 7).

In poplars, willows, alders, and birches the female flowers are borne in many-flowered catkins, each capable of producing several dozen or a few hundred seed. Furthermore, a single branch may bear several catkins. Thus a few days work in a moderate-sized greenhouse can produce a few hundred thousand control-pollinated seed, and practically all modern breeding work in these genera is done indoors (Figs. 6.9

FIG. 6.9. Indoor controlled pollination of birch (*Betula verrucosa*) in Finland. The worker on the left is pouring pollen into the suction end of a vacuum cleaner, and the worker on the right is directing a pollen-filled airstream from the exhaust end around the entire greenhouse. There was no natural pollen being shed at the time. The bagged branches will be subjected to another type of pollination. (Photograph courtesy of Lauri Kärki of the Finnish Forest Tree Breeding Foundation.)

FIG. 6.10. Heavy crop of birch seed produced indoors in Finland. By maintaining a CO_2-enriched (2000 to 2500 ppm) atmosphere in a greenhouse, crops of 700,000 seeds per tree can be obtained from 4- to 5-year-old seedlings at low cost of time and money. (Photograph courtesy of Lauri Kärki of the Finnish Forest Tree Breeding Foundation.)

and 6.10). Similar techniques could be used with larger seeded species but the cost per seed is usually too high to be justified.

Fruit Collection and Seed Handling

Protection against Insects, Animals, and Premature Drop

Insects may cause heavy damage to developing fruits, especially of pines and oaks. If the life history is known, it is possible to work out a spray schedule to minimize losses. For example, Coyne (1957) was able to reduce cone losses in southern pine by 60% by applying a water emulsion of benzene hexachloride in March of the second year. If the time of egg deposition is not known, or if several different insects are involved, it is safest to apply a cloth protective bag around each branch as soon as the pollination bag is removed.

Rodents and birds are fond of many tree seeds. Their depradations can be controlled by placing a strong canvas or stapled wire screen bag around the developing fruit a few weeks before ripening. Only experience with a species in a given locality will show whether such precautions are necessary.

Most tree seeds can be collected from one to several days before normal ripening and dispersal. If so, they can be allowed to ripen normally in the open without fear of loss. However, a breeder may be so busy at harvest time that he is unable to collect all fruits before the seeds are dispersed. If that is the case, it is wise to place cloth bags over the pollinated branches a month or two prior to harvest.

Seed Collection, Cleaning, and Sowing

Detailed procedures vary so much with species that only a few general remarks are appropriate here. The book "Seeds of Woody Plants in the United States" (U.S. Forest Service, 1974) is a complete treatise on seed handling and should be consulted for individual species.

It is desirable to assign each female × male combination a permanent seedlot number soon after collection and to record pertinent information about that seedlot in a special accession record. After that is done, it is normally desirable to extract, clean, sort into full and empty, and store each seedlot separately until the proper time for sowing. Then the seed is sown in a nursery or greenhouse according to an experimental design worked out in advance. These steps can usually be done in a leisurely and orderly manner in spruces, pines, and eucalypts.

Many species produce dormant seed which requires scarification or roughening of the seed coat (seed coat dormancy), prolonged storage under cool and moist conditions (embryo dormancy), or moist–warm followed by moist–cool storage (immature embryos) in order to germinate promptly. If so, the appropriate treatment must be added to the above schedule.

Most spring ripening (e.g., poplars and willows) and some fall ripening trees (e.g., magnolias and white oaks) produce nondormant seeds which are short-lived and should be sown soon after ripening.

Some are so sensitive that they lose viability after 1 to 2 days of improper storage. In these cases, haste in sowing is essential, and it may be necessary to dispense with cleaning and sorting and sow in the order of collection rather than according to a specified design. To this list may be added some conifers (e.g., golden larch and some true firs) whose cones shatter upon ripening and whose seeds do not withstand prolonged storage well. Success is apt to be greatest if the seeds are sown immediately, even though mostly empty and contaminated with a very great amount of cone scales and other debris.

A breeder must usually deal with many small rather than a few large seedlots. Hence relatively unsophisticated cleaning equipment (mostly small screens of varying size mesh) usually suffices.

Seed Sorting

If time permits, it is desirable to sort each seedlot into fulls and empties, and sow only the fulls. Among the sorting techniques used are the following.

Visual examination. In many pines full seeds are darker in color than empties. Birches and elms produce thin samaras in which the presence of a full seed is readily apparent.

Flotation in water or ethyl alcohol. With the proper liquid, pine and spruce seeds and various types of nuts will float if empty and sink if full. Full seeds should be completely dried after immersion and should be sown the same year.

Air sorting. An air sorter is a vacuum cleaner in reverse, fitted with a vertical shaft containing baffles which trap chaff and empty seeds. Full seeds remain at the bottom of the shaft.

X-rays. Gustaffson and Simak (1956) and Kriebel (1965) have used X-rays for sorting as well as to determine degree of embryo development in full seeds.

Oak acorns need to be inspected for signs of insect damage but not otherwise sorted into fulls and empties; acorns do not develop to mature size unless they contain a well-developed seed. Nor are true fir and yellow-poplar seeds sorted; the seed coats or pericarps (ovarial tissue) surrounding the embryos are so thick in relation to the size of the seed itself as to make separation extremely difficult.

Laboratory testing of the type normally done with commercial seed is rarely attempted with control-pollinated seed. The seedlots are ordinarily too small to permit the sacrifice of the 100 to 200 seeds necessary for a germination or cutting test. Instead, breeders tend to rely on tested treatment and storage schedules and on reasonably prompt sowing.

Success Expected from Controlled Pollination

The success of controlled pollination work depends on factors such as flower structure and arrangement, seeds per fruit, pollination mechanism, physiological ability of a tree to mature fruit, size of tree, difficulties encountered in climbing, reaction to bagging. The following estimates of numbers of sound seeds per man-day breeding effort (combined bagging, pollination, and harvesting) are based on the assumption that the work is done under favorable conditions (small, conveniently located trees, favorable weather, light insect damage, etc.) and that the crosses are made between trees of the same species. Seed yields are generally much lower when trees of different species are crossed (see Chapters 16 and 17).

Indoor pollination of poplars, willows and birches can yield many thousand seeds per man per day.

Grafted southern pines yield up to a few thousand seeds per man per day; seed yields are generally slightly lower than with open pollination.

Some northern pines yield up to several thousand seeds per man-day but jack pine suffers frequent cone abortion and generally lower seed yields.

Grafted white spruce can yield up to 50,000 seeds per man-day.

Eucalypts and acacias yield up to a few hundred seeds per man-day; the flowers are scattered and must be brush pollinated but each flower can yield several seeds.

Maples yield 100 to 200 seeds per man-day; seed sets per flower pollinated can be high but the pollination work is time-consuming and each flower can yield only 1 to 2 seeds.

Walnuts and chestnuts can yield 50 to 100 seeds per man-day; the female flowers are borne a few to a branch and each flower can produce only 1 nut (walnut) or to 1 to 3 nuts (chestnut).

Yellow-poplar can yield 25 to 500 seeds per man-day; the trees are difficult to climb, and the percentage of sound seed is usually 2–10%.

Ashes may yield up to a few thousand seeds per man-day; if trees flower they are apt to flower very heavily; of those trees which flower many mature no seeds, and others produce heavy seed crops.

Oaks are very recalcitrant. About 6000 control-pollinated acorns have been produced as the result of 40 years effort by American and Russian breeders.

Study Questions

1. Name three types of climbing apparatus, with advantages and disadvantages for each.

2. How may climbing be minimized? List four safety precautions for climbing.

3. Give general rules for identification of flower buds. Observe three local species and describe flower and vegetative buds in each.

4. What are the general characteristics of a bagging material? Name two satisfactory and two unsatisfactory bagging materials.

5. Describe the placement of a pollination bag.

6. Describe optimal conditions for pollen storage. Which species have long-lived pollen? Short-lived pollen?

7. List the general rules used to determine when stigmas are receptive. How long does receptivity last in a flower? Tree? Locality?

8. Describe two different types of pollen applicators and the method of use for each.

9. Describe the labeling and record keeping processes used in controlled pollination work.

10. Select five genera representing contrasts in emasculation method and describe the method used in each. Tell why the methods differ.

11. Do the same for pollen collection methods.

12. Do the same for pollination technique.

13. Do the same for seed sorting methods.

14. Do the same for seed yields per man per day of controlled pollination effort.

7

Useful Techniques—
Vegetative Propagation

Introduction

Many trees can be propagated from vegetative shoots, thus reproducing their genotypes exactly. In a few species shoot cuttings can be
rooted easily enough that vegetative propagation is the accepted method
in commercial practice. In more species, however, its use is limited
to the preservation of breeding material or the establishment of clonal
seed orchards. Those are usually established with grafts.

Before proceeding with the detailed description of methods, four
words should be defined, as they will be used several times in this
and later chapters. An "ortet" is the original tree from which a cutting
or graft is taken. A "ramet" is any vegetative offspring of such an ortet.
A "clone" consists of the ortet and all ramets considered collectively;
all members of the same clone have the same genotype. A "vegetative
propagule" is any type of vegetatively propagated tree.

A good general reference is the book "Plant Propagation, Principles
and Practice" by Hartmann and Kester (1968).

Grafting and Budding

Description of Methods

A "graft" is a severed small branch (scion) attached to a rooted
tree (rootstock or understock) in such a way that the cambial layers

111

are in close contact and grow together. This is accomplished by making one or more slanting cuts in both scion and rootstock and binding the cut portions together. Usually the completed graft union is covered with a thin rubber grafting tape and coated with a suitable waterproof compound to preserve moisture and help support the scion until the cut surfaces grow together. In earlier years the sealing substance was primarily grafting wax based upon beeswax, but that requires heating and causes trouble to the tender young tissues; there are newer water-miscible emulsions on the market which can be applied cold and dry to become permanent and waterproof. The most common types of grafts are illustrated in Fig. 7.1 and described below.

In a "veneer graft," the base of the scion is cut diagonally to expose a large oval of cambium. A wedge-shaped patch of bark of the same size is removed from the side of the understock and the scion and understock are bound together with cambiums matching. Or, a slit may be made in the side of the understock, into which the base of the scion is inserted. After the graft union has healed, the upper portion of the understock is removed (Fig. 7.2).

In a "cleft graft" the base of the scion is cut to form a wedge-shaped

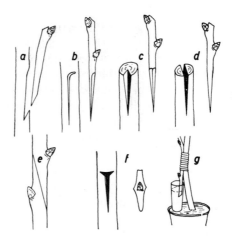

Fig. 7.1. Types of grafts used with forest trees. (a) A type of veneer graft. (b) Another type of veneer graft in which the bark of the understock is not removed and the scion is cut to a very sharp point. (c) "Incrustation in head" graft, not commonly used. (d) Cleft graft. (e) Splice graft. A tongue graft is similar except that extra cuts are made in both scion and stock. (f) Budding. (g) Bottle graft, a variation of an approach graft. (After Nanson, 1974a, originally from P. Gathy in 1958.)

Fig. 7.2. Veneer-grafting white spruce. (A) Making the initial cut. (B) Fitting the scion into the cut in the rootstock. (C) Wrapping the completed graft with thin rubber tape. (D) Brushing the union with a waterproof compound. (Photograph courtesy of Hans Nienstaedt of the United States Forest Service at Rhinelander, Wisconsin.)

point. The understock is cut horizontally and then split. It is preferable that the top of the understock and the base of the scion be of the same diameter so that the cambiums can be matched on both sides. Sometimes, however, the understock may be larger than the scion, in which case the cambiums are matched on one side only.

In a "tongue graft" the base of the scion is cut diagonally, then split. The top of the understock is also cut diagonally and split. Scion and understock are then fitted together in such a manner that the cut ends dovetail. A "splice graft" is a variation in which single diagonal cuts are made in scion and rootstock and the cut ends are bound together.

In an "approach graft" the scion is usually an unsevered stem of a growing seedling. An oval patch of bark is removed from the scion, a similar shaped patch of bark is removed from the understock, then the two are bound together at the point of wounding. A "bottle graft" is a variation on this technique in which the scion is $1\frac{1}{2}$ to $2\frac{1}{2}$ ft long, severed, and kept with its base in water. "Inarching," as used sometimes with chestnut, is another variation in which the scion is a basal sprout of the rootstock and is approach grafted to it at some distance from the ground, to provide insurance in case the main stem should die.

In all except approach grafts, the scions are usually a few inches long and consist of the current year's growth. Most grafting is done in winter or spring.

"Budding" is a variation of grafting in which the scion is a dormant bud with a small patch of attached bark. A small cut (usually T-shaped) is made on the side of the understock, the bark is peeled back, and the bud is inserted in the cut and bound. Budding is usually done in the summer, late enough for buds to have been formed, yet early enough that the bark can be peeled readily.

Grafting work can be further categorized as "topworking" (grafting scions into the tops of old trees), "bench grafting" (grafting inside at a bench), "bare-root grafting" (grafting on rootstocks which have been lifted from the ground), "nursery grafting" (grafting outdoors on seedlings growing in the soil), or "field grafting" (grafting on seedlings already planted in their permanent locations).

Factors Affecting Success

Exposed cambial layers should match as closely as possible, at least on one side of the graft. With veneer or cleft grafts on young seedlings, this is most easily accomplished if the rootstock is the same or slightly larger in diameter than the scion.

The grafting knife should be razor sharp and all cuts must be made speedily with minimum damage to the cambium. It is also desirable to keep the wound areas moist on both scion and understock. At best, a sharp cut kills all cambial cells which are cut. If the knife is dull or a worker is slow, additional cambial cells die and graft union is inhibited.

The point of union should be protected from desiccation for several weeks after grafting. That is accomplished by wrapping with specially made thin rubber tape, and sometimes with a grafting compound in addition. The rubber tape is preferably of a type which will disintegrate naturally in a few weeks. If not, it must be cut after the graft union is complete to prevent girdling.

In cool climates, much grafting is done in a greenhouse, and the newly made grafts are kept several weeks at cool temperatures (e.g., 65°F night and 70°F day), without shade but often with additional light (to maintain an 18 to 20 hr photoperiod) and with daily or twice daily watering. In warm, humid climates much grafting is done in the field. Nearly always in such cases it is necessary to protect the newly made grafts from desiccation by covering scion and understock with a plastic bag and covering that with a paper bag to prevent overheating.

Much fruit tree grafting is done indoors, with bare rootstocks. The work can be done rapidly at a bench. That is also possible with several hardwood forest trees. Bare root grafting has not, however, proved successful with most conifers. With conifers, the understocks should be grown in place or should be potted approximately 1 year prior to grafting.

Fruit tree growers have had some success by purposely using different species as scions and understocks, and in this way have discovered understocks that promote heavy fruit production while the trees are still small. As yet no such understocks have been developed in forest trees, and success is usually greatest if scion and understock are genetically similar.

Although it is possible to graft almost any tree species by almost any method, the success ratio varies considerably among species. Within a species it also varies with the type of graft used and with the season at which the work is done.

Recommended Methods for Different Species

The following partial list is based on reviews by Nienstaedt *et al.* (1958) and Wright (1962) and other papers as indicated. German grafting methods are summarized in a paper by Weisberger and Kohnert (1974).

Alder. Veneer grafts on established rootstocks are satisfactory.

Ash. Most ashes are easily grafted on established or bare root understocks, using tongue or splice grafts.

Aspen. Success is good with almost any method, but bottle grafts are most common.

Beech. Cleft or veneer grafts on established rootstocks have proved best, but results are usually poor.

Birch. Veneer, bottle, and bud grafts on established rootstocks are satisfactory.

Douglas-fir. Veneer grafts on established rootstocks are most common. Initial success is high but a high percentage of grafts fail after a few years (Silen and Copes, 1972; Radu and Blada, 1966).

Eucalypt. Cleft grafts are best.

Larch. Veneer and cleft grafts are successful.

Oak. Many oaks are difficult to graft. Veneer grafts in early spring on established rootstocks are best.

Pine. Most pines are easy to graft, using established rootstocks. Veneer grafts are most common. Most grafting has been done on 1-year-old wood in late winter or early spring, but early summer grafts on new growth are better in some species (Parks, 1974). Loblolly pine in southeastern United States should be grafted after February 1 (Webb, 1961). Incompatibility is often a serious problem in Virginia and jack pines and a minor problem in loblolly pine. Many century-old grafts between Corsican and Scotch pine were still alive in France in 1960 (Bouvarel, 1960).

Spruce. Veneer grafts from late fall to early spring are good. Most grafting is done in a greenhouse, on prepotted understock seedlings which have been brought indoors 2 to 4 weeks prior to grafting.

Sugar maple. Bud grafts are best but the success ratio is low.

Walnut. Cleft and tongue grafts are satisfactory, using bare root understocks. To inhibit later suckering, sever the rootstock slightly below the former ground line and thus graft the scion onto a portion of the root rather than onto the stem (Shreve, 1974).

Effects of Rootstock on Scion and of Scion on Scion

After a graft union heals there is no exchange of genetic material between scion and understock, each maintaining its own genotype. Thus in general each tends to grow in its own way, as if it were connected to a root or top of the same genetic constitution. That can be demonstrated easily by observing the bark on either side of a graft union, particularly in grafts between species. The point of union is easily visible, as the two types of bark remain in strong contrast to each other (Fig. 7.3). There is, however, the possibility that plant foods elaborated one place can be transferred across the union and thus influence the other part of the tree. There is also the possibility that exceptionally slow or rapid growth on the part of the rootstock may influence the growth of the scion by limiting food or providing it in excess.

Research on this subject has been prompted by experience with apples and other fruit trees. Particularly in apples a number of dwarfing rootstocks are known which result in trees that bear fruit heavily when only 4 to 6 ft tall. The flavor, size, color, etc., of the apples are not changed, however.

Ahlgren (1962) was able to hasten flowering of seedling white pines by grafting them into the crowns of old trees of the same species. He

Fig. 7.3. White poplar (*Populus alba*) grafted on largetooth aspen (*P. grandi-dentata*) near Maple, Ontario. Scion and stock tend to retain their own identities, as is easily visible from the contrasting barks at the graft union. (Photograph courtesy of Peter W. Garrett of the United States Forest Service, Durham, New Hampshire.)

was not able, however, to hasten flowering by grafting scions on root-stocks of a different species.

In Mississippi slash pine grows fastest, loblolly pine next fastest, and shortleaf pine slowest. Allen (1967) grafted them in various combinations, e.g., slash on slash, slash on loblolly, etc., and measured the trees 5 years later when they were 15 to 20 ft tall. He found that growth of the scion was not affected by the understock on which it was grown. He also grafted roots on entire seedlings in such a way that he had a slash pine top on a slash + loblolly root system and a slash pine top on a slash + shortleaf root system. The presence of this additional root system did affect growth. Thus, slash pine tops did not grow as fast on slash + shortleaf as on slash + slash or on slash + loblolly roots.

Schröck (1966) made reciprocal grafts among several clones of Scotch pine. He found that diameter of the understock was correlated very strongly with diameter of the scion and less strongly with height and leaf length of the scion.

Clones of rubber trees differ in rubber-producing ability, high-yielding clones having many latex-yielding tubes in the phloem. They also differ in resistance to a leaf disease. Three-part trees have been obtained by grafting scions of high-yielding clones on ordinary rootstocks and then by grafting (at a height of 8 to 10 ft) scions of disease-resistant clones at the tops of the high-yielding stems. There were no reciprocal effects, yields being determined by genotype of the trunk and resistance by genotype of the leaves.

In forest trees there are many examples in which grafted trees are more crooked or have other special characteristics not found in seedlings of the same size. Thus a grafted seed orchard is nearly always identifiable as such. Generally, this phenomenon is related more to the part of the crown from which the scion was taken or the age of the scion-producing tree than to an interaction between rootstock and scion (see later section on topophysis).

Scion–Rootstock Incompatibilities

In a successful graft the actual union tends to be confined to one or two tiers of cells on either side of the graft. These tiers of cells overlap each other longitudinally. If scion and rootstock are fully compatible the union is strong and plant materials can be transported successfully from root to stem and vice versa. With a slight degree of incompatibility, cells of scion and understock live in harmony with each other, but do not interlace in such a manner as to form a mechanically strong joint. The scion may grow normally until some sudden strain causes the stem to break at the point of grafting; if this happens the break often resembles a smooth saw cut more than the ragged edge with is common when a normal tree breaks in two. With a greater degree of incompatibility, cells of scion and understock are antagonistic to each other, with consequent difficulties in nutrient transport and ultimate failure of the graft union. The problem is similar to that encountered with organ transplants in humans. Severe incompatibilities may be apparent immediately as the graft union fails to heal; or they may not be apparent for several years (Figs. 7.4 and 7.5).

Fowler (1967) found that red pine scions on Scotch pine understocks succeeded initially but usually failed after a few years unless the grafted trees were planted so deeply that the scions could produce roots. Ahlgren (1962) found that eastern white pine scions could live indefinitely on eastern white pine understocks, 5+ years on mugo pine understocks, and only 1 to 5 years on understocks of balsam fir and some hard pine species; red and eastern white pines were incompatible from the start.

FIG. 7.4. Incompatibility in a graft of Douglas-fir. Transport of elaborated food material from crown to root is inhibited, resulting in overgrowth of the scion. Many grafts of this type will die in time, and incompatibility is a major problem in Douglas-fir breeding. (Photograph courtesy of Don Cope of the United States Forest Service, Corvallis, Oregon.)

In contrast, Hans Nienstaedt and others working with spruce have found few incompatibilities, being able to graft practically any species on white spruce understocks.

FIG. 7.5. Graft incompatibility in jack pine. The grafts were made in 1965 and photographed in 1972. Mortality sets in at different ages, depending on the clone. Mortality was complete in 1969–1970 for the clone in the center. (Photograph courtesy of Hans Nienstaedt of the United States Forest Service at Rhinelander, Wisconsin.)

In California, where Persian walnut is grown commercially for its nuts, it is sometimes grafted on understocks of another species. The grafts usually live many years, but many then succumb as a black line develops along the graft union.

Scion–rootstock incompatibilities pose a major problem in the development of grafted seed orchards of Douglas fir, even when scions and rootstocks are of similar genotypes. The problem is especially serious because the grafts usually live many years, only to die as flowering is about to start. It has not been possible to identify particular clones or rootstocks or particular clone–rootstock combinations which succeed better than others (Silen and Copes, 1972). It has been possible, by microscopic examination of the graft union 12 to 18 months after grafting, to predict which graft unions are most likely to fail later.

Similar incompatibility problems are serious in jack and Virginia pines. Loblolly pine breeders usually plant an excess of grafted trees in order to have the desired number remain alive and vigorous for the duration of a seed orchard.

Propagation by Stem Cuttings

Description of Methods

A stem cutting is a severed twig whose base is placed in a moist rooting medium and permitted to develop adventitious roots (Figs. 7.6–7.8). The most successful general procedure is as follows. (1) In spring or early summer cut twigs 6 to 10 in. long from vigorously growing shoots, remove most but not all of the leaves, and keep the cuttings moist. It is preferable that the cuttings be of the current year's growth. (2) Dip bottom of each cutting in a powder containing a root-inducing hormone. (3) Place cuttings vertically (buried to $\frac{2}{3}$ their depth) in a rooting medium, such as sand, sand–peat, vermiculite, etc., in a shaded greenhouse bench. (4) By means of a buried insulated electric cable, maintain the temperature of the rooting medium at 80°–85°F, slightly warmer than the surrounding air. (5) Keep the tops of the cuttings moist and cool by means of an intermittent spray; overwatering can be as bad as underwatering. Various modifications of this general schedule are needed for individual species. For example, with some easy to root species it is necessary only to place 1 ft long cuttings in the soil. Some species root best if cuttings are made in the fall or winter; some species are almost impossible to root with any treatment. The reviews by McAlpine (1965) and Nienstaedt et al. (1958) summarize experience in several forest trees. Recent experience with Norway spruce is summarized by Lepistö (1974) (Figs. 7.6–7.8).

Factors Affecting Rooting—Species

The kind of tree is the single most important factor governing the success of rooting. Roughly, forest trees can be categorized as follows.

VERY EASY TO ROOT

Most willows and non-aspen poplars are so easily rooted from dormant-season cuttings that vegetative propagation is the normal commercial practice.

EASY TO ROOT

Monterey pine, cryptomeria, and various species of yew and juniper root readily if the cuttings are maintained under favorable conditions

Fig. 7.6. Rooted cuttings of Norway spruce at the Haapastensyrjä Nursery in Finland. The cuttings are made in November–December, stored at −3° to −5°C, and placed in the rooting medium (usually peat moss and peat mixed) in an unheated greenhouse in April. The relative humidity is usually maintained above 80%. Rooting takes place by late May or June. Most of the cuttings are made from 2- to 4-year-old trees selected from progeny tests. (Photograph courtesy Lauri Kärki of the Finnish Forest Tree Breeding Foundation.)

in a shaded nursery bed or greenhouse. Costs are usually 5 to 10 times as great as for seedlings. Cuttings are used frequently for experimental purposes and uncommonly to establish commerical plantations.

MODERATELY DIFFICULT TO ROOT

Aspens, red maple, some birches, and some hemlocks fall into this category. Under favorable rooting conditions a reasonable percentage

Fig. 7.7. Roll of rooted Norway spruce cuttings at the Haapastensyrjä Nursery in Finland. The rooting medium is spread on a long strip of waterproof material which is then rolled. The rolls are placed vertically as shown and unrooted cuttings are inserted into the exposed peat–peat moss mixture. After rooting the entire roll is transported to an open nursery bed. (Photograph courtesy Lauri Kärki of the Finnish Forest Tree Breeding Foundation.)

FIG. 7.8. A 2-year-old rooted cutting of Norway spruce ready for field planting. This cutting was started in the greenhouse shown in Fig. 7.5, then grown two summers in an outdoor nursery bed. In 1972 approximately 140,000 rooted cuttings of Norway spruce were grown in Finland for reforestation purposes. (Photograph courtesy of Lauri Kärki of the Finnish Forest Tree Breeding Foundation.)

of success can be expected but per cutting costs are usually many times as great as per seedling costs.

DIFFICULT TO ROOT

Most eucalypts, dalbergias, pines, spruces, larches, birches, maples, and black locusts fall into this category. Most reported successes are with cuttings made from young trees, and the rooting of cuttings is not generally practiced even for experimental purposes. See Hare (1973) for details with some pines.

VERY DIFFICULT TO ROOT

Most oaks, chestnuts, beeches, ashes, and walnuts fall into this category. Where vegetative propagation is needed, it is usually accomplished by grafting.

Other Factors Affecting Rooting

Among the other factors, age of the ortet is most important—the younger the tree from which a cutting is taken, the easier it is to root. This is true in nearly all species. In eastern cottonwood, which falls into the very easy to root category, cuttings from the tops of trees up to 20 years old can be rooted with ease, but cuttings taken from 100-year-old trees root only if they are given optimum conditions in a greenhouse rooting bench. At the other extreme, cuttings from a 1-year-old oak tree can be rooted, even though oaks are normally regarded as impossible to root.

Age and size of cutting are also important. Preferably, a cutting should consist of the current year's growth only, and should be longer than 4 in. It should contain buds from which aerial shoots can arise. Only in very easy to root species are cuttings as long as 2 ft practical.

In a hardwood which sprouts from the base, it is usually best to cut the tree near the ground line, let it sprout, and collect cuttings from the vigorous sprouts which develop. This is the usual practice with poplars and willows and has also met with some success in elm, walnut, and yellow-poplar. Apparently this results in a partial return to the juvenile easy to root state. The return is limited, however, as success is usually not as good as if cuttings were to be made from 1- or 2-year-old seedlings.

A variation of the above technique has been used successfully with the conifer Monterey pine by Libby and Conkle (1966). They sheared trees annually and found that cuttings made from the resulting adventitious shoots rooted more easily than those from normal branches. With this technique they were able to obtain reasonable rooting percentages from trees more than 15 years old.

The best season for rooting varies among species. In most cases cuttings are made in the spring before the start of growth or in early summer before the new growth becomes hardened, but there are species in which other times of year are best.

IBA   *IAA*   *NAA*

Plant hormones (principally indoleacetic acid, naphthaleneacetic acid, indolebutyric acid, and their derivatives) are usually beneficial, increasing the percentage and speed of rooting. Exact concentration usually does not matter, and the hormones are usually applied in powder form, using a commercial preparation consisting of a carrier (e.g., talcum powder) containing a low concentration of active chemical. Bases of cuttings are dipped in this powder. Hormone treatment is not a panacea and does not change a difficult to root species into one which is easy to root.

The rooting medium should be moist, well aerated, and sterile, and preferably should be kept at a warmer temperature than the air. Ordinary soil and water do not work well except for the species that are classified as very easy to root. Fungicides applied at time of planting are often beneficial. With loblolly and slash pines it is helpful to water the cuttings with a nutrient solution and to grow them in a CO_2-enriched atmosphere. With the relative humidity conditions prevailing in most temperate countries, rooting is often done best in a greenhouse or cold frame where high humidity can be maintained; in warm and humid countries, such as Taiwan, rooting can be done as easily outdoors as indoors.

In most species there are pronounced clonal differences in rooting ability so that selection for this ability is desirable.

Commercial Propagation of Willows and Cottonwoods

Most species in these genera root so easily that unrooted cuttings are usually planted for commerical reforestation. The general procedure is as follows (Schreiner, 1959; Chmelar, 1967). (1) If starting with old trees, take small cuttings from upper branches, root them under optimum conditions, and plant the rooted cuttings outdoors. (2) Shortly before the start of growth in the second year, mow the trees and cut the stems into 1 to 2 ft lengths. (3) Allow the root systems to resprout and produce multiple stems 5 to 20 ft tall which are the source of the next year's cuttings. Annually thereafter in the spring mow the sprouts and make into cuttings. (4) Treat the newly made cuttings with a fungicide and store them under cool moist conditions until time for planting. (5) Punch holes in the ground, in the places in which the trees are to be grown permanently. The holes should be $\frac{2}{3}$ as deep as the cuttings are long. (6) Drop the cuttings into the holes, basal end down, with 2 to 4 in. exposed to the air, and firm the soil around the cuttings. (7) Maintain a weed-free condition around each tree for 1 to 2 years.

Some selection may be necessary to find clones that root satisfactorily. Ordinarily, 100% rooting can be expected from those chosen for commercial production. The species of willow with which this method is practiced often grow 6 to 8 ft per year in cool temperate regions and 10 to 15 ft per year in subtropical regions.

Propagation by Root and Root–Stem Cuttings

Some species, particularly those which spread naturally by means of root suckers, can be propagated more easily by root than by stem cuttings. Such cuttings are usually $\frac{1}{4}$ to $\frac{1}{2}$ in. in diameter and 4 to 6 in. long. They are planted horizontally at depths of $\frac{1}{2}$ to 1 in. and treated the same as stem cuttings. The method is uncommon because of the labor involved and the difficulty encountered in making sure that the cuttings are from the desired tree.

A variation of this method is used in aspen. First, root cuttings are made as described above. These are allowed to produce aerial sprouts, and these sprouts are then treated as ordinary stem cuttings. They root more easily than cuttings taken from normal trees.

Propagation by Leaf Cuttings

Single leaves, with accompanying basal buds, are preferable to stem cuttings in some horticultural plants such as rhododendrons.

In some pine species, single leaf fascicles have been found to root more easily than stem cuttings from trees of the same age. There is a difficulty, however, in that the rooted fascicles usually fail to produce a bud which can grow into a new tree. In jack pine, but not in some other species, this difficulty can be countered by using fascicles with preformed buds.

Root Suckers

Among the trees which propagate themselves naturally by means of root suckers are the aspens, beech, black locust, ailanthus, black tupelo, and sassafras. In these species, roots produce buds from which aerial stems arise, frequently several feet from the main stem. As these new trees grow, the roots spread laterally and give rise to yet more aerial stems, so that a single clone may occupy an acre or more. Suckering can often be increased by deep cultivation.

This characteristic is of great silvicultural importance. If it were not for their suckering ability, most species listed above would be rare, as they do not ordinarily reproduce well in nature by seed. The characteristic is also of genetic interest, because it provides naturally established clonal material for special studies.

Other Methods of Occasional Interest

"Layering" is the production of roots from a lower branch that comes in contact with the soil and as a result produces roots. It occurs commonly in certain shrubs (e.g., forsythia and black raspberry). It is a rare occurrence in trees, but can happen in almost any species growing in a moist habitat where there is a deep accumulation of litter which can keep the branch moist where it is in contact with the soil.

"Air layering" is essentially the same phenomenon but the soil (or other rooting medium) is brought to the branch rather than the branch to the soil. Rooting is facilitated by wounding the branch, surrounding it with moist sphagnum moss or peat, and covering the whole with a waterproof wrapping. If one waits long enough (rooting may require 24 months), roots can be produced on a great variety of tree species that are otherwise difficult to propagate vegetatively. After roots develop, the ·branch is severed and planted in a greenhouse before transferral to its permanent location.

A limited amount of clonal material (2 to 4 ramets per clone) for special experiments can be obtained by "splitting seedlings" longitudinally in such a manner that each piece has roots, stem and leaves or leaf buds. The success ratio is highest if the cut surfaces are waxed to prevent desiccation. After being split, the parts of seedlings are planted in a greenhouse or nursery, as would be the case for ordinary seedlings. The method is easy to practice in some hardwoods that normally sprout readily; in these hardwoods it is necessary only to start with a sturdy seedling, remove all but 2 in. of the top and 2 in. of the root, and split the 4-in. seedling longitudinally. It is much more time-consuming in conifers, where an entire seedling must be split lengthwise.

Apomixis

"Apomixis" is the development of seed without fertilization. It occurs naturally in several herbaceous plants and among trees possibly occurs in the hawthorns. Unfortunately, apomixis can not be induced at will (at least not readily) as can other forms of vegetative propagation. Where it does occur, it is of great interest genetically because all seedlings are of the same genetic constitution.

Solbrig (1971) described the mechanism in the common dandelion, which appears to be entirely or primarily apomictic. Early stages of ovule development are normal, as is the first half of the meiotic division. Then the process is interrupted and a "restitution nucleus" is formed. That nucleus has the same chromosome number and genotype as the female parent and from it develops the embryo, the seed, and finally the new plant. Thus a heavy seed crop producing identical seedlings is possible. Also, if a mutation occurs which can produce a visible effect, it is passed on to all offspring.

Topophysis or Persistent Nongenetic Effects

Cellular Growth Capabilities

A fertilized egg cell in a female flower divides to become two cells, these divide to become four cells, these divide to become eight cells, etc. Thus the embryo is produced. Some cells in the embryo become differentiated to form the root primordium, some to form the primordia of the cotyledons, and others to form the primordium from which the stem and leaves will arise. As a tree grows older the cambial cells divide longitudinally, the inner ones becoming differentiated into xylem and the outer ones into phloem. Still later, the cells in a flower primordium become differentiated into those producing sepals, petals, stigma, and anthers.

Much of this differentiation is related to a cell's position within a tree. Thus a normal dividing cell in the cambial layer can be expected to develop into xylem or phloem but not into a root or a leaf. A normal dividing cell in a flower bud can be expected to develop into some part of a flower but not into a leaf or root. Some of the differentiation is related to happenings in other parts of a tree. Thus a hardwood tree can be cut and some of the cambial cells may develop into vegetative buds which produce new shoots. Some of the differentiation is related to a tree's age. In most species the leaf primordia in a young seedling produce very different types of leaves than those produced on mature trees. A cell in a vegetative bud is much more apt to become differentiated into a flower primordium in an old than in a young tree.

Sometimes the differentiation is reversible, as when a cutting is removed from a tree and some cells in the cambium are induced to form roots rather than xylem or phloem. Or, single pollen grains can be placed on an appropriate culture medium and in some species can be induced to grow into normal plants instead of fulfilling their usual function of fertilizing egg cells. Sometimes the differentiation seems irreversible, or reversible only with great difficulty. Root formation cannot be induced from the cambial cells of cuttings taken from old trees of many species. In aspen, single cambial cells have been excised and placed on agar where they could be induced to divide and grow into a mass of undifferentiated "callous" tissue, but only in rare cases into complete seedlings having roots, stems, and leaves.

Presumably, such differentiation does not occur as the result of any change in the mechanism of DNA replication, as the chromosome and gene complements remain the same in cells throughout a tree. Possibly, differentiation involves a change in the mechanism by which DNA is transcribed to RNA, so that at a certain stage in the development of a tissue some chemical within a cell inhibits the transcription of certain portions of the DNA molecule, thus blocking the action of some genes. Or, at a certain stage in tissue development, a chemical may be present which permits the transcription of some RNA which was previously blocked.

Definition and Examples of Topophysis

Up to now, we have discussed differentiation within an intact tree, and mentioned the fact that some is difficult to reverse. This can be true even when a cutting or scion is removed from a tree and grown by itself. Differentiation may have progressed to the stage where the cutting growing on its own roots for many years retains the characteristics it has as a branch on an intact tree.

This phenomenon is called "topophysis" and can be defined briefly as the long-time persistence of age or position effects in trees grown from cuttings or grafts. I use the term to include "periphysis" or the long-time persistence of position effects and "cyclophysis" or the long-time persistence of age effects. The phenomenon was investigated extensively by Schaffalitsky de Muckadell (1959). Among the numerous examples are the following.

TREE FORM

Scions or cuttings taken from lower, horizontal branches continue to grow horizontally for seveal years in many conifers. Even when scions are taken from upper branches, grafted pines are often more crooked than seedlings (Fig. 7.9).

FIG. 7.9. Topophysis in 5-year-old grafted Sakhalin fir growing in Hokkaido, Japan. The scions were taken from lower branches and continue the growth habit of branches rather than terminal shoots.

LEAF SHAPE

English ivy produces palmately lobed leaves before flowering, and cordate leaves after flowering. An ivy cutting will produce a palmately lobed, nonflowering plant or a cordate-leaved flowering plant if taken from a nonflowering or flowering portion of an ivy plant, respectively. Some junipers produce mixtures of awl-shaped and scalelike leaves, the proportion of scalelike leaves increasing with a tree's age. Sometimes cuttings taken from branches with awl-shaped leaves will produce trees with predominantly awl-shaped leaves. Scions taken from the branches of old pine and spruce trees produce mature needles after grafting, not the juvenile-type foliage found on a seedling of the same size.

FRUITING

In many conifers cones are borne predominantly on the upper branches of old trees. Scions taken from these upper branches are prone to retain their flower-producing ability after grafting, or at least to fruit a few years earlier than seedlings. This phenomenon seems reversed in jack pine, in which grafted trees may fruit later than seedlings.

THORNINESS

In honeylocust, vigorous outer branches are thornless, and less vigorous inner branches become thorny. Eight years after grafting, scions from thornless branches remained thornless, whereas scions from thorny branches remained thorny.

LEAF RETENTION

In European beech, the leaves drop earlier in the autumn from outer than from inner branches. Grafts made with scions taken from outer and inner branches tend to drop their leaves early or retain them late, respectively.

GROWTH RATE

In pine, the terminal shoots on upper branches are longer than those on lower branches. In a South African experiment, measured after 8 years, pines grown from cuttings taken from the uppermost whorl were taller than pines grown from cuttings taken from the second whorl down, which were taller than trees grown from cuttings taken from the third whorl down, etc. In an Illinois experiment, cuttings and seeds were collected from two geographically separated populations of eastern cottonwood; after 8 years seedlings from the two populations differed in height by more than 10%, vegetative offspring by much less.

RESPONSE TO PHOTOPERIOD

Seedlings of several spruce species respond dramatically to being grown under continuous light year long, sometimes making the equivalent of 4 years' growth in 1 year. Grafted trees made with scions collected from old ortets do not respond and grow at normal rates if provided with continuous light.

RESISTANCE TO DISEASE

Grafted trees produced with scions taken from old eastern white pines are more resistant to white pine blister rust than are seedlings.

Practical Importance of Topophysis

Tree breeders make practical use of topophysis when they establish grafted seed orchards with scions collected from upper, female-flowering parts of the crown. The ability to produce female flowers is often retained in the young grafted trees.

It is probable that many cultivars used in horticulture are of topophytic origin. Thornless honeylocust has already been mentioned. (Some thornless cultivars may be genetically thornless, however.) Some of the spreading types of yew and juniper and some of the foliage types of juniper may have originated as cuttings from particular branches of otherwise normal trees. Conversely, some horticultural cultivars do not come up to expectations because of topophysis. One recently distributed clone of dawn redwood, probably reproduced by cuttings taken from the lower branches of old trees, has undesirable form and growth rate, unlike those of the ortet.

From the many examples quoted above it is evident that data from clonally propagated trees are not good indicators of those trees' ability to transmit desirable form, growth, etc., characteristics by seed. Thus the behavior of trees in a clonal seed orchard is largely ignored in deciding which clones to favor. Clonal testing is limited to those few species in which commercial clonal propagation is feasible.

Study Questions

1. Define clone, ortet, ramet, stem cutting, scion, understock, graft, air layer, rooting medium, and topophysis.
2. Describe three methods of grafting.
3. List five factors affecting the success of grafting, and tell how each works.
4. How is it possible to demonstrate an effect of rootstock on scion?
5. Describe a graft union. How and when do graft incompatibilities result?
6. Describe instances in which understocks have noticeable effects on scions. In which they do not.
7. Describe the procedures necessary to root cuttings in a species in which rooting is difficult. In willow and poplar.
8. Describe how species, age of ortet, age of cutting, hormones, season, and rooting conditions affect the success of rooting.
9. What vegetative propagation method is most successful with beech? Eucalypt? Most pines? Monterey pine? Willow? Cottonwood? Ash? Sassafras? Aspen? Douglas-fir? Oak? Spruce? Fir? Larch? Redwood? Black tupelo? What degree of success can be expected in each?
10. Describe the procedure followed to air-layer a tree.
11. Describe three examples, other than those given in the book, of tissue differentiation in trees. How is that related to topophysis?
12. Describe five different examples of topophysis in trees and tell the practical importance of each.

8

Useful Techniques—
Establishment and Measurement
of Test Plantations

Introduction

This chapter is devoted to another important aspect of tree breeding technique—testing in the nursery or greenhouse and in the field. It includes a preliminary section on definitions, then a section describing the simplest and most commonly used experimental designs. There are discussions of the nonstatistical and statistical aspects of testing, followed by sample schedules which can be followed to establish statistically adequate nursery or greenhouse and field tests. Finally are a few hints on measurement techniques.

Definitions

As defined here, a "seedlot" is a group of related trees (a single clone, a single half-sib family, a single full-sib family, etc.) given a number and identified as a unit throughout the course of an experiment.

A "plot" is a group of 1 to 100+ trees belonging to a single seedlot and planted adjacent to each other in the nursery or plantation. A plot may be linear, rectangular, or square.

"Replication" is the practice of planting different plots of the same seedlot at different locations within a plantation, in different plantations, or in different years. Replication is necessary because site conditions are always somewhat variable, and therefore the only way to determine the true genetic potential of a seedlot is to test it at several places.

True replication exists only if plots representing a single seedlot are scattered, separate by intervening plots of other seedlots.

A "block" is a portion of a plantation containing one plot of each of several different seedlots. A complete block contains one plot of each seedlot. Blocks may have regular or irregular outlines.

A "plantation" is a group of blocks planted close to each other (often contiguous).

An "experiment" consists of the nursery test and the one or more plantations devoted to the testing of a particular group of seedlots. Most modern experiments involve many seedlots and include several scattered plantations, often in different states or countries.

"Random" means without apparent order and usually applies to the distribution of plots within a block. Seedlots are randomly distributed within blocks to prevent any particular seedlot from being planted exclusively on poor or good sites in all blocks, to prevent the same two seedlots from always being adjacent to each other, and to prevent bias during measurement.

"Statistical precision" is the ability of an experiment to differentiate among seedlots and is often measured by the "least significant difference" (often denoted as LSD), which is the smallest difference which can be shown to be statistically significant at a given confidence level. With regard to growth rate, the LSD varies from as little as 3% of the mean in very precise experiments to as high as 50% in poorly executed experiments.

"Statistical efficiency" is the ability of an experiment to yield the greatest amount of useful information per unit of cost, as measured by number of trees planted, measurement time, etc.

Commonly Used Experimental Designs

Randomized Complete Block Design

In this design, a plantation is divided into a number of equal-sized blocks, each containing one and only one plot of every seedlot. The sequence of plots within blocks is random. The randomized complete block design is the most commonly used, is simple to understand, and is relatively easy to analyze statistically. The statistical analysis is usually done in such a manner as to recognize differences due to seedlots and to blocks, the remainder of the variation being regarded as error.

The statistical precision is higher than can be obtained with a completely randomized experiment but lower (especially if the number of seedlots is very large) than can be obtained with lattice and incomplete block experiments. However, the analysis can be performed in such a way as to make the precision equal to that obtained from more sophisticated designs; this is true if data from each plot are analyzed in terms of superiority (or inferiority) to the two or three plots adjacent on either side.

Randomized Complete Block Design with Some Blocks Incomplete

It often happens that one desires a randomized complete block design but has insufficient trees of some seedlots to include all seedlots in all blocks. If so, it is most convenient (and probably best from the statistical standpoint) to include every seedlot in every block as far as the supply of planting stock permits and include in subsequent blocks all seedlots of which stock is still available. Opinion varies as to whether it is better to make some plantations complete and others incomplete or to distribute complete and incomplete blocks equally to all plantations. Statistical precision varies, being high for seedlots represented in all blocks and lower for incompletely represented seedlots.

Complete Randomization

All seedlots may be completely randomized within a plantation. This design is not recommended, as good randomization is relatively difficult to achieve when working with bundles of trees which cannot be shuffled as can cards. Without good randomization, statistical efficiency is lowered. Also, a completely randomized experiment is no easier to install or analyze than a randomized complete block experiment.

Compact Family Design

This design is meant for the testing of a number of groups of seedlots, seedlots of each group being more closely related to each other than to seedlots of other groups. Thus, it is applicable when testing several half-sib families from each of several stands, several clones belonging to each of many families, etc.

In this design, the seedlots belonging to one group are planted compactly in adjacent subplots, the adjacent subplots representing a group of seedlots comprising a plot. The plots should be randomized within blocks; subplots may or may not be randomized within plots.

The compact family design is ideal if expected intragroup differences are small as compared with expected intergroup differences, as statistical precision is greatest for testing the intragroup differences. Also, if intra-group differences are small, they can sometimes be ignored during measurement and analysis, with a large saving in time.

Latin Square

In a Latin square, a plantation is divided into equal numbers of rows and columns, and one plot of each seedlot is included in each row and column. Thus the number of replications equals the number of seedlots. Therefore, the design, although capable of yielding statistically precise results, is not often used because it is limited to experiments having small numbers of seedlots.

Balanced Incomplete Block and Lattice Designs

There are special designs in which blocks are of equal size and contain much fewer than the total number of seedlots, with each block containing a different but representative sample of the entire experiment. The number of blocks is much greater than the number of replications. The designs are relatively inflexible in that the number of seedlots is usually fixed for given numbers of blocks and replications.

The blocks are smaller in size and therefore contain less site variation than in a complete block experiment. Thus, balanced incomplete block designs can give greater statistical precision than complete block designs. To give great precision they must be planned and analyzed very carefully.

Nonstatistical Considerations

A precise experiment is one in which the error variance is low. This is most apt to be the case if every tree is planted well on a good uniform site and receives good uniform care. The goal should be a plantation with at least 90% initial survival (98% is often achieved), with all trees growing vigorously, and with all trees easily locatable. All too often, results from poorly managed plantations are of little value.

Care and Quality of Planting Stock

Most experimental plantations are established with seedlings grown in a nursery or greenhouse for an appropriate period (Fig. 8.1). In tropical climates the seedlings are often grown outdoors in soil-filled

FIG. 8.1. The Haapastensyrjä Tree Breeding Center of the Finnish Forest Tree Breeding Association. The carefully maintained greenhouses and outdoor seedbeds of this 10 hectare (25 acre) nursery provide uniformly favorable conditions for progeny testing, indoor controlled pollination, and vegetative propagation. All seedlings raised in this nursery are parts of genetic experiments or are of genetically improved stock. (Photograph courtesy of Lauri Kärki of the Finnish Forest Tree Breeding Foundation.)

containers made of tarpaper, bamboo, paper, etc. In temperate climates they are usually grown in outdoor seedbeds, although the practice of growing them (or at least starting them) in containers inside a greenhouse is growing in popularity. The choice is one of cost and uniformity—choose a method which assures uniformly good planting stock at a reasonable cost.

Since uniformity and high quality are important, it is best to sow the seeds at a uniform spacing wide enough to grow without severe competition until time for field planting. This usually means a wider spacing than would be used in an ordinary commercial nursery for the same species. Extra care in maintaining uniformly favorable growth conditions through careful soil preparation, watering, fertilization, and weed control is usually warranted. Slow-growing nursery grown species usually benefit from root pruning or transplanting within the nursery prior to field planting.

Even with the greatest care, some variability in growth conditions can be expected, with the result that some seedlings grow three times as tall as others in the same seedlot. In some experiments, such nursery-induced size differences have been maintained for 15 years, with a consequent decrease in experimental precision. Therefore it is usually good practice to cull the poorest seedlings prior to field planting. Also, it is good practice to replicate within the nursery or greenhouse, so that no one seedlot can receive extra favorable treatment prior to field planting.

During the lifting and packing operations it is necessary to maintain the identity of every seedling even when several hundred seedlots are involved. It is also necessary to work quickly and carefully so that the planting stock does not suffer from prolonged exposure or storage. With careful planning and supervision both things are possible.

Site Selection

A planting site should have the following prerequisites.

Average or above average in quality to ensure high survival and good growth

Uniform, at least within blocks

Prepared beforehand to ensure high survival and lack of competition for the newly planted trees

Open and relatively free of obstructions, as attainment of a uniform partial cover is difficult to obtain

Large enough to accommodate at least one complete block; it is desirable but not necessary that the site be large enough to accommodate several contiguous complete blocks

Owned by the public or a large company to ensure permanence

Accessible enough to ensure adequate follow-up

Care after Planting

High initial survival and good growth are essential to good results. Also, silvicultural practices will probably become more than less intensive with time. Therefore, better than average care is frequently warranted.

If weeds are a problem, as they are for most species planted in north central United States, they must be controlled. In areas with high labor costs, this is usually accomplished most satisfactorily by means of chemicals. In Michigan, for example, good results have been obtained by applying aminotriazole to strips the season before planting and simazine to the same strips after planting (Figs. 8.2–8.4).

Invading woody vegetation, especially of the same species, should be controlled periodically to prevent confusion, to prevent unwanted competition, and to facilitate measurement.

FIG. 8.2. Chemical weed control used prior to the establishment of a test planta-tion in southern Michigan. The strips were sprayed with aminotriazole the previous autumn to kill existing vegetation and will be sprayed immediately after planting with simazine to kill germinating weed seeds. Good cross-alignment was obtained easily by planting the tree rows perpendicular to the sprayed strips. The black cherry planted in these strips averaged 16 ft (maximum 29 ft) after 8 years.

Artificial pruning is unnecessary in most eucalypts and southern pines which drop their shaded dead branches quickly. It is desirable, however, in trees, such as the spruces and many northern pines, which retain dead branches for 40 to 50 years or in trees, such as walnut, which are grown primarily for high-value sawlogs or veneer logs. In such cases the cost of pruning is more than balanced by the decreased cost of measurement. Also, in high-value trees one can obtain data on response to artificial pruning which may be of more value than data on rate of natural pruning.

Control of Pest Damage

In areas with dense populations of herbivores, such as deer and rabbits, fencing may be necessary, but is a very expensive adjunct to plantation establishment.

Data on resistance to insects or diseases may be very desirable, and can be obtained only from plantations suffering a high incidence of attack. Therefore, heavy infestations are often welcomed rather than

FIG. 8.3. A 1-year-old (from seed) black walnut plantation in Illinois. Weed control is very important in the establishment of most hardwood plantations, whether experimental or commercial. Chemicals were used to maintain weed-free 5-ft strips in this progeny test. (Photograph courtesy of Calvin Bey of the United State Forest Service, Carbondale, Illinois.)

controlled. Light infestations of nondamaging pests are usually ignored.

On the other hand, it is often best to control light infestations of pests which cause growth loss or growth deformities. If the attack rate is less than 10%, one is not likely to obtain statistically reliable data on genetic differences in resistance, but likely to suffer undesirable variation in other growth traits.

Spacing

Most experimental plantations are designed to be used as breeding arboreta or seed orchards as well as to provide data about growth. Trees flower most heavily if their crowns do not quite touch.

Most experimental plantations will be thinned one or more times. It is well to delay such thinnings until one is reasonably confident of not removing any trees that might ultimately be among the best if they had been retained.

FIG. 8.4. A well-designed full-sib progeny test of white spruce in northern Wisconsin. The rows are in straight lines in both directions and the weeds were controlled around each tree. The result is a high-survival plantation which is easy to manage and measure. The plantation was established with 4-year-old seedlings and photographed 3 years later. (Photograph courtesy Hans Nienstaedt of the United States Forest Service at Rhinelander, Wisconsin.)

The thinning intensity may be such as to call for the removal of a very high percentage of the trees, in which case an initial close spacing may become a rather wide final spacing. Also, representative growth data are more apt to be forthcoming if the spacing approximates that used in commercial practice.

All these factors must be considered in determining the proper spacing. Most investigators have resolved the question in favor of a spacing similar to or slightly greater than that used in commercial plantations of the same species.

Thinning

In a test plantation to be used as a seed orchard, maximum improvement results if thinning removes all trees of the poorest seedlots and the poorest trees of the best seedlots. If data on genetic differences are the objective, the data are of most value if based on retention of the maximum number of trees of the best seedlots until maturity. Only if data on juvenile–mature correlations are the main objective is it best to retain equal numbers of trees in all seedlots during thinning.

Removal of all trees of the poorest seedlots is feasible only in small-plot experiments. In large-plot experiments it is necessary to remove some trees of all seedlots or to leave undesirably large gaps.

Thinning is usually accomplished in several stages, removing the poorest trees first, the next poorest trees or seedlots next, etc. Thinning should be started early enough to leave the remaining trees with adequate growing space. Yet it should not be started so early that there is a strong possibility of removing trees that might have proved to be among the best if kept until maturity.

While absolute uniformity of spacing in the mature test plantation would be desirable, it is difficult to achieve. In most experiments, however, it is desirable to make a few compromises and leave a few trees of mediocre seedlots or cut a few exceptionally good trees to obtain moderately uniform final spacing.

Alignment

Both maintenance and measurement are made much easier if the trees are check planted so that one can easily follow a row in an up and down, crosswise, or diagonal direction. On level open land it is easy to maintain proper alignment within ±3 in.; on rough land it may be necessary to have some trees out of line by 2 ft. If straight lines are marked on the ground in one direction (by a weed-control strip, wheel track, scratch mark, etc.), good alignment can be obtained most easily if the planting crews proceed in straight lines perpendicular to the on the ground lines.

Border Rows

It is good practice to surround each experimental plantation with 1 to 2 border rows of the same species. All trees in such border rows should be mapped but not necessarily measured.

Labeling and Mapping

There is no truly satisfactory way to label trees permanently at the time they are planted. Therefore, maps are prepared to enable future workers to find and measure the trees. It is most convenient to plant first and map immediately afterward, thus avoiding the necessity of looking for the proper trees to plant in a particular plot.

Among the various types of temporary labels used, gummed masking tape inscribed with a glass-marking pencil is used by several breeders. The labels are applied to the tops of seedlings before lifting from the nursery, and remain legible for several months after planting. Being attached to the trees, they are not apt to become lost during the planting operation. They rot before they can cause girdling; wired labels are unsatisfactory in this respect.

A plantation map should be accurate and complete enough to enable a person 20 years hence to locate the plantation and every tree in it, and to be able to measure and report intelligently on the plantation without reference to another person. To accomplish this it should contain the following.

Plantation number
Statement of species, type, and design of experiment
Location data, including exact distance and direction to a permanent landmark from some point in the plantation
Description of site, vegetative cover, site treatment, date, personnel, method of planting
Directional arrow
Statement of plot size, orientation, and spacing
Statement giving further source of information about the identities of the seedlots, which can be identified only by number on the map
Actual map of the plantation, showing the location of each plot and of the border trees and showing the demarcation of blocks

As for the map proper, it is most convenient to recognize rows and columns of single trees and to designate rows by letters and columns by numbers (or vice versa) so that single trees may be identified as Z-12, AA-106, etc. But if the plots are of equal size (as they should be), only one seedlot number need be entered for any one plot. Thus, for four-tree linear plots the column headings might be A (for the border column), B to E (for columns B, C, D, and E), F to I, etc. Thus the map can be compact.

The map, with its accompanying information, should be made in the field soon after the planting work is done (preferably on the same day). If neatly done, this field map can be photocopied and serve as a permanent map. Many breeders prefer, however, to prepare the final map on a typewriter and reproduce several copies to ensure against loss and to supply copies to all interested persons.

Large maps which must be unfolded are unwieldy to use in the field. Therefore it is most convenient to make each plantation map page-size or (if there are a few thousand plots in a plantation) to sectionalize the map.

Breeding and Demonstrational Aspects of Test Plantations

As tree breeders progress from the first to the second and third tree generations, more and more test plantations are destined to be used as breeding arboreta. The breeding value of a test plantation is enhanced by wide spacing, heavy flowering, good maintenance, large enough size to permit selection within as well as between seedlots, and maintenance of the maximum numbers of trees of the best seedlots until maturity.

Tree breeding results are accepted by the general tree planting public less through scientific publications than through popular publications and demonstration tours through test plantations. The demonstration value of a test plantation is enhanced by good alignment, good maintenance, good growth, small plots size (one-tree plots are less good than plots containing 2+ trees, however), and accessibility. Several small plantations can be visited by more people than one large plantation.

Statistical Considerations

Plot Size and Number of Blocks per Plantation

The smaller the plots, the smaller the blocks, the less the amount of intrablock site variability, and the more reliable the comparisons among seedlots planted in the same block. Thus, statistical precision is greater in a plantation containing small plots and many blocks than in one containing larger plots and fewer blocks. Measurements made on actual plantations in Michigan and North Carolina by Wright and Freeland (1960) and Conkle (1963) indicate that the amount of information gained per tree planted is 20–30% less with four-tree than with one-tree plots, and 80–90% less with 100-tree than with one-tree plots. Calculations made by these same investigators showed that one- to ten-tree plots were statistically most efficient, giving the most information per unit of time spent on measurement and analysis.

There can be a difference in type of information supplied by small and large plots. With small plots (or with linear plots of any size), every tree grows in competition with some others of unlike genotype, and therefore may not grow at the same rate as in a commercial plantation composed of trees similar in genotype. In large square plots interior trees are adjacent to trees of the same seedlot, and therefore grow as if they were planted in a commercial plantation. To give this type of information on performance under stand conditions, however, a "large" plot must be at least 7×7 (= 49 trees total) or preferably 11×11 (= 121 trees total) trees in size to contain several interior trees after one or two thinnings.

The difference in type of information is probably not present during the first third of a test rotation, and may be slight after that if a small-plot plantation is thinned to remove all trees of the poorest seedlots. In view of the great sacrifice in statistical precision and efficiency accompanying the use of large plots, the majority of recent experiments in the United States employ four- to 10-tree plots with 8 to 10 replications per plantations. The extremes of which I am aware are 100 one-tree plots and three 625-tree plots per seedlot.

Orientation of Plots and Blocks

An experimental plantation yields the most precise results (i.e., the unexplained error variance is lower) if laid out so that the amount of site variation among plots within any one block is the smallest. This is true if the plots are parallel to and the blocks are vertical to any site gradient which may be present. On a hillside plots should run up and down the hill, and the blocks should be elongated parallel to the contours.

With plots containing six or fewer trees, it is usually not possible to demonstrate a difference in statistical efficiency between 1×6, 2×3, or 3×2 shapes. As row plots one tree wide are easiest to plant and measure, they are most commonly used.

Confusion is almost certain to result if all plots in a plantation are not of exactly the same size and orientation. At planting time, leave an empty space if a tree is missing from a plot bundle, or throw away an extra tree included in a plot bundle. Block boundaries do not need to be regular in outline, and it is better to make them irregular than to leave plots empty.

Calculating the Desired Number of Blocks per Plantation

Rather than following a model applicable to another species and a different set of conditions, it is best to make some calculations applicable to one's specific conditions. The philosophy and procedures are described here briefly, in as nonstatistical terms as possible.

For a given set of measurements, the error mean square is a measure of the amount of variation which cannot be explained as being due to differences among seedlots or blocks. The smaller the error mean square, the better the experiment. This statistic is governed very largely by the care given the plantation and is small if care is excellent. It is also governed by the design used and is smallest if the design is good and the plots are small.

The standard error of a seedlot mean is a measure of the reliability of that mean and is calculated as the square root of the error mean square divided by the number of plots per seedlot. In other words it varies inversely as the square root of the number of blocks.

The LSD (least significant difference) is approximately equal to the standard error of a seedlot mean multiplied by a constant of about 3.2 (5% level) or 4.2 (1% level). To determine the desired number of blocks, one has first to decide what will be an acceptable LSD. This is to some extent an arbitrary decision but there are some guidelines. In some experiments the LSD is as low as 6% of the mean, and that could be the goal. Or, one can make a preliminary guess as to the probable amount of genetic variability in the material being tested and decide that the experiment is not worth doing unless a certain level of precision is reached.

Having decided the desired plot size and an acceptable LSD, the next step is to determine the probable size of the error mean square. This varies so much with site and cultural conditions that it can be determined only by actual measurement of existing plantations similar to the contemplated experimental plantation. The final step is to divide the error mean square by the square of the desired standard error of a seedlot mean and thus obtain the desired number of blocks per plantation.

Judging from much recently published data, the answer is often so large as to be totally impractical. In this case, the best solution is to refine technique and thus reduce the size of the error mean square.

Number of Blocks per Plantation versus Number of Plantations

Recent reviews of multiplantation experiments by Squillace (1970) and Wright (1973) indicate that most such experiments contain numerous instances in which seedlot A grows 5% faster than B at one place but 5% slower than B at another place, and a few instances in which seedlot A grows 10–15% faster than B at one place and 10–15% slower than B at another place. Statistically these differences in relative performance are called interaction. Interaction may also be present in other characteristics or in plantations established in the same locality in different years. Because of interaction, a person can never put as much trust in data from a single plantation, no matter how precise, as in data from several plantations. To obtain good data on average performance it is better to distribute a given number of blocks among several than among a few plantations.

No matter how large and statistically significant an interaction, it is best regarded as "error" and is of little value in making practical recommendations unless it can be interpreted. This is mainly a matter of planting every seedlot on a large number of site conditions, so that the relative performances of seedlots A and B can be compared on several hilly versus several level sites, several warm versus several cold sites, several high-nitrogen versus several low-nitrogen sites, etc. If that is done, one can learn the precise site conditions under which A will grow better than B or vice versa and make valid practical recommendations as to where to plant A rather than B or vice versa. Interpretation of interaction based upon data from only a few plantations is usually a matter of guessing and all too often has been erroneous.

Even when data from many plantations were available, interpretation of interactions has been a difficult matter. Even with a sophisticated statistical analysis, Hattemer (1968, 1969) was unable to identify a particular combination of site characteristics responsible for exceptional growth of any 1 of 22 Japanese larch seedlots tested in 13 plantations in West Germany. In northern Argentina (Barrett, 1969) and north central United States (Wright, 1973) there were several interesting interactions involving plantations on different soil types, different aspects, etc., but the few which could be interpreted involved seedlots differing considerably in cold hardiness and planted on sites differing appreciable in average winter temperature. In those particular regions it seems best to establish a few plantations each in the warmer and colder parts of the region, thus obtaining data on average performance on warm and cold sites, respectively.

Missing Plot Calculations

The statistically most efficient experiments are those employing small plots and many replications, with a few replications assigned to each of several plantations. The ultimate, and one favored by the late R. A. Fisher, is an experiment consisting of many plantations, each consisting of 1 to 2 complete blocks, with very small (one- or two-tree) plots. Use of such a design would inevitably lead to a large number of missing plots. This is not an insurmountable problem, however.

The solution is very simple if only 2–3% of the plots are missing and differences among blocks are small. In this case the seedlot mean can be substituted for the missing plot mean. Or a more complicated formula involving seedlot and block means can be used (see a statistical text for details).

If many plots are missing, a least-squares routine requiring the use of a computer can be used (see an advanced statistical text for details). Two relatively simple procedures yielding moderate to good results are also available. Both involve preliminary calculation of the block means. In one, the plot means are calculated as deviations above or below the block means; the seedlots totals and means are calculated (with due regard to sign) in terms of those deviations. In the other calculation, the plot means are expressed as percentages of the block means; seedlot means are then calculated as percentages of the overall mean.

Purposeful Division of Seedlots into Groups to Be Planted in Different Plantations

A recent work plan calls for the testing of 55 seedlots, to be divided into two groups of 30 each, with 5 seedlots common to each group. One group is to be planted in plantations A and B, the other in plantations C and D. The grouping was arbitrary, there being no *a priori* reasons· to believe any particular seedlot incapable of growing at a particular place. Plots were to be large and blocks per plantation few. This practice was planned to conserve space and ease the measurement problem.

Similar schemes have been used in the past and appear necessary if several hundred seedlots are involved or if seeds must be collected and sown in different years. However, analysis and interpretation are never as straightforward as if all seedlots are represented in all plantations. If the groups are so small as not to be representative of the entire experiment and if no seedlots are common to all plantations, it is extremely difficult to make comparisons between seedlots in the A–B and C–D groups. If some seedlots are tested at all places, they serve as standards by which to compare an A–B with a C–D seedlot but the comparison is a two-step one. In such a case the LSD between an A–B and a C–D seedlot is approximately $\sqrt{2}$ as great as the LSD for an intragroup comparison. Thus there is an inevitable loss in precision.

In other cases, seedlots have been grouped according to their supposed ability to survive and grow well on certain test sites. In some provenance tests, for example, only seedlots from warm regions have been tested in warm regions, and only seedlots from cold regions have been tested in cold regions. The wisdom of this practice can be judged from comparable experiments in which an effort was made to include all seedlots in all plantations. Nearly always the *a priori* reasoning has been only partially correct and there have been surprises, i.e., seedlots which should have died grew very well.

The advisability of including near certain failures in a test plantation depends somewhat upon the plot size used. With small plots the failures constitute the first thinning and the experiment may actually proceed better than if all trees were to live. With large plots, large bare spaces are undesirable.

Sample Schedules for Experiment Establishment

The following sample schedules are based largely upon experience at Michigan State University and are meant to give the reader some idea as to how the work can be done. Actually a great variety of procedures is in use, and there is probably no one procedure which is best for all conditions. To simplify the presentation, I assume the use of a randomized complete block design and assume exact numbers of seedlots, trees per seedlot, etc.

Outdoor Nursery Experiments

To establish a 100-seedlot, 20-tree per plot, 5-block experiment in outdoor nursery seedbeds, proceed as follows. (1) Place the 100 seed containers in a row with 5 envelopes near each, inscribe the seedlot number on each envelope and place 20 (more if seed is variable in quality) sound seeds in each envelope. (2) Place 1 envelope per seedlot in each of 5 piles (to become blocks 1–5) and randomize envelopes in each pile; (3) With another color ink renumber the envelopes in each pile from 1 to 100. (4) Prepare a nursery map placing blocks 1–5 in seedbeds 1–5, respectively, and including both sequence and seedlot number for each envelope. (5) Prepare nursery stakes, including sequence and seedlot number on each. (5) Carefully prepare nursery seedbeds and mark equidistant (exact distance to be obtained by a steel tape) crosswise rows. (6) Place nursery stakes at ends of rows and make a final check against the nursery map. (7) Distribute envelopes and plant seeds equidistant in rows, checking each envelope against nursery stake. (8) Cover the seed.

Containerized Experiments in Nursery or Greenhouse

The schedule can be nearly the same as given above, with one seed planted per container. Sometimes the seeds are sown in flats and the seedlings transferred to containers at an early age. Extra care is needed in labeling, as the containers are apt to be moved (Fig. 8.5).

Preparation of Seedlings for Field Planting

Assume that stock from the above-mentioned nursery experiment is being prepared for planting in four plantations, each to include 100 seedlots, 5 trees per plot, and 4 blocks. The sample schedule is as follows. (1) Make a final check of nursery stakes against nursery map. (2) Label (with seedlot number) 1 of every 5 trees in the seedbeds. (3) Loosen all trees so that they may be pulled easily. (4) Proceeding down one seedbed pull 5 trees per seedlot (including a labeled one), tie the 5 in a bundle, and keep the bundle moist. (5) When all 100 five-tree

Fig. 8.5. Containerized seedlings of Caribbean pine (Honduran variety) ready for field planting. These seedlings, grown in the Fiji Islands, are 8 months old. Container-grown seedlings are used commonly in subtropical experiments and are becoming more common in temperate regions. The containers should be removed before planting. (Photograph courtesy of Bill Dvorak of the Fiji Pine Scheme.)

bundles are ready, randomize them and pack them in a 500-tree bundle destined for block 1 in plantation 1. (6) Repeat the process in seedbed 2 to obtain the stock for block 2 in plantation 1. (7) Repeat the process in seedbeds 3 and 4. (8) On the same or another day prepare the stock to plant blocks 1–4 in plantations 2–4.

One important feature of the schedule is prelabeling of trees before lifting. This prevents confusion and saves time during the planting operation, as only a labeled tree rather than a tree and a label need be planted. Another important feature is the skipping from one nursery block to the next when obtaining stock for a single plantation. This ensures against the inclusion of extra large or extra small trees of any one seedlot in any one plantation.

A similar schedule can be used when handling containerized seedlings. However, extra care is needed to prevent confusion. The 5 seedlings destined for a single plot can be tied in a 5-container bundle or placed in a 5-container box. Or they may be kept carefully in sequence in larger boxes, in which case some randomization procedure must be adopted at the time of planting.

Establishment of Field Plantations

Assume that the planting stock has been delivered to a prepared planting site, and that one wishes to establish a 100-seedlot 4-block plantation with 5-tree row plots. A sample schedule is as follows. (1) Ascertain the site variation pattern and lay out approximate block boundaries to make each block as uniform as possible. (2) Establish alignment procedure to assure straight rows in both directions. (3) From the 500-tree block-1 bundle choose at random a 5-tree plot bundle. (4) Plant the 5 trees in a row, placing the labeled tree in the first position. (5) Proceed down the row and plant the second, third, etc., plots. (6) Upon reaching a block boundary, reverse direction and plant in the adjacent row, placing the labeled tree last in each plot. (7) Make sure that there are 5 and only 5 tree-spaces per plot. (8) Upon completion of planting prepare a plantation map by walking along the front (i.e., labeled) tree of each plot and making occasional cross-checks with plots in adjacent tiers.

Placement of plots randomly while planting rather than in a predetermined sequence is an important feature of the schedule. It is difficult to keep bundles of trees or containers in a predetermined order, and the other alternative would be to search for a plot-bundle to be planted at a particular space. That can be very time consuming. Also, with the schedule given above the supervisor can devote full attention to quality of planting and to alignment, as paper work is delayed until later.

Hints on Measurement and Data Recording

A plantation established in 1 to 2 days may exist for 50 years and be measured several times during that period. Ultimately the time spent on measurement and analysis far exceeds that spent on establishment. Hence extra time spent on establishment to facilitate later measurement is well worthwhile. Also, improvement in measurement and analysis techniques are worthwhile. The problems encountered are so varied that no complete treatment of techniques is possible within a few pages, so I shall attempt here to give only a few hints as to how to save time or get better results. A more detailed account is given in a paper by Wright (1970b).

Hints on Measurement

If row plots are used, walk across the centers of the plots rather than along the plots. It is easier to keep oriented and also saves considerable time when measuring traits that do not require a visit to every tree.

Measure to an accuracy of 1/20 to 1/25 of the range between extremes. If the shortest and tallest trees are 20 and 45 ft tall, respectively (difference = 25 ft), measure to an accuracy of 1 ft. If judging color, try to establish 20 color grades (difficult but not impossible). Lesser accuracy usually results in some loss of information. Greater accuracy is rarely warranted because of the inherently large variability of biological material.

Having established the accuracy of measurement, use no decimals or fractions when expressing plot means. They require extra time to calculate in the field and much extra computation time in the office, and the extra accuracy is not warranted.

Mortality must be measured as a yes–no trait. Generally, however, statistical precision is much greater if a trait is measured in terms of quantity per tree (i.e., 25% of foliage eaten by insect A) rather than in terms of presence or absence (i.e., tree attacked by insect A).

Accuracy is always greatest if one trait is measured at a time. This is especially true for things such as foliage color or number of fruit. Judgments are more consistent if one can observe the same trait on different plots with little time lag.

Measurement of one trait at a time may also be quicker than measurement of all traits while visiting a given plot. This can be judged by computing the walking versus the measuring and recording time. For example, in a 660-plot plantation, spaced 8 × 8 ft, a complete across plot traverse requires 1 mile or approximately 15–25 min of walking time. Is that less than or greater than the time spent on switching from the measurement of one to measurement of another trait on each of 660 plots? The answer varies.

Postulate from start to finish the statistical analysis and use to be made of any particular set of data, thus avoiding the accumulation of data that cannot be analyzed or turn out to have no value. At best, little use is made of 50–75% of most sets of measurements, and without forethought the amount of low-value data can approach 95% of the total gathered. As an example, there is little possibility of detecting statistically significant differences in pest resistance if the incidence of attack is low, so the attack data are of little value. As another example, height and diameter are usually so strongly correlated that there is little purpose in measuring both unless one goes further and determines whether differences in the ratio of the two differ among seedlots.

Height can be measured at periodic intervals. Damage from a pest or from cold weather may be of much greater importance but only measurable soon after the damage occurs. Thus a measurement schedule should be somewhat flexible.

A trait such as foliage color is best measured in terms of live-tree standards defined while one is in the plantation. It is more accurate to make tree–tree than tree–color card–tree comparisons. It is also faster. It is well, however, to translate the live-tree grades into standard grades understandable to others.

If intraplot variation is normally distributed (as is usually the case) there is apt to be a very high correlation between plot means based upon measurement of the 25–50% tallest trees per plot and plot means based upon measurement of all trees. Use of this shortcut results in a uniform increase in all seedlot means, but this is not important where comparative rather than absolute data are needed. Measurement of a randomly chosen 25–50% of the trees per plot does, however, result in a decrease in statistical precision.

Even if several trees per plot are measured, it is desirable to record plot means only. Mental calculation of such means is relatively easy with a small amount of practice. One way is to assume a mean, measure and remember the deviations from that mean, divide the sum of deviations by the number of trees, and add the answer to the assumed mean. Another way is to observe and grade individual trees and accumulate the individual tree grades for the total plot grade. Barring mistakes in arithmetic, the answers are as accurate as if individual tree data were recorded. Usually plot means only are used in subsequent calculations.

Most growth traits must be measured (but not necessarily recorded) by individual tree. There are better methods, however, when working with traits (foliar chemistry, wood density, etc.) which require complex equipment and considerable time per sample. In such cases it is most economical to collect material tree by tree, bulk the samples from all trees in a plot (or even from all trees in a seedlot), and make one analysis per bulked sample. Only enough duplicate analyses need be made to provide an appropriate error term for statistical analysis.

It is most pleasant and often most efficient to work in teams of 2 to 3 persons, particularly if the plots or trees are large. If the plots and trees are small, skilled workers often find it most efficient to work alone.

A Sample Data Recording Folder

A type of data recording folder in common use in north central United States consists of the following.

Manila folder giving the plantation number, species, and approximate location

Plantation map glued to the back inside cover of the folder

Set of data recording pages, prepared soon after plantation establishment; the pages are made of ruled paper, with one line devoted to each plot; down the left-hand side of each page the plots are listed in the sequence by which they occur in the plantation and identified by row and column number, seedlot number, and block number; the remaining columns on each page are left blank for future measurements

Set of blank pages on which to record the number, date, personnel, type of measurement, and units for each set of measurement data

Set of previously prepared ruled-paper pages with seedlot numbers across the top on which to record trait by trait the plot means by seedlot and block; appropriate columns are labeled in which to insert seedlot sums and means; space at bottom to insert block sums and means, overall sum and mean, and analysis of variance

List of seedlot numbers and descriptions

The folders are prepared soon after planting is completed, a separate folder for each plantation. Duplicate or triplicate folders are prepared for those plantations in which two or more people located at different places are seriously interested. By insertion of photocopies of the seed-lot $\times$ block tabulations for each trait all folders can be maintained complete with little extra effort.

The pages are temporarily bound in the folder for convenience of handling in the field. Statistical analyses are inserted below or to the side of the trait to which they apply. With proper care each folder becomes a complete and easily intelligible record of all maintenance, measurements and statistical analyses made for the plantation.

Hints on Data Recording

Whether or not a computer is actually used in data analysis, it is well to proceed as if this were to be the case. For computer analysis, all data must be recorded legibly and neatly in straight rows and columns. Data should be recorded only in terms of numbers (no decimals, slant lines, etc.). An entry must be recorded in every space provided, and all data applying to a single item (tree or plot) must be on a single line. If these precautions are diligently observed, one can come in from a few days' measurement work and proceed directly to computer-card preparation without any recopying. Or, if one prefers hand analysis, the same precautions considerably facilitate the work.

In experiments with trees, one frequently desires to use data gathered 15 to 20 years previously. Conversely, data gathered today may be useful 15 to 20 years hence. It is essential to bear this in mind and to provide enough descriptive information about each set of measurements to make the data intelligible to any interested person.

Study Questions

1. Define plot, block, random, randomized complete block, replication, statistical precision, statistical efficiency, LSD, Latin square, and balanced incomplete block.

2. Compare the statistical precision of seedlot intragroup and intergroup comparisons for the randomized complete block and compact family designs. How do the two designs differ?

3. How is statistical precision affected by care during planting, large amount of site variation within blocks, large amount of site variation among blocks, presence of herbaceous weeds, presence of competing woody vegetation, plot size, number of blocks, and deliberate exclusion of some seedlots from some plantations?

4. What is the purpose of replication? Of randomization?

5. Under what conditions should pests not be controlled? Be controlled?

6. Under what conditions is artificial pruning a desirable practice? An unnecessary practice?

7. Describe the important features of a good test site. Think of possible test sites in your vicinity and rank them in order of desirability for each of three species.

8. How do breeding plans, proposed thinning intensity, proposed thinning time, and plot size affect decisions as to the most desirable initial spacing?

9. What are the advantages and disadvantages of thinning by removal of trees within plots? Of complete removal of poor seedlots?

10. How do plans for thinning vary with plot size used?

11. Outline the procedures used to determine the desired number of blocks per plantation. What can be done if the answer is impractical?

12. Devise three methods of obtaining good tree alignment and for each tell how to check the adequacy of alignment while planting is in progress.

13. Describe the essential features of a good plantation map.

14. Describe optimum experimental procedure from the standpoint of breeding needs, demonstrational needs, and statistical efficiency.

15. How does the type of information vary with 1-tree, 2-tree, 5-tree, 16-tree, 49-tree, and 121-tree plots before thinning? After removal of 25% of the trees or plots? After removal of 75% of the trees or plots?

16. How does statistical efficiency vary for these same plot sizes?

17. Give advantages and disadvantages of one many-block plantation. Of a few few-block plantations. Of many one-block plantations.

18. Describe a sample schedule for the establishment of a replicated nursery or greenhouse experiment.

19. Describe a sample schedule for the establishment of a replicated test plantation.

20. Describe a sample record folder for a plantation.

21. If possible, obtain actual establishment schedules or examples of records and decide which features are better than, worse than, or equal to the features mentioned in the samples.

22. What aspects of measurement and recording procedure affect statistical precision? Usefulness of information? Speed? Convenience in later years?

9

Selective Breeding—General Principles and Methods

Role of Selective Breeding in Tree Improvement

Selective breeding is the selection and crossing of individual trees. One form, mass selection, is an extension of natural selection, with choice of trees left to man instead of nature. Other forms, family and clonal selection, involve establishment and measurement of test plantations.

Generally, productive selection programs involve the expenditure of considerable effort. For that reason, intensive selective breeding work is usually confined to those species that are planted commercially on a large scale.

Theoretically, genetic improvement can be expected in any tree characteristic. Actually, the possibility of obtaining improvement in a specific trait varies widely among species and does not become known until after considerable experimentation. Also, costs vary among traits. Compare stem form and wood density, for example. Improvement in either will require measurement of many hundreds or thousands of trees. The stem form measurements (most often ocular estimates) may be made in a few days, the wood density measurements in a few months.

Most widely distributed trees are geographically variable, and the differences among races may be several times as large as the improvement expected from one or two generations of selective breeding. Thus it is wise to learn first the geographic variation in a species and delay intensive selective breeding until the areas generally having the best trees can be identified. Scotch pine and Douglas-fir for use in north central United States are examples. Both are native to mountainous regions and have well-defined geographic races that differ in many respects. In each case large amounts of improvement were obtained by geographic origin tests, and selective breeding within the best races may now be profitable.

There are numerous cases in which simultaneous experiments on geographic and individual tree variation are desirable. Experiments of this type are under way in several parts of the world. Usually seeds are collected from several trees in each of many stands scattered over a large area; sometimes the parents are carefully selected in each stand, more often not. Then the seedlings are kept separate by both parent and stand of origin. Results vary. In some cases differences among the offspring of different stands are large, and differences among the offspring of different trees in the same stand are small so that the selection and crossing of individual trees seems unwarranted. In other cases the variation among the offspring of individual trees is considerable, and intensive breeding work is indicated.

Growth rate, and most other traits, can be increased by selective breeding or by intensive culture. However, it is usually not a question of one or the other. The genetic improvement obtained through breeding can be added to the improvements obtained through culture. Often there is a mutual reinforcement between the genetic and cultural improvements. A genetically faster growing variety may justify more intensive culture, which in turn justify more planting.

General Principles

Books by Lerner (1958) and Falconer (1960) are good accounts of the theory of selection. They form the basis for much of this chapter. A recent paper by Nanson (1974b) gives tables comparing the efficiency of four types of selection practiced in forestry.

Genetic Gain, Selection Differential, and Heritability

The general formula for genetic gain is

$$\text{Gain} = \text{selection differential} \times \text{heritability}$$

To increase gain, a breeder must manipulate one or both components.

Selection differential is the difference between average and selected trees, families, or clones. If a stand has an average height of 50 ft and selected trees are 55 ft tall, the selection differential is $55 - 50 = 5$ ft or 10%. If a progeny test has an average height of 20 ft and selected families are 21 ft tall, the selection differential is $21 - 20 = 1$ ft = 5%. Selection differential, which is easy to measure, is often expressed in terms of the standard deviation (σ).

Heritability is that portion of the total variance (a statistical term translatable roughly as variability) which is genetic. Only the additive genetic variance is transmissible by seed. Hence, when calculating gain in a seed-propagated species, heritability is defined as that portion of the total variance due to genes with additive effects. Estimation of heritability requires data from progeny tests, and even with good experimental data exact heritability calculations may be difficult.

Increasing Selection Differential

Selection differential is a function of the numbers of trees (or families or clones) selected and the numbers of trees (or families or clones) from which selections are made. The relationship is shown in Table 9.1, which is expressed in standard deviation units.

The greater the ratio of trees observed to trees selected, the higher the selection differential. The relationship is not a direct one, however. That ratio must increase tenfold to achieve a selection differential of 2σ rather than 1σ; it must increase 15-fold to achieve 3σ rather than 2σ; it must increase 40-fold to achieve 4σ rather than 3σ.

The idea in selection work is to choose trees which are genetically

TABLE 9.1

Relation between Selection Differential and Number of Trees Observed When Making a Selection

Selection differential (standard deviation units)	No. of trees observed
1.0	4
1.5	13
2.0	42
2.5	159
3.0	739
3.5	4,298
4.0	31,540
5.0	3,588,000
6.0	100,000,000

superior. The numbers in Table 9.1 are realistic only if selections are made under similar conditions. Thus there is a practical limit to the size of the selection differential. Generally one can make valid comparisons only among trees that grow within a few hundred feet of each other and consider that he has selected the best one in a few hundred trees. There is also an economic limit—too much money is needed to scout thousands of acres for every selected tree.

Selection differentials of 3σ to 4σ are realistic when practicing mass selection in a nursery or when selecting for disease resistance under epidemic conditions. Selection differentials of 2σ to 3σ are realistic when practicing mass selection in a uniform wild forest or unimproved plantation. Selection differentials are most apt to range from 1σ to 2.5σ when practicing family selection or selecting within families in a progeny test.

Increasing Heritability

Heritability can be expressed mathematically as follows:

$$\text{Heritability} = \frac{\text{additive genetic variance}}{\text{environmental variance} + \text{nonadditive genetic variance} + \text{additive genetic variance}}$$

The numerator additive genetic variance is fixed for any given trait in any given population (as is nonadditive genetic variance). Additive genetic variance can be increased, however, in one of three ways.

1. Make selections over a wide geographic area, not in one stand only.

2. Confine selection work to traits with a large amount of additive genetic variance. This usually does not become known until after some progeny testing. There are now sufficient inheritance data for loblolly and slash pine, wattle, and a few other species to permit rational planning of selection work.

3. Hybridize diverse types (species, races, trees from different stands, etc.) prior to the start of selection work. This is commonly done in agricultural crop plant breeding work. It seems to be the only solution for the improvement in resistance of white pines to the white pine weevil. Eastern white pine seems to be uniformly susceptible, western white pine to be much more resistant. Hybridization may engender enough variability to permit selection for resistance and the many desirable growth traits of eastern white pine.

The denominator can be decreased by decreasing the environmental variance. That is, when practicing mass selection, make the selections in pure, evenly spaced, even-aged stands growing on uniform sites, and make comparisons among trees in the dominant crown class. In family selection programs, make test conditions uniform.

Direct versus Indirect Selection

Direct selection for height (or for insect resistance) means selecting the tallest (or least damaged) trees as parents. Indirect selection for height might mean selecting trees with the highest photosynthetic rate, in the hope that photosynthetic rate and growth rate are causally related. Indirect selection for insect resistance might mean selecting trees with the highest content of a particular chemical in the hope that concentration of that chemical is related to resistance.

One type of indirect selection is commonly practiced. That is selection for growth rate at relatively young ages in the belief that trees tallest then will also be tallest at maturity. Juvenile selection is almost a necessity. Otherwise, no trees could be discarded in the nursery, test plantations could not be thinned, and breeding generations would be impossibly long. An increasing amount of data indicate that such selection is effective, if not carried to extremes.

Otherwise, however, direct selection is generally recommended for two reasons: (1) It is usually most economical. Direct measurement of height, wood density, insect damage, etc., is usually much simpler than measurement of chemical and internal traits. (2) It is more effective. For indirect selection to be effective at all there must be a demonstrable cause and effect relationship between the two traits. When such a relationship is present, the gain from indirect selection for trait B is to the gain from direct selection for trait A as r_{AB}^2 is to 1, where r is the (product moment) correlation between traits A and B. In most studies, correlations have been such that $r = 0.5$ or less and $r^2 = 0.25$ or less. This indicates that gains from indirect selection will usually be less than 25% as fast as gains from direct selection.

Concentration on a Few Traits

Most trees need improvement in many ways, but progress will be fastest if selection work is concentrated on a few traits. To show why assume a population of 1000 trees, from which the best tree is to be chosen on the basis of superiority in 1, 2, 3, 4, or 5 traits. That best tree will have an average per trait and a total superiority as shown in the tabulation on p. 164.

No. of traits selected	Selected tree/ No. of trees observed	Superiority of selected trees	
		Per trait	Total
1	1/1000	3.29	3.29
2	1/31.6	2.53	4.55
3	1/10	1.65	4.94
4	1/5.62	1.35	5.38
5	1/3.98	1.14	5.72

Total superiority of the selected tree increases as the number of traits increases, but the value of that superiority may decrease because of differences in value of the traits. Assume the first two traits to be equally important, the next three to be only half as important. The value of the gain is less after selection for five than for only two traits.

Concentration on Unitary Traits

It is easy to fix a trait governed by one gene pair, less easy to fix a trait governed by two gene pairs, and still less easy to fix a trait governed by three gene pairs, etc. At best, most economically important traits seem to be governed by several gene pairs. The problem can be minimized by breaking traits into components and breeding for these components individually. For example, growth rate may be broken down into components such as seasonal duration of growth, juvenile growth rate, mature growth rate, etc. Insect resistance might be broken down into components such as resistance to insect A, resistance to insect B, etc.

Avoidance of Inbreeding

At the same time as he is trying to maintain a high selection differential by selecting the best few of very many trees, a breeder must avoid inbreeding. Otherwise inbreeding depression may result, and excessive uniformity may prevent further progress after a few generations.

The effects of inbreeding can be minimized in two ways: (1) by working on a large scale or (2) relaxation of selection standards. Usually, both methods are used. Many tree breeders select hundreds of plus trees or include hundreds of families in their progeny tests. Commonly, 10 to 15 or more clones or families are retained after thinning in clonal orchards or progeny tests.

For calculation of probable inbreeding rate, N is the number of trees selected in mass selection programs. With family selection, N_{female} and N_{male} are the numbers of female and male parents of the families retained for the next generation, not of all families in the progeny test (see Chapter 3).

Breeding Methods Described

Mass or Phenotypic Selection

Mass selection is the choosing of trees based on their phenotypic performance. For growth rate, it means choosing the tallest trees. Mass selection and natural selection are similar except that man rather than nature makes the choice of parents (Fig. 9.1).

In spite of the utmost care, selection conditions are never uniform. Any particular selected tree may owe its superiority to the site on which it is growing rather than to the genes which it contains. However, if a trait is under genetic control, offspring of selected trees should surpass offspring of average trees.

In its simplest form, mass selection can be inexpensive. The silvicultural operation known as "thinning from below" is an example. In that operation a forester quickly marks the smallest and most crooked trees for removal. On the other hand, mass selection can be expensive if a breeder attempts to maintain a very high selection differential and to select many hundred trees. In that case it involves searching hundreds of stands scattered over a wide area, looking for the 1 to 2 best trees in each stand.

The effectiveness of mass selection becomes known only after offspring of selected and average trees (preferably from the same stands) have been grown together in replicated experiments. Once the effectiveness is known, a breeder can decide whether to (1) practice mass selection at all, (2) use mass selection in conjunction with family or clonal selection, or (3) use mass selection as the only improvement method. In most breeding programs, mass selection is used as a prelude to family selection, or in choosing the best trees within the best families.

Mass selection has been effective for growth rate and form in loblolly pine (Zobel, 1971). In slash pine, it was effective for growth rate and fusiform rust resistance, if the selections were made in heavily infested stands (Goddard et al., 1973). It has been effective for blister rust resistance in western white pine (Bingham et al., 1969) but not in eastern white pine (Riker et al., 1943). It has been relatively ineffective for growth rate in longleaf pine, jack pine, red pine, and eastern white pine (Goddard et al., 1973; Kuo et al., 1971; Yao et al., 1971; Canavera, 1969).

Fig. 9.1. Selected or "plus" tree of birch (*Betula verrucosa*) in southern Finland. By the end of 1971 some 1500 such plus trees had been selected on the basis of their growth rate, form, and freedom from pests. The initial phenotypic selection work on birch has nearly been completed in Finland, and the next step is progeny testing of such superior trees. (Photograph courtesy of Lauri Kärki of the Finnish Foundation for Forest Tree Breeding.)

Mass Selection and Bulk Progeny Tests

A bulk progeny test is a simple way to learn the effectiveness of mass selection. To perform such a test, select 25 to 50 trees for mediocrity in the same trait. Collect seed from them and bulk the seed into two samples, "excellent" and "average." Establish a replicated progeny test and measure it later for the trait under consideration. Tests of this type have been commonly used to determine genetic quality of seed produced by clonal seed orchards. They can also be valuable for species for which there are no inheritance data and in which future intensive breeding work is contemplated.

Fig. 9.2. A half-sib progeny test of Scotch pine in southern Finland. The seedlings will be grown in place an additional 2 to 3 years at which time the best seedlings of the best families will be moved to permanent test plantations. Each labeled row represents the offspring of a different, carefully selected female parent and the entire experiment is replicated several times. Careful records are necessary to keep track of the thousands of plots in such an experiment. (Photograph courtesy of Lauri Kärki of the Finnish Foundation for Forest Tree Breeding.)

Half-Sib Progeny Tests

The word "sib" is a contraction of "sibling," which is a general term meaning brother or sister. Half-brothers or half-sisters share the same mother or father, but not both. In trees, half-sibs have the same seed parent or (more rarely) the same pollen parent, but not both. A "half-sib family" or "half-sib progeny" is a group of trees sharing one parent (usually the seed parent) in common (Figs. 9.2 and 9.3).[1]

To perform a half-sib progeny test, collect seeds from a large num-

[1] For purposes of simplicity I treated "open-pollinated" and "half-sib" families as synonymous. This is not strictly true, as an open-pollinated family may contain a mixture of selfs, crosses with related trees, and crosses with unrelated trees. With large interstand differences, open-pollinated families from the same stand may approach full-sib families with regard to genetic correlation among the offspring.

FIG. 9.3. First step in a half-sib progeny test of birch (*Betula verrucosa*) in Finland. The seedlings are grown 2½ months under plastic, which is then removed. After 1 year the seedlings are moved to field plantations which will grow for 10 to 20 years. Already in the first year differences among families are visible. (Photograph courtesy of Lauri Kärki of the Finnish Forest Tree Breeding Foundation.)

ber of parents, keep the seeds separate by parent, sow the seeds in a replicated nursery test, and later outplant the seedlings in a replicated permanent test plantation. In later years, measure them and calculate the differences among families.

In mass selection, a breeder looks at single trees and judges their genotypes from their phenotypes. There is always doubt as to the genotype of any particular tree—its excellent performance may be due solely to its being grown on an exceptional microsite. In a half-sib progeny test, the breeder judges the genetic quality of any parent by measuring a number of its offspring growing under slightly different conditions. Thus the genetic quality of the parents can be judged with much greater certainty than in mass selection.

The genetic quality of a parent, as judged from a half-sib progeny test is known as "general combining ability." That is the ability of a parent to cross with ("combine" with) other parents in general to produce superior offspring. A parent with high general combining ability for height is one which produces tall offspring.

The difference between mass and family selection can be shown by the following hypothetical example. Assume that seeds were collected from 10 parent trees (A–J) and grown in a replicated progeny test, and that the relative heights of parents and offspring were as indicated in the following tabulation.

Generation	Relative height of parent or offspring									
	A	B	C	D	E	F	G	H	I	J
Parent	120	116	112	108	104	96	92	88	84	80
Offspring	105	103	106	95	96	104	97	102	98	94

Note that there is a small correlation between parent and offspring. In mass selection, Parents A–E would be chosen, and the selection differential would be $(120 + 116 + 112 + 108 + 104)/5 - 100 = 12$. Those parents would produce offspring with an average relative height of $(105 + 103 + 106 + 95 + 96)/5 = 101$. In other words, the offspring would be $101 - 100 = 1\%$ superior to average.

With data from the progeny test, parents A, B, C, F, and H would be chosen, even though F and H were below average in height. The offspring of those five trees would have an average height of $(105 + 103 + 106 + 104 + 102)/5 = 104$. This time the offspring would be $104 - 100 = 4\%$ superior to average.

The difference between mass selection and selection on the basis of progeny performance can also be illustrated graphically, as in Fig. 9.4. Figure 9.4A illustrates a situation in which heritability is low. There is a small correlation between parental performance (X axis) and offspring performance (Y axis). If a breeder practiced mass selection he would choose parents A, D, G, I, and J, as being to the right of the dashed line marked standard set for mass selection. He would obtain slight improvement, as the offspring of those parents are slightly above average. If he selected on the basis of offspring performance, he would choose parents A, B, E, F, and K, as being above the line marked standard set for family selection. The gain would be much greater.

Figure 9.4B illustrates a very different situation. It is assumed that heritability is high and there is a very strong correlation between parent

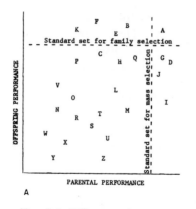
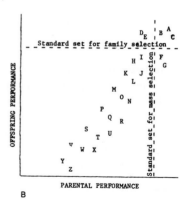

A   B

FIG. 9.4. Differences between mass and family selection with low heritability (A) and high heritability (B). With mass selection, parents to the right of the dashed lines marked standard set for mass selection are chosen, and with family selection, families above the lines marked standard set for family selection are chosen. With low heritability, the offspring of selected parents are little better than average; with high heritability, the offspring of selected parents are much above average and nearly as good as if family selection was practiced.

and offspring. A breeder practicing mass selection would choose parents A, B, C, F, and G. If he selected on the basis of offspring performance he would choose parents A, B, C, D, and E. There would be relatively little difference in performance between the two selection methods.

Breeders use what is called "family heritability" to calculate the gain from a half-sib progeny test. Family heritability is calculated from the plot means and family means in the progeny test. There is a definite mathematical relationship between "family heritability" and the "true heritability" or "single tree heritability" that is applicable to single trees. Formulas for conversion from one to the other are given by Lerner (1958).

Table 9.2 shows the relationship between family heritability and single tree heritability for two sizes of half-sib families. Note that family heritability is 12 to 21 times as large as single-tree heritability if the latter is as low as 0.01, and only 3 times as large if the latter is 0.3.

From Table 9.2 it is possible to make rough comparisons between the gains possible from mass selection versus selection on the basis of offspring performance. The comparisons can be only approximate, however, because they do not take into account differences in selection differential. There is always more variation among single trees (hence greater selection differential) than among family means. Hence, the difference in gain between mass selection and selection based upon offspring perfor-

TABLE 9.2

Heritability of Single Trees and of Half-Sib
Family Averages for Two Sizes of Families

Heritability of a single tree	Heritability of a half-sib family average	
	50-tree families	100-tree families
0.01	0.12	0.21
0.02	0.21	0.34
0.05	0.41	0.56
0.10	0.60	0.74
0.20	0.76	0.87
0.30	0.86	0.93

mance is not as great as indicated by the differences in heritability. Actually, at low heritabilities, gains are 5 to 10 times as great if based upon offspring performance than from mass selection only. With single tree heritabilities of 0.3 or greater there is frequently little difference between the methods.

Costs of half-sib progeny tests are moderate. The extra cost is usually that of record keeping and measurement; seed collection and actual planting are often no more difficult than for commercial plantations of comparable size. Frequently, progeny tests cost no more than intensive mass selection.

The work described by Goddard et al. (1973) furnishes an example of an actual half-sib progeny test and the uses of the data from it. They selected 442 slash pine trees for superior growth rate and resistance to fusiform rust. With scions collected from those trees they established a clonal seed orchard. At the same time they collected open-pollinated seed and established a half-sib progeny test. In the progeny test they included the offspring of average trees from the same stands. They measured the progeny test at age 5.

Of the 442 families of selected-tree offspring, 362 (82%) grew faster than the controls. This indicates the mass selection for growth rate had been effective. However, selection on the basis of offspring performance was even more effective, because 80 of the selected parents had offspring which grew more slowly than the controls.

With regard to rust resistance, only 200 (46%) of the selected tree families had less rust than the controls, but the differences among families were statistically significant. This indicates that mass selection for rust resistance had been ineffective, probably because there had been too little rust in the parental stands.

Goddard *et al.* set arbitrary standards by which to rate the acceptability of the parents or families for continued inclusion in the clonal seed orchards or for further breeding work. Of the 442 parents which were progeny tested, 151 (34%) met these standards as regards growth rate, and 37 (8%) met these standards as regards both growth rate and rust resistance. They will regraft the 151 clones that have good general combining ability for growth rate, and establish new clonal orchards to produce seed for use in areas with little fusiform rust. They will also regraft the 37 clones that have good general combining ability for growth rate and rust resistance, and establish other new clonal orchards to produce seed for use in areas with much fusiform rust. They will also make crosses among those 37 clones as a prelude to further improvement.

Goddard *et al.* could also, if they wished, use the data from their half-sib progeny tests to thin their present clonal orchards. If so, they would eliminate all but the 151 clones with good general combining ability for growth rate, or all but the 37 clones with good general combining ability for growth rate and rust resistance. Many breeders use such data in similar ways.

Half-Sib Family Selection

Half-sib progeny tests can also be considered as part of a breeding method called "half-sib family selection." In this method, progeny tests are established and measured, and are then thinned to leave only the best families. After thinning, the progeny tests are considered as seed orchards or as breeding arboreta. The original parents may or may not be retained.

In a half-sib progeny test, a superior family is the result of crossing a genetically superior seed parent with several (presumably average) pollen parents. That family possesses only half the genetic superiority it would have if both seed and pollen parents were genetically superior, as would be the case in the clonal seed orchard of Goddard *et al.* Actually, gain from half-sib family selection may be slightly more than half as much as in the clonal orchard because there is opportunity for some mass selection within superior families.

For cost comparisons it is necessary to remember that half-sib family selection involves only half-sib progeny tests; the twofold gain from clonal orchards involves half-sib progeny tests and grafted orchards.

A paper by Yao *et al.* (1971) describes the first step in a half-sib selection program for red pine in Michigan. Seeds were collected from 272 trees in the fall of 1960; no preliminary mass selection was practiced because of the scarcity of suitable pure stands for selection work. The seeds were sown in the spring of 1961, and the resultant seedlings were field planted in the spring of 1963 in four permanent plantations. These were measured when the trees were 10 years old and will be remeasured at age 15, when fruiting is expected to start. The families varied significantly in growth rate only. The 136 shortest families will be removed at age 15, and subsequent thinnings will remove all but the 25 tallest families. According to statistical analysis of the age 10 data, those will produce offspring capable of 5% greater volume growth than would be obtained with seed from average trees.

In this study by Yao *et al.* gains from half-sib family selection were estimated to be somewhat greater than half as much as if all parents with superior combining ability had been cloned and placed in a grafted seed orchard. This is because the superior parents were nonrandomly distributed. Most parents that produced superior offspring were from a few stands. Probably other trees (the pollen parents) in those same stands were above average genetically also. Therefore, the superior half-sib progenies were not the result of crosses between superior seed and average pollen parents, but of crosses between superior seed parents and above-average pollen parents.

Wells and Switzer's (1971) study of loblolly pine is another example of half-sib family selection. They collected seeds from 5 trees in each of 115 stands in Mississippi and Louisiana and grew the seedlings in replicated progeny tests. Seedlings from an area in southeastern Louisiana grew rapidly and were resistant to fusiform rust. Seedlings from an area in northern Mississippi grew rapidly but were susceptible to fusiform rust, as were seedlings from most other areas. Data from this experiment can be used to designate entire natural stands in southeastern Louisiana as *pro tem* seed orchards consisting mostly of parents superior in both traits. When the half-sib progeny test flowers, it can be thinned to become a seed orchard to replace the natural stands. Because of the nonrandom distribution of the parents in nature, gain will be nearly as great as if the superior parents had been cloned and placed in a clonal seed orchard.

Full-Sib Progeny Tests and Full-Sib Family Selection

Full brothers and full sisters share the same mother and father. Similarly, the members of a full-sib tree family share the same seed

Fig. 9.5. A full-sib progeny test of Douglas-fir in western Oregon. The younger trees, 5 years after planting, represent the offspring of many matings between selected females and selected males. The objective of such a full-sib progeny test is to provide data on specific combining ability. (Photograph courtesy of Kim K. Ching of Oregon State University.)

Fig. 9.6. Part of a full-sib progeny test of Virginia pine. A commercial seedlot (left) is compared with the control-pollinated offspring of two selected trees (right) on lands of the Hiwassee Land Company in Tennessee. (Photograph courtesy of North Carolina State University at Raleigh.)

parent and pollen parent. A full-sib progeny test is a replicated experiment containing many full-sib families that are the result of crossing various seed parents with various pollen parents. In full-sib selection, trees are selected on the basis of average performance of the full-sib families (Figs. 9.5–9.7).

For reasons of convenience, most crossing work is done either in clonal seed orchards or in young half-sib progeny tests. At least in most pines, the costs of such work are moderate. Often the crossing work is spread out over two to three seasons, during which time a crew of workers can produce 300 to 500 full-sib families. The costs are prohibitive in some genera such as oak.

Fig. 9.7. Two 7-year-old loblolly pine full-sib families growing in northern Florida. Selection for stem straightness has been very effective in loblolly pine. All trees in the family on the left are straight, whereas nearly all in the family on the right are crooked (straight tree in foreground at right is part of the border row.) (Photograph courtesy of Ray E. Goddard of the University of Florida.)

Various mating schemes are possible. Some are shown in Table 9.3. The principal types are as follows.

Diallele. With a complete diallele crossing pattern, each parent is crossed with every other parent; reciprocals and selfs are usually excluded. With N parents there are $N(N-1)/2$ possible combinations. This becomes very cumbersome if there are more than 15 parents.

Systematic, progressive. There are various schemes, one of them illustrated in Table 9.3B, by which many pollen parents can be crossed with many seed parents according to some regular mating scheme in such a way that no particular pollen or seed parent is used in more than a few combinations.

Tester. The same few pollen parents are crossed with each of many seed parents, as shown in Table 9.3C.

Irregular. Each pollen parent is crossed with a few seed parents, not according to any regular scheme (Table 9.3D). In the example shown, each seed parent is crossed with four pollen parents, but the number of crosses made with any particular female parent or with any particular pollen can vary. This can be carried to the extreme of using each female and male parent only once.

Full-sib progeny tests yield information about general combining ability and also about "specific combining ability," which is a tree's ability to transmit superiority to its offspring when crossed with a specific tree rather than with trees in general. General and specific combining abilities result from additive and nonadditive genetic variance, respectively. In most experiments, general combining ability has been most important.

A good full-sib family owes its superiority to a mating between a genetically superior female and a genetically superior male. Therefore mating schemes which use large numbers of female and male parents (e.g., those illustrated in Table 9.3B and D) give most improvement. Another way to obtain much improvement is to learn general combining ability from many parents from a half-sib progeny test, then make crosses only among those parents with proven high general combining ability.

With a complete diallele crossing scheme (Table 9.3A), genetic gain is approximately twice as great as with half-sib family selection; one unit of gain is contributed from the female parent and one unit from the male parent.[2] That is true if general combining ability is not

[2] Note in the previous section exceptions to this statement if superior parents are distributed nonrandomly.

TABLE 9.3

Examples of Four Different Mating Schemes Used in Full-Sib Progeny Testing[a]

A. Diallele, no reciprocals

Female	Male									
	A	B	C	D	E	F	G	H	I	J
A	—	—	—	—	—	—	—	—	—	—
B	×	—	—	—	—	—	—	—	—	—
C	×	×	—	—	—	—	—	—	—	—
D	×	×	×	—	—	—	—	—	—	—
E	×	×	×	×	—	—	—	—	—	—
F	×	×	×	×	×	—	—	—	—	—
G	×	×	×	×	×	×	—	—	—	—
H	×	×	×	×	×	×	×	—	—	—
I	×	×	×	×	×	×	×	×	—	—
J	×	×	×	×	×	×	×	×	×	—

B. Progressive, 4 pollen parents

Female	Male													
	A	B	C	D	E	F	G	H	I	J	K	L	M	N
A	×	×	×	×	—	—	—	—	—	—	—	—	—	—
B	—	—	×	×	×	×	—	—	—	—	—	—	—	—
C	—	—	—	—	×	×	×	×	—	—	—	—	—	—
D	—	—	—	—	—	—	×	×	×	×	—	—	—	—
E	—	—	—	—	—	—	—	—	×	×	×	×	—	—
F	—	—	—	—	—	—	—	—	—	—	×	×	×	×
G	×	×	×	×	—	—	—	—	—	—	—	—	—	—
H	—	—	×	×	×	×	—	—	—	—	—	—	—	—
I	—	—	—	—	×	×	×	×	—	—	—	—	—	—
J	—	—	—	—	—	—	×	×	×	×	—	—	—	—

C. Tester, 4 pollen parents

Female	Male			
	N	O	P	Q
A	×	×	×	×
B	×	×	×	×
C	×	×	×	×
D	×	×	×	×
E	×	×	×	×
F	×	×	×	×
G	×	×	×	×
H	×	×	×	×
I	×	×	×	×
J	×	×	×	×

D. Irregular, 4 pollen parents

Female	Male									
	A	B	C	D	E	F	G	H	I	J
A	—	×	—	—	—	—	×	×	—	×
B	×	—	—	×	×	—	—	—	×	—
C	—	—	×	×	—	×	—	×	—	—
D	×	—	—	—	×	—	×	—	×	—
E	—	×	—	×	×	—	—	×	—	—
F	×	×	—	—	—	×	—	—	×	—
G	×	—	×	×	—	—	×	—	—	×
H	—	—	—	×	—	×	—	×	—	×
I	—	×	—	—	×	×	—	—	×	—
J	×	—	×	—	—	—	×	—	×	×

[a] Many other schemes are possible. Crosses made and not made are denoted by × and —, respectively. In part A, diallele, with every tree crossed with every other parent. B, Progressive, in which each pollen parent is used the same number of times, but with slightly different parents each time; crosses are planned systematically. C, Tester, in which the same few pollen ("tester") parents are crossed with all females. D, Irregular, in which each female is crossed with a few males, but not according to any regular scheme.

TABLE 9.4

Size of Half-Sib and Full-Sib Progeny Tests
Needed to Produce Equal Gain

	Full-sib families		
Half-sib families	Female × Male		= Total
15	4	5	20
44	8	5	40
161	21	5	105
741	70	5	350
4300	250	5	1250

known beforehand and if the half-sib and full-sib diallele progeny tests involve equal numbers of parents.

That is not apt to be the case, and therefore it is often more useful to consider the sizes of half-sib and full-sib progeny tests which produce equal genetic gain. Such comparisons are made in Table 9.4 which was prepared on the assumption that each female is crossed with five males. The numbers were calculated (from a table of areas under the normal curve) on the assumptions that (1) the best half-sib family results from crossing the 1 female with the highest GCA (general combining ability) of X many females with average males, and (2) the best full-sib family results from crossing the 1 highest GCA female of X many females with the 1 highest GCA male of 5 males.

As can be seen, small (i.e., 44 families or fewer) half-sib progeny tests are about equal to full-sib progeny tests of the same size. However, many modern progeny tests include 100 or more families. With increasing size, the advantages of the full-sib progeny tests become more clear-cut.

Goddard *et al.* will practice full-sib family selection in slash pine. They have already determined general combining ability of 442 parents from half-sib progeny tests. They will cross those parents with the best general combining ability to obtain a large number of full-sib families. They will then select the best trees within the best of those families.

Bingham (1968) and Bingham *et al.* (1969) practiced full-sib family selection for blister rust resistance in western white pine. They started with mass selection and chose 200 rust-free parents from very heavily infected stands. In the first experiment, they used an irregular mating scheme and made 77 crosses among 21 of those parents. They also grew open-pollinated offspring of the selected trees and of average trees. The proportions of resistant trees from various types of matings are shown in tabulation on p. 179.

Average × wind (= controls)	5%
Selected × wind	9%
Selected × selected (all families)	18%
Selected × selected (5 best half-sib families)	27%
Selected × selected (5 best full-sib families)	41%

The heritability of full-sib family averages was 0.69. Thus there was a genetic gain of 0.69 (9 — 5) = 3% from mass selection if seed were to be collected from selected trees surrounded by average trees or of 0.69 (18 — 5) = 9% from mass selection if the selected trees were to be placed in a seed orchard where they could pollinate each other. There was a genetic gain of 0.69(41 — 5) = 25% from full-sib family selection. In this, as in other experiments, Bingham and his co-workers found that trees with high specific combining ability also had high general combining ability. That is, data from a half-sib progeny test could be used to indicate which full-sib families would perform best.

In some sections of northern Idaho where Bingham worked, blister rust has taken a very heavy toll, frequently killing 99+% of the western white pines in the seedling stage. That is, natural selection has been operative. Presumably the gain from this natural selection will be 9%, i.e., the mass selection gain. In other words, 5 + 9 = 14% of the offspring of the natural survivors should be resistant. As of 1974, natural selection has nearly run its course in some areas, leaving only resistant trees to pollinate other resistant trees. If an open-pollinated, half-sib progeny test should be conducted now, there should be a gain of 0.69 (27 — 5) = 15% by selecting the five best half-sib families.

The original full-sib progeny tests of western white pine were 20 years old and fruiting in 1971. If thinned to leave only the best families (remember that blister rust has already done much thinning), they are seed orchards.

Most of Bingham's data were collected from nursery experiments in which currants and gooseberries (alternate hosts for white pine blister rust) were interplanted to provide exceptionally high levels of the disease. There was undoubtedly overkill. The expected frequencies of survivors are higher than shown above if the seedlings are planted under natural conditions where conditions for the disease are less favorable.

Combined Selection

By combined selection is meant selection of the best families and best trees within families. In actual practice this is the most common type of selection. A paper by Nanson (1974) includes tables showing the relative merits of combined selection, mass selection, family selection, and intrafamily selection for varying heritabilities and family sizes.

Optimum Family Size

In combined selection some gain results from selecting the best families and some from selecting within the best families. Optimum family size is that which results in the maximum combined gain. Three factors must be considered in calculating optimum family size. The first is total size of progeny test; the larger the experiment the greater the size of each family.

The second is heritability. The lower the heritability the greater the family size needed to detect statistically significant interfamily differences and hence to achieve gain from between family selection. The lower the heritability, the greater the gain from interfamily versus intrafamily selection. The greater the family size, the fewer the families, the smaller the selection differential among families, and the smaller the gain from interfamily selection. The net result of these conflicting trends is to cause optimum family size to be greater for low than for high heritabilities.

Third, intensity of selection must be considered. Inbreeding results if a progeny test is thinned to too few families, and a breeder must decide in advance how many families are to be retained when the final selections are made. The more families to be retained, the less accurate the performance estimate of a particular family needs to be, and the smaller the family needed to provide such an estimate.

Robertson (1957) worked out an approximate formula for half-sib families incorporating these three variables. His formula is

$$\text{Optimum half-sib family size} = 0.56(T/Nh^2)^{1/2}$$

where T, N, and h^2 are the total number of trees in test, number of families to be retained, and single tree heritability, respectively.

Note that in using this formula it is necessary to fix the total size of experiment and number of families to be retained and to estimate heritability. Robertson's formula was used to prepare Table 9.5, applicable to a half-sib progeny test originally including 5000 trees.

Robertson's formula is strictly true only if a progeny test is conducted with one-tree plots. Multiple-tree plots are usually used, and they are not as efficient statistically as one-tree plots. Hence, with multiple-tree plots somewhat larger family sizes are optimal.

With full-sib progeny tests there is less intrafamily variation and greater need to retain a large number of families than half-sib progeny

TABLE 9.5

Optimum Family Size in a 5000-Tree Progeny Test

Single tree heritability	Optimum number of trees per family if test is to be reduced to	
	5 families	10 families
0.1	56	40
0.2	40	28
0.3	32	23
0.4	28	20
0.5	25	18

tests. Hence optimal family size tends to be somewhat smaller with a full-sib than with a half-sib progeny test.

Actual family sizes in forestry experiments have often been much larger than the optimum family sizes from Robertson's formula. Thus, greater gain can be expected by increasing the number of families tested and decreasing the number of trees included in each family.

Clonal Selection

A "clonal test" is an experiment in which rooted cuttings (willows and poplars) or grafted trees (some horticultural shrubs and trees) are planted according to a replicated design and measured after a period of years. In clonal selection, clones are chosen on the basis of their average performance in such an experiment.

According to some results obtained in E. J. Schreiner's experiments with poplars, clones may grow differently if planted in different years. This would indicate that clonal tests should be repeated in each of several years for the best results.

Clonal selection should not be confused with clonal propagation. A particular willow or cottonwood tree may be chosen on the basis of its phenotypic excellence (i.e., be mass selected) and then be clonally propagated without having been clonally tested. Or, different species could be hybridized to produce hybrid full-sib families; various clones in those families could be propagated commercially without having been clonally tested.

Because of topophysis, clonal selection is most useful in those groups which are propagated commercially by clonal means. For timber production, clonal selection is effectively limited to willows and cottonwoods.

It is also warranted in some nut-producing trees propagated by grafting and in a great many horticultural plants.

In horticulture, clonal selection is most frequently used in those groups in which the parents are themselves very complex hybrids which produce exceedingly variable offspring (apples, tea roses, etc.). Crosses are made among promising parents, and promising seedlings are mass-selected and clonally tested. Seedlings of such crosses are so variable that differences among them can be detected by planting a few ramets per clone.

It is probable that clonal selection should be limited to such situations in forestry. At the very least, clonal selection should follow family selection, and should be limited to clones of the best families.

Stevens (1949) pointed out the disease and insect hazards inherent in vegetatively propagated crops. He studied a large number of different crops and classified them as to whether they were clonally propagated (absolute genetic uniformity within a variety), selfed (relatively great genetic uniformity within a variety), or cross-pollinated (relatively variable within a variety). Seriousness of disease problems, as measured by numbers of diseases, money spent on control, etc., was about three times as great for the clonally propagated as for the cross-pollinated crops. That problem is magnified if clonal selection is carried to the ultimate of selecting only one or a few best clones to be propagated on a large scale. Poplar clone "I-214" selected in Italy is an example. It was chosen for its excellent growth traits and freedom from pests and has been widely planted. Its commonness in pure stands has generated new disease problems. Elm clone "belgica" is another example. Originally chosen for its form and growth rate, trees of that clone comprised about 300,000 of the million elm trees in the Netherlands. It happened to be susceptible to Dutch elm disease and therefore was almost a total loss when that disease became serious.

The planting of selected clones results in great uniformity. In itself, uniformity is desirable in many horticultural crops. A peach grower, for example, wishes a variety which will ripen uniformly, ship uniformly well, and be uniform in taste after freezing or canning. A grower of black walnuts for nut production wishes a variety with uniform (good) taste, uniform shell thickness, etc. Uniformity per se is less necessary and sometimes is undesirable in a tree grown for wood. To escape from the disease and other problems associated with uniformity, Schreiner (1966) recommended the planting of clonal mixtures.

It should also be realized that clonal selection involves an extra test rotation per generation. Assume that many parents are crossed in year 0 and that full-sib families are raised and measured at age 10 (a common harvesting age in fast-growing willows and cottonwoods). Selected families could then be recommended for planting. But with clonal selection, seedlings from the best families are mass-selected and clonally tested, and the selected clones can only be recommended at age 20. (Alternatively, every seedling in every family can be clonally tested from year 1 to year 10, but that is very expensive.) To justify this extra rotation, intrafamily differences should be as large as inter-family differences.

In hybrid tea roses, crosses are made among good existing varieties or open-pollinated seed is collected from plants with good color, hardiness, etc. The resulting seedlings are then grown until flowering age when they are mass-selected, and the best are chosen for clonal testing. These are grafted (they grow better if not on their own roots) and tested a few years in the breeder's own test plots. Then the most promising clones are regrafted and 2 to 3 plants per clone are distributed to specially designated "rose test gardens" situated in many parts of the country. After 2 to 3 years the local person in charge of a rose test garden grades the clones for color, flower size, hardiness, freedom from pests, etc. Those results are communicated back to the breeder and clones with the best overall ratings are released to the public.

Poplar breeders in southern Europe generally mass-select natural seedlings from stands of natural hybrids between European and eastern (America) cottonwood. They clone these but usually do not undertake formal, replicated clonal tests. They maintain a high selection differential and release only a few clones for commercial propagation. The data are not usually analyzed in such a way as to distinguish between intra-family and interfamily selection gains. Many further details of poplar breeding and culture in Europe are contained in a book by Schreiner (1959).

Schreiner and others in northeastern United States artificially crossed 33 different trees (mostly belonging to different species) in the late 1920's grew the offspring in Maine, and made mass selections among them. Starting in the late 1940's, 250 trees were mass-selected from many different full-sib families. These were cloned and tested in a series of replicated clonal tests. There have been large differences among clones in growth rate, disease resistance, form, etc., but the extent to which these represent variation within families and between families is not known. After 10 to 15 years of clonal testing, mixtures of 25 to 50 clones, often belonging to several families, have been released for commercial use.

Poplar breeders in southern United States have concentrated on native populations of eastern cottonwood. They have performed half-sib and full-sib progeny tests and have done a limited amount of clonal testing to determine the additional gain possible from clonal selection. In a 2-year experiment, Farmer (1970) found much greater differences among than within families in date of foliation and resistance to *Melampsora* rust, smaller differences among than within families in growth rate and specific gravity.

Barrett and Rial Alberti (1972) performed a special experiment to determine the most suitable breeding method for the improvement of willow in Argentina. By artificially crossing various species and selected hybrids, they produced 15 full-sib families and grew these 4 years. The tallest family averaged 6.8 m (range 5.0 to 8.2 m); the second tallest family averaged 5.5 m (range 3.7 to 6.5 m); and the third tallest family averaged 4.6 m (range 3.7 to 5.1 m). Note the relatively small overlap between the tallest and next tallest families. From these data they concluded that family selection should have highest priority. They also established clonal tests including several clones per family, but results from these were not available as of 1974.

Thulin (1969) described a proposed clonal selection program for Monterey pine in New Zealand. That is one of the few pines which can be grown with moderate ease from rooted cuttings. The program will start with 600 mass-selected clones (each the best tree on a 3-acre block) and will involve the planting of 8,000,000 cuttings by 1981. In the early years the cuttings will be planted in replicated clonal tests and selection in these will ultimately reduce the number of selected clones to 50. The per tree cost is estimated to be about 3 times as great as if seedlings were used, but potential genetic gains are expected to be great. For future generations, the clonal selection program will be integrated with family selection programs, and new clones will be used in the second and third generations.

Study Questions

1. Give the formula for genetic gain and define each of its two components.

2. How can selection differential be increased? Is selection differential higher in mass selection or family selection, and how would it be computed in each case?

3. How do total gain, gain per trait, and economically valuable gain vary with the number of traits selected for?

4. Describe three ways to increase heritability. Why is family heritability higher than single tree heritability?

5. Differentiate between mass selection, half-sib family selection, full-sib family selection, and clonal selection.

6. Describe a bulk progeny test and tell how it can be used to test the effectiveness of mass selection.

7. Assume that parents A through L have been selected and measured, and

that their open-pollinated offspring have been grown and measured, with the following relative heights (the mean is 100 for all parents and for all progenies).

Generation	Relative height of parent or offspring											
	A	B	C	D	E	F	G	H	I	J	K	L
Parent	110	108	106	104	102	100	100	98	96	94	92	90
Offspring	101	99	105	97	103	102	99	101	96	100	97	100

Assume you will select the 4 best parents or offspring.

(a) Compute the selection differential with mass selection.

(b) Compute the average performance of the offspring with mass selection.

(c) Compute the average performance of the offspring with family selection.

(d) Compute the single-tree heritability with mass selection.

8. Draw a diagram showing the difference between mass and family selection with low heritability. With high heritability.

9. Describe the relations between family and single tree heritability with varying heritability.

10. Was Goddard's mass selection for growth rate and rust resistance equally effective. Describe how he can use data from the half-sib progeny test to increase the genetic gain from his seed orchards. Why does he wish to retain at least 37 clones in his seed orchard?

11. How does the red pine half-sib progeny test of Yao differ in design from the slash pine half-sib progeny test of Goddard? In what ways will the experiments be treated differently?

12. Differentiate between half-sib and full-sib progeny tests.

13. Under what conditions is the gain from a full-sib progeny test approximately twice that from a half-sib progeny test? Under what conditions are the gains equal? Under what conditions are the full-sib gains greater but not twice as great as the half-sib gains?

14. Describe four different mating systems for full-sib testing.

15. Differentiate between general and specific combining abilities.

16. How do the gains from an unthinned clonal seed orchard compare with the gains from a clonal seed orchard thinned in accordance with results from a half-sib progeny test? With results from a full-sib progeny test?

17. Assume that all clones have been progeny tested and the clonal orchard has been thinned to leave only those with high general combining ability. Assume that those selected clones are crossed with each other to produce full-sib families. Compare the genetic gains.

18. Give the formula for optimum family size and tell in words how it is affected by T, N, and h^2.

19. Describe clonal selection.

20. In clonal selection, why is it desirable to differentiate between differences within and between families?

21. Discuss the advantages and disadvantages of the uniformity engendered by clonal selection.

22. Describe how to propagate superior clones of willow without practicing clonal selection.

10

Choice of Breeding Method and Type of Seed Orchard

Factors Affecting Choice

There is no one "best" procedure for all situations. Choice depends on several factors. The principal ones are discussed here.

①Genetic gain is important. If heritability is very low, family selection may result in some gain whereas mass selection may result in no demonstrable gain. With moderate heritabilities, family selection may give 2 to 4 times as much gain as mass selection.

②The amount of available inheritance data must also be considered. If none are available, they must be obtained by means of progeny tests, preferably simple ones. If many are available, there is an opportunity to choose between mass selection and various types of family selection.

③ Costs are also important. Half-sib family selection seems the only practical approach to improvement of the oaks, in which controlled pollination and vegetative propagation are costly. But full-sib family selection and clonal seed orchards are reasonable alternatives to other methods in the pines and spruces, which can be control-pollinated and grafted relatively easily.

The choice of seedling versus clonal seed orchard depends primarily on three factors: earliness of flowering of grafted trees and of seedlings, relative difficulty and speed of establishing grafted trees and seedlings, and selection differential possible with clones and families. Grafted trees may or may not flower appreciably earlier than seedlings; if they do clonal orchards have a time advantage. Pines require 2 years to produce seed and must be grown at least 1 year in a nursery prior to field planting. Thus there must be at least a 3-year lag between selection and controlled pollination and the field planting of full-sib families; grafted trees in contrast could start fruiting very soon after selection. Poplars, on the other hand, mature their seeds so quickly and grow so rapidly that one can have tall full-sib families or clones within a few months of doing selection work. In a half-sib or full-sib progeny test which will become a seedling seed orchard there is a very definite limit to the size of the selection differential, as one cannot realistically plant hundreds or thousands of trees of each of hundreds or thousands of families in order to be able to select the best 1% of the trees in the best 1% of the families. However, if heritability is reasonably high, one can realistically practice mass selection and include only the best 1% of the trees as clones in a clonal orchard.

Presumably most improvement programs will continue for several generations. Therefore, first-generation work must set the stage for future work by providing inheritance data and guarding against severe inbreeding as the work progresses. Regardless of the type of seed orchard desired, large-scale progeny testing (and probably some reliance on family selection) are required during the first generation.

Need should be considered. Each year, slash and loblolly pines are planted by the hundred million. In these species, full-sib family selection promises about as much gain as clonal orchards accompanied by progeny tests, and possibly at a lower cost. But the clonal orchards can produce several hundred million improved seed during the waiting period for the full-sib progeny tests to flower; that improved seed is more than worth the extra cost. This might not be true in a species planted at the rate of only one million trees per year.

The following sections include more detailed information to help in the very complex problem of deciding the best procedures for particular situations.

Thinned Natural Stands versus Planted Seed Orchards

A natural stand or plantation can be thinned and managed to increase seed production. The usual benefit is cheaper seed, providing the trees have large enough crowns to respond to thinning and are not so large as to make seed collection expensive. Genetic gain is usually small because of the low selection differentials. Costs higher than those sustained in an ordinary silvicultural thinning are rarely justifiable. Thinned natural stands can be recommended in place of planted seed orchards in the following situations:

1. Minor species, in which intensive breeding work is unwarranted

2. After a combination half-sib progeny test-provenance test, if intra-stand differences are small and minimum fruiting age is large. Ponderosa pine in Idaho is an example (see Chapter 11). Wang and Pattee's progeny tests will fruit and become seed orchards in 15–20 years. The progeny tests have already pinpointed genetically superior stands, which are the best seed orchards for the next 15 to 20 years

3. After provenance tests (see Chapter 13) if inter-provenance dif-ferences are large and the genetically superior stands are readily avail-able to seed collectors. Douglas-fir for use in Michigan is an example. Trees from parts of Arizona and New Mexico grow fastest. Pending the start of intensive breeding work in Michigan on the Arizona–New Mexico race, Michigan growers can obtain the cheapest and best seed from designated natural stands in Arizona and New Mexico

4. After a heavy epidemic of a damaging disease. R. E. Goddard found the average incidence of fusiform rust in 117 open-pollinated slash pine families obtained from a heavily rusted site in Florida to be only 28%, in contrast to an incidence of 48% for 126 families selected in an area of low rust incidence. The heavily rusted natural stand can be considered as an improved seed orchard.

Mass versus Family Selection

Family heritabilities are always higher than single tree heritabilities (see Chapter 9). However, the differences among family means are not so high as those among individual trees. Thus, selection differential is always higher in mass than in family selection. The total result is greater gain from family than from mass selection, but not by such a great margin as indicated by the differences in heritabilities.

As already indicated in Chapter 9, computation of expected gains is a complex matter, depending very much on the numbers of trees involved. The possible assumptions are infinite. Table 10.1 (from Lerner, 1958) was prepared under the assumption that the total number of trees involved, parents plus offspring, remains the same under all types of selection. It was prepared for animal breeders, who usually deal with herds or flocks of constant total size. In trees it would be more realistic to assume that the number of offspring could be expanded to many times the numbers of parents. If calculated in that manner, the relative gains from family selection would be increased. Nevertheless, Table 10.1 is a good approximation of relative gains.

In Table 10.1 the genetic effectiveness of mass and three types

TABLE 10.1

Gains from Various Types of Family Selection, Relative to Those from Mass Selection, for Two Family Sizes and Varying Heritabilities[a]

| Single tree heritability | Gains (%), relative to those from mass selection, from | | | | | |
| | Half-sib family selection with family size of | | Combined half-sib + mass selection with family size of | | Full-sib family selection with family size of | |
	50 trees	100 trees	50 trees	100 trees	50 trees	100 trees
0.01	177	230	191	242	324	413
0.02	168	210	183	222	296	358
0.05	148	172	164	187	243	302
0.10	125	134	148	157	193	206
0.20	101	102	124	131	149	153
0.30	87	89	116	118	125	126

[a] From Lerner, 1958.

of family selection are compared. Note that the advantages of family selection are much greater at low than at high heritabilities. For a realistic comparison, then, it is essential to have some data on heritability, which means that some progeny testing is necessary before making any final decision.

Decisions should be made on the basis of cost and effectiveness. Mass selection can be inexpensive if it consists only of choosing the best of each of two or three trees as a probable parent. It can be very expensive, however, if it consists of searching hundreds of stands for the one best tree in each stand.

Costs of half-sib progeny tests vary from moderate to high. Many pines and spruces fall into the "moderate" category. In those groups, 1 to 2 man-months may suffice for the establishment of a 500-family, open-pollinated progeny test, provided planting conditions are favorable and there is a reasonably high juvenile–mature correlation (permitting elimination of half to two-thirds of the families prior to field planting). Another 1 to 2 man-months usually suffice for its measurement through age 15.

A breeder usually wishes to achieve a balance between gain and expenditure. With a single tree heritability of 0.01, gain from half-sib family selection is approximately 2 to $2\frac{1}{2}$ times as great as from mass selection. In that case a breeder would be justified in spending 1 to $1\frac{1}{2}$ man-months on establishment and measurement of progeny tests for every man-month spent on phenotypic selection of parents. With a single tree heritability of 0.20, gain from half-sib family and mass selection is 20 to 30% greater than from mass selection alone. In this case a breeder could advantageously spend most of his time on phenotypic selection of parents.

Half-Sib versus Full-Sib Progeny Testing

Sometimes preliminary progeny tests are conducted to screen a large number of parents for inclusion in a clonal seed orchard or a full-sib family selection program. Goddard et al. (1973) did this in slash pine to learn which of the 442 clones in their present seed orchards merit inclusion in special seed orchards or in the full-sib selection program. For that purpose they needed information on general combining ability of each clone. This is so for two reasons. First, most experiments have shown little enough difference between general and specific combining ability as to permit the latter to be ignored. Second, they cannot use information on specific combining ability when selecting clones for a clonal seed orchard unless they wish to reduce their orchard to only 2 clones ultimately. They actually desire to include at least 37 clones. There is no way in which each of 37 clones can be mated with every other clone to take advantage of high specific combining ability as opposed to high general combining ability.

Full-sib progeny tests, especially those made with a "tester" mating scheme, provide statistically more precise information on general combining ability than do open-pollinated progeny tests. However, the increase in precision is rarely sufficient to justify the additional cost. For example, in Bingham's work with western white pine, there were only slight differences in the ranking of parents for general combining ability according to whether the ranking was based on data from open-pollinated or control-pollinated progeny tests.

Thus, for screening purposes, half-sib and full-sib progeny tests are nearly equal.

Half-Sib versus Full-Sib Family Selection

Half-sib and full-sib progeny tests are not the same in all respects, however. Gains possible from full-sib family selection (i.e., selection of the best families in a full-sib progeny test) are always greater than those possible from half-sib family selection. The exact amount of the difference depends on the geographic distribution of genetic variation in natural stands, on heritability, and on the numbers of parents and families tested.

Two very different situations must be considered. First is the situation in which all parents in particular stands produced superior offspring. That was true for ponderosa pine in Idaho, Maximowicz birch in Hokkaido, and loblolly pine in Mississippi. In these cases parents with superior general combining ability were surrounded by other trees with superior general combining ability. Even open-pollinated offspring tended to be the result of superior × superior matings. If a diallele mating scheme (every parent mated with every other parent) had been used, little additional gain would have resulted from full-sib family selection.

The second situation is one in which genetic superiority is distributed more or less randomly, as seems to be the case with blister rust resistance in western white pine, growth rate of slash pine in Florida, and growth rate of loblolly pine in parts of its range. In this situation the approximate gain comparisons shown in Table 10.1 are applicable.

It is possible to make decisions purely on a cost basis and to practice full-sib instead of half-sib family selection only if the gain ratio exceeds the cost ratio. However, it is usually better to consider whether the entire benefits (as influenced by gain, acreage planted, etc.) outweigh prospective costs and not to calculate costs too finely.

Species is a major factor influencing probable costs of controlled pollination work and full-sib family selection. At one extreme are the poplars and willows. In them, controlled pollinations are so easily made that full-sib family selection is nearly always preferable to half-sib family selection. Seedlings of those groups need much care during the first 2 to 3 years, and the cost of such care is usually a very large percentage of the total cost of progeny testing. At the other extreme are oaks and beeches, in which controlled pollination is so expensive that full-sib family selection cannot be considered realistic.

Most pines and spruces are intermediate. Under favorable conditions, one man can produce 50 to 200 full-sib families in a single breeding season as the result of 2 to 3 weeks work. A small group of breeders can expect to produce enough control-pollinated seedlings for a good full-sib family selection program in 1 to 3 seasons. If the species is a commonly planted one, the additional cost of full-sib family selection can usually be justified easily.

Generation is another factor which influences costs of controlled pollination work. In most examples quoted in Chapters 9 and 11, the parents were tall wild trees scattered over large areas. They would have been difficult to climb and pollinate artificially. Working under such conditions, a breeder might spend one tree generation producing the numbers of control-pollinated families necessary for full-sib family selection. He might better achieve less gain with half-sib family selection the first generation. Then he can make controlled pollinations on the relatively small and conveniently located trees in open-pollinated progeny tests, and practice full-sib family selection in the second generation.

This problem can be partially avoided by cloning all parents. Then controlled pollinations can be made on relatively small, conveniently located grafted trees. This lengthens generation time because it is necessary to do the grafting and wait for the grafted trees to flower before undertaking controlled pollination work. Therefore its advisability depends on the ages at which seedlings and grafted trees produce flowers. To explain more fully, it is desirable to symbolize these ages as follows.

YS = years required for seedlings to flower
YG = years required for grafted trees to flower
YE = years required to evaluate seedling progenies (often assumed to be the same as YS)

With this symbolization, YS or YE years are needed to accomplish half-sib family selection in an open-pollinated progeny test. $YS + YG$ years are needed to clone the parents, practice controlled pollination on them, and practice full-sib family selection in the progeny tests. If $YG = \frac{1}{2}YS$ (probably the case in slash and loblolly pines) the extra gain from ful-sib family selection can be achieved in 1.5YS years. If $YG = YS$ (probably the case with some races of Scotch pine), full-sib family selection requires about twice as long as half-sib family selection. If YG exceeds YS (possibly true in black cherry and jack pine), full-sib family selection requires more than twice as long as half-sib family selection.

Many data indicate that YE is equal to or slightly less than YS, in which case YS should be used to calculate generation length. However, there are probably instances in which YE is appreciably greater than YS, in which case YE should be used to calculate generation length.

Considering probable gains, costs, and time, half-sib family selection is usually preferable in the first generation if YG equals or exceeds YS; full-sib family selection is preferable if YG is 25–50% less than YS.

Avoidance of inbreeding is another factor involved in the choice between half-sib and full-sib family selection. Any open-pollinated family is the result of a mating between one female and several male parents. Also, the establishment of a large progeny test is usually much more economical with open-pollinated than with control-pollinated progenies.

Both factors combine to make the avoidance of inbreeding easier to accomplish with half-sib than with full-sib family selection.

Progeny Test Seed Orchards versus Clonal Seed Orchards

In family selection, the progeny test itself may serve as a seed orchard. To accomplish this, the test is measured and thinned to leave only superior families. There is usually some additional intrafamily selection. Those superior families will transmit that part of their superiority which is due to additive genetic variance to their offspring. In this case, progeny testing and seed orchard development are one and the same thing (Figs. 10.1–10.3).

Clonal seed orchards may yield somewhat improved seed due to mass selection or greatly improved seed due to family selection. However, the higher gains are possible only if there are accompanying pro-

FIG. 10.1. A seedling progeny test seed orchard of birch (*Betula verrucosa*) in Finland. The greenhouse (100 × 20 × 7.5 m) contains 125 seedlings moved indoors from outside progeny tests at ages 1 to 4, approximately 4 months prior to the picture. The greenhouse is unheated but flowering is stimulated by a CO_2-enriched (2000 to 2500 ppm) atmosphere. In 1974 (the third year inside) the 125 trees produced 40 kg (88 lb) or 80 million seeds; production for 1975 is estimated as 50% greater. (Photograph courtesy of Lauri Kärki of the Finnish Forest Tree Breeding Association.)

FIG. 10.2. A progeny test seed orchard of black walnut in southern Illinois. The seed orchard is actually part of a family selection program. By the use of continued weed control, Dr. Bey stimulated nut production by age 4 from seed. (Photograph courtesy Calvin Bey of the United States Forest Service, Carbondale, Illinois.)

FIG. 10.3. Practicing selection within a seedling seed orchard of black walnut. The 3-year-old half-sib families are growing in multiple-tree row plots, with closer spacing within than between plots. Thinning will start quickly to leave the best trees in the best families. (Photograph courtesy of Calvin Bey of the United States Forest Service, Carbondale, Illinois.)

geny tests showing which clones to favor as parents. Thus, gain and cost comparisons should always include clonal orchards, progeny tests, and clonal orchards + progeny tests. In addition to cost and gain, relative fruiting ages of clones and seedlings must be considered.

These factors are considered in greater detail in the following sections.

Unaccompanied Clonal Orchards versus Half-Sib Progeny Tests

An "unaccompanied clonal orchard" is defined as one for which there is no accompanying progeny test, or in which thinning to the clones with the best combining ability has not yet occurred (Fig. 10.4). In such an orchard the genetic gain results entirely from mass selection and, judging by most examples quoted in Chapters 9–11, is apt to be less than could be obtained by thinning a half-sib progeny test to leave only the superior families. The cost is ordinarily greater for a clonal orchard than for a progeny test, and a clonal orchard does not contain within itself proof of its effectiveness as does a progeny test.

Therefore, considering both gains and costs, open-pollinated half-sib progeny tests are usually preferable to unaccompanied clonal seed orchards.

Accompanied Clonal Orchards versus Half-Sib Progeny Tests

An "accompanied clonal orchard" is one which has an accompanying progeny test, either half-sib or full-sib. Data from the progeny test are used to select those clones to be included in the orchard ("delayed clonal orchard") or to be retained after thinning (in a "simultaneous clonal orchard") (Fig. 10.5).

In such a clonal orchard, genetic gain results from full-sib family selection. The final orchard contains only clones with proved high general combining ability that can produce seed with approximately twice the genetic superiority of that produced by the thinned half-sib progeny test.

For cost comparisons, it is necessary to distinguish between "simultaneous" and "delayed" clonal orchards. The original 442-clone orchard of Goddard et al. may be considered as "simultaneous" in that the grafts were made at about the same time the half-sib progeny tests were established. They also plan to establish "delayed" orchards, which will include only those clones with proved general combining ability.

The costs for simultaneous clonal orchards and progeny tests are usually much more than twice as great as for progeny tests alone. Such clonal orchards involve a large amount of grafting because at the outset they must contain several ramets of every selected clone in order to produce large amounts of seed after removal of the clones with poor general combining ability. The costs of delayed clonal orchards and progeny tests need not be much greater than for progeny tests alone. For example, Goddard et al. can establish their delayed orchards with only 37 clones known to produce fast growing offspring that are resistant to fusiform rust.

FIG. 10.4. Grafted seed orchard of Japanese larch in Japan at age 12. In 1968 the trees were girdled half-way around on the north side at a height of 4 ft and half-way around on the south side at a slightly greater height. That treatment induced heavy cone production the following (twelfth) year. This seed orchard is not accompanied by a progeny test.

FIG. 10.5. A 55-hectare (135 acre) clonal seed orchard of Scotch pine in Finland. Reasonable amounts of seed can be expected at age 10 to 15 in such an orchard versus age 30-40 for seedlings. The trees are permitted to grow naturally, as shearing to reduce height decreases cone production considerably. An extensive progeny testing program is under way to evaluate the combining ability of all clones included in such orchards. (Photograph courtesy of Lauri Kärki of the Finnish Forest Tree Breeding Foundation.)

Repeating from a previous section, let YS, YG, and YE represent years required for seedlings to fruit, for grafts to fruit, and for seedlings to be evaluated, respectively. A simultaneous clonal orchard can produce improved seed at YG or YE years, whichever is greater. A delayed clonal orchard can produce improved seed at $YE + YG$ years. A progeny test can produce seed at YS or YE years, whichever is greater.

To summarize the gain, cost, and time comparisons: (1) accompanied, simultaneous clonal orchards produce seed with about twice the genetic superiority of that produced by a half-sib progeny test, at more than twice the cost, in less time (if YG is less than YS), or equal time (if $YG = YS$), or more time (if YG is more than YS); (2) accompanied, delayed clonal orchards produce seed with about twice the genetic superiority of that produced by a half-sib progeny test, at somewhat greater cost, in equal time (if $YG = YE = \frac{1}{2}YS$), or twice the time (if $YG = YE = YS$), or more than twice the time (if YG is greater than YS).

Accompanied Clonal Orchards versus Full-Sib Progeny Tests

An accompanied clonal orchard is thinned to those clones with proved high general combining ability. The clones are parental generation and produce F_1 offspring for commercial planting.

A full-sib progeny test can be thinned to the best F_1 families. To the extent that the parents possessed high general combining ability, those F_1 families should be equal in genetic quality to the F_1 offspring of the clonal orchard. To the extent that there was additional specific combining ability, some F_1 families will be superior to the F_1 offspring of the clonal orchard. If the families in the progeny test are large enough there is an opportunity for intrafamily selection. The selected F_1 families will produce F_2 offspring for commercial planting. They will transmit to these F_2 offspring all their genetic superiority due to general combining ability plus some additional gain due to specific combining ability and intrafamily selection. Thus the seed produced by the full-sib progeny test should be equal to or slightly superior to the seed produced by the clonal orchard.

The above is true only if the full-sib progeny test results from a diallele mating scheme in which each parent is crossed with every other parent. However, this is not practical. Usually the mating pattern is such that each of numerous female parents is crossed with each of a few male parents so that only one sex contributes fully to genetic gain. On the other hand, a clonal orchard can be made to contain only superior male and female clones so that the F_1 offspring produced are as good as if a complete diallele full-sib progeny test had been thinned to the best families. From this standpoint an accompanied clonal orchard can produce higher quality seed than a full-sib progeny test.

A clonal orchard and progeny test can be "simultaneous," that is established immediately after the parents are phenotypically selected. Or it may be "delayed," i.e., established after general combining ability of the parents becomes known as the result of a half-sib progeny test. The former are much the most expensive; the latter involve a delay of YE years in the availability of improved seed.

For cost comparisons it is best to compare only "simultaneous" or "delayed" clonal orchards and progeny tests. In that way, the comparisons can be made tree for tree. In choosing parents to be included in a clonal orchard, fruitfulness is nearly always considered, and one grafted tree is apt to have 2 to 4 times the seed-producing capability of one seedling. One trained man can usually produce 100 to 200 living grafts per day in easily grafted species and several thousand control-pollinated seeds per day in easily pollinated species. When judged this way, clonal orchards are probably much more expensive than full-sib progeny tests in many pine, spruce, and birch species, and cheaper than full-sib progeny tests in many maple, walnut, and oak species.

Another aspect of cost must be considered. A clonal orchard does not produce highest quality seed unless accompanied by a half-sib or full-sib progeny test. Therefore cost comparisons must be between clonal orchard + progeny test and a full-sib progeny test.

Clonal orchards possess a time advantage if YG is appreciably less than YS. Simultaneous and delayed clonal orchards produce seed in YG and $YE + YG$ years, respectively. Simultaneous and delayed progeny tests produce seed in YS and $YE + YS$ years, respectively.

Earliness of Flowering on Grafts and Seedlings

Much of the choice between clonal and progeny test seed orchards depends on earliness of flowering of grafted trees and seedlings. Often progeny tests are preferable from cost and/or gain standpoints if grafts and seedlings flower at about the same time, and clonal orchards are preferable if grafts flower many years earlier than seedlings.

Direct experimental evidence is scarce, as few investigators have grown both types of trees under the same conditions. Hence most of the following comparisons must be considered as approximate. For brevity, YS and YG have been used to denote years required for seedlings and grafts, respectively, to produce reasonably large numbers of flowers.

Slash pine $YG = 6$ to 7 and $YS = 12$ to 14

Loblolly pine $YG = 6$ to 7 and $YS = 12$ to 14 in most of the North Carolina State-Industry experiments; in one Texas experiment, YG and YS were about equal, (7 years)

Scotch pine   In Sweden, $YG = 12$ to 15 and $YS = 40$; in north central United States, $YS = 6$ to 8 for fast-growing races from central Europe.

Shortleaf pine $YS = 6$ near Gulfport, Mississippi

Virginia pine $YS = 4$ to 5 years commonly

Eastern white pine   According to Ahlgren's (1962) experience in northern Minnesota and Patton and Riker's (1966) experience in Wisconsin, $YG = 15$ to 20; a reasonable percentage of grafted trees bore cones at age 5 to 6 but did not produce many flowers thereafter; $YS = 15$ to 20

Jack pine $YS = 4$ to 5, YG is probably greater. Much heavier flowering has been observed on 8 ft seedlings than on 8 ft grafts

Red pine $YG = 4$ to 8 in Wisconsin; $YS = 5$ in Nebraska; 9 on a widely spaced plantation in southern Michigan; commonly 15+

Japanese red pine $YS = 5$ to 6 in the United States, a little more in Japan

Japanese black pine   Similar to Japanese red pine

Monterey pine   J. M. Fielding found $YG = 6$ and $YS = 7$ in Australia

Norway spruce   In Sweden, $YG = 12$ and $YS = 30$ to 40

White spruce   In Wisconsin, $YG = 3$ to 4 and $YS = 15+$

Blue spruce YS and YG seem similar, both probably $20+$

European beech $YG = 8$, YS is much larger

Japanese and European larches $YG = 3$ to 4 in European grafted orchards; $YS = 10$ to 15 in Europe; for some races of Japanese larch, $YS = 9$ in parts of north central United States

American tamarack $YS = 4$ to 5

Douglas-fir   In British Columbia, $YG = 3$ to 4, $YS = 7$

Peach $YG = 3$ to 4, $YS = 4$ to 8

Sour cherry and sweet cherry $YG = 5$, $YS = 6$ to 7

Apple $YG = 5$ to 6 and $YS = 6$ to 7 for precocious varieties; for varieties not so precocious, $YG = 10$ to 12 and $YS = 15$ to 20

Cryptomeria $YS = 3$ to 4 if seedlings are sprayed with gibberellin

Sugar maple $YG = 15+$ in Ohio; YS probably $= 25+$

Black wattle $YS = 5$ or less

Eastern cottonwood and hybrid poplars $YS = YG = 6$ to 7

Many eucalypts $YS = 5$ to 10

As the above examples show, YG frequently is smaller than YS, but it can be larger. Often YS is small enough that the time advantage of grafted trees is negligible. In a few species the time advantage of grafted trees is large enough to justify considerable effort on grafting.

Study Questions

1. Why is there no one best solution for all situations? How does choice of breeding method and type of seed orchard vary with heritability? With amount of genetic data available? With cost? With earliness of fruiting? With need?

2. Under what circumstances are thinned natural stands useful as seed orchards? What genetic gain is expected in such a seed orchard?

3. What are the approximate genetic gains from mass selection, half-sib family, and full-sib family selection if single tree heritability is 0.01? If single tree heritability is 0.20?

4. Under what situations are the gains from full-sib family selection much greater than those from half-sib family selection? Under what situations are the gains apt to be nearly equal?

5. What is meant by accompanied and unaccompanied clonal orchards? Why are the gains from the two types so different?

6. What is meant by simultaneous and delayed clonal orchards? How do they compare in cost? How do they compare with respect to time at which improved seed becomes available?

7. Compare thinned full-sib progeny tests and thinned accompanied clonal orchards as to gain and time of seed availability.

8. Assume a 5000 ramet (200 clones × 25 ramets per clone) clonal orchard with an accompanying half-sib progeny test of 50 seedlings per family. Assume also a 200 female × 4 male full-sib progeny test with 50 seedlings per family. Use the symbolism CHS, CG, and CFS as the per seedling or per graft costs for half-sib progeny tests, grafting, and full-sib progeny tests, respectively. Develop statements of the total costs of the accompanied clonal orchard and the full-sib progeny test.

9. Assume that one grafted tree produces as much seed as 3 seedlings. Assume also that the plantations mentioned in question 8 will be thinned to the 25 clones with best general combining ability or to the 25 best full-sib families. Using the statements developed in question 8, develop statements as of the final per seed producing equivalents for the thinned clonal orchard and the thinned full-sib progeny test.

10. Give 1 to 3 examples each of instances where YS is appreciably greater than, equal to, or less than YG.

11

Examples of Progress through Selective Breeding and Seed Orchard Management

Introduction

Chapters 9 and 10 contain descriptions of the different breeding methods used to take advantage of the genetic variability among individual trees and discussions of the factors involved in choice of the best breeding method for a particular situation. In this chapter, I give further examples of progress through selective breeding and descriptions of types of seed orchards which are used or proposed.

The first section includes a few examples of genetic differences which are presumably under the control of a few genes. The next section includes examples, species by species, of progress made in the improvement of traits which are presumably under the control of many genes. The last sections cover the problems met in the actual management of seed orchards, for which I have relied heavily on the extensive experience with loblolly and slash pines in southern United States, with shorter treatments of the problems encountered in other species.

Examples of Simple Inheritance

Table 11.1 contains examples of traits that are believed to be under the control of single dominant–recessive gene pairs. These data are from control-pollinated experiments.

Also worth mentioning are a few additional cases of simple inheri-

TABLE 11.1

Some Tree Characteristics Controlled by Genes with Large Effects[a]

Species	Characteristic	Phenotype		
		AA	Aa	aa
Siberian pea tree	Seedling color	Green	Green	Yellow
Siberian pea tree	Branches	Upright	Upright	Drooping
Norway spruce	Seedling color	Green	Yellow	White
Cryptomeria	Autumnal seedling color	—	Red	Green
European beech	Leaf color	—	Copper	Green
Black wattle	Presence of the flavonoid "mearnsitrin"	Present	Present	Absent
Western white pine	Presence of more than 2.7 % 3-carene in cortical oleoresin	Present	Present	Absent
Slash pine	β-Pinene content	21–74 %	21–74 %	2–8 %
Slash pine	Myrcene content	6–45 %	6–45 %	0–5 %

[a] After Wright, 1962; Hanover, 1966; Zeijlemaker and Mackenzie, 1966; Squillace, 1971.

tance, the evidence for which is based upon open-pollinated progeny tests. Presumably the abnormal expression of each trait is controlled by one or a few dominant genes, as 10–50% of the offspring are abnormal. The "Schwedler" clone of Norway maple has leaves which remain purplish-red throughout the summer. Open-pollinated offspring tend to be either red or green, with few intermediates. The "fastigiate" clone of English oak has steeply ascending branches. In small tests, most open-pollinated offspring were either normal or had steeply ascending branches like the seed parent. In central Michigan there is a wild red pine with a semi-fastigiate growth habit and branches ascending at an approximate 45° angle. Surrounding it, and presumably some of its open-pollinated offspring, are a few trees with a similar branching habit.

The Schwedler Norway maple and the fastigiate English oak are grown commercially for landscape purposes. They are usually grafted, but could also be grown from open-pollinated seed.

Polk (1974) assembled a collection of jack pine having branches borne at different angles from the trunk, some being nearly fastigiate. Nearly all control-pollinated offspring resulting from crosses between parents with the same branch angle were similar to the parents in branch angle.

Variation and Improvement in Quantitative Traits

Loblolly Pine

Loblolly pine is an important tree of southeastern United States. Several agencies are engaged in its improvement. Among these, the North Carolina State–Industry Program deserves special mention. That program involves cooperation between North Carolina State University and several forest industries. Many hundreds of acres of clonal seed orchards have been established with grafts taken from trees phenotypically selected, usually for stem form and growth rate. Many of the following data are taken from publications by Woessner (1965), Zobel *et al.* (1973), Zobel (1971), and North Carolina State (1970) pertaining to that program.

In one experiment conducted in Tennessee, North Carolina, and South Carolina, open-pollinated offspring of selected trees were grown in comparison with offspring of average dominant trees in the same stands. Offspring of selected parents were tallest in 5 of 9 test plantations. The results are summarized in the tabulation below.

Trait	Size of offspring of average trees	Superiority of offspring of selected parents (%)
Height	14.7 ft.	1.2
Diameter	2.5 in.	1.2

In another experiment, seeds were collected from trees in grafted seed orchards; such seed thus represented crosses among phenotypically selected parents. Commercially collected seeds, from different stands than those in which the parents were selected, was used for comparison. The seedlings were grown in three test plantations and were measured at age 5, with results as shown in the following tabulation.

Height of commercial check (ft)	Superiority of offspring of selected parents (%)
17	4
13	5
7	6

Clones in two seed orchards were crossed with "tester" male clones and the offspring were grown 4 years to determine the relative gains from mass selection and full-sib family selection. Commercial seed was used as a control for comparison purposes. Mass selection was effective, but full-sib family selection was 3 to 4.5 times more effective, as shown in the following tabulation.

Location of experiment	Average height of controls (ft)	Height superiority of	
		Offspring of all selected (%)	The 25% best families (%)
Georgetown, South Carolina	16	4	18
Rincon, Georgia	15	4	11

Stem form of loblolly pine is also highly heritable. Some crooked parents produce uniformly crooked offspring. Selected parents usually produce straighter offspring than if commercial seed is used. In this trait, as in growth rate, full-sib family selection can result in larger genetic gain than mass selection.

Wood density varies considerably within a tree, from center to outside and from bottom to top. It also varies genetically among trees. In South Carolina, 34 trees were selected for high and low density, and their open-pollinated offspring were grown for 3 years. Offspring of the light highest-density parents produced wood weighing 22.4 lb/ft^3, 6% heavier than the wood produced by the offspring of the lowest-density parents.

Wells and Switzer's (1971) work in Mississippi and Louisiana was mentioned in Chapter 9. They combined half-sib progeny and provenance testing. There were large genetic differences in growth rate and fusiform rust resistance. The largest differences were associated with stand of origin rather than individual trees within stands. Thus they concluded that mass selection within stands is not as effective as testing the offspring of a great many stands.

Slash Pine

Slash pine is also a commercially important species of southeastern United States. As in loblolly pine, several agencies are engaged in its improvement. Programs of the Georgia Forestry Commission and the University of Florida involve cooperation with forest industry, the phenotypic selection of parents, the establishment of clonal seed orchards, and the progeny testing of many of the selections. The United States Forest Service also cooperates on these programs.

The recent results of Goddard *et al.* (1973) were mentioned in Chapter 9. They found that mass selection for growth rate was effective, as 82% of the selected parents produced superior offspring. They found that mass selection for resistance to fusiform rust was ineffective if done in lightly infected stands, but effective if practiced in heavily infected stands.

As part of the North Carolina State–Industry Program, Zobel *et al.* (1973) collected seed from grafted trees in clonal seed orchards established with phenotypically selected parents, and tested the seedlings in comparison with commercial seedlings. In three separate experiments, the commercial seedlings averaged 11, 11, and 13 ft tall at age 5; offspring of the seed orchard trees were 6% taller.

Squillace (1966) collected seeds from a few average trees in each of many stands and grew the offspring for 2 years. He measured height and diameter as well as several needle characters. In growth rate, differences among the offspring of different trees in the same stand were of about the same magnitude as the differences among the offspring of different stands. In other words, for maximum improvement in growth rate, it is desirable to select offspring of superior trees in superior stands.

Squillace (1971) also studied the inheritance of certain monoterpenes found in the oleoresin of slash pine in an experiment which included 34 full-sib families. The parent–offspring correlations are shown in the following tabulation.

Monoterpene	Correlation (r)
α-Pinene	0.79 ± 0.29
β-Pinene	0.49 ± 0.29
Myrcene	0.63 ± 0.16
β-Phellandrene	0.55 ± 0.17

Each of these correlations is high enough to indicate that considerable gain would result from mass selection. Note also that the error terms associated with the correlation coefficients are relatively high. This is commonly the case in experiments such as this and shows that it is often difficult to estimate genetic gain with an accuracy much greater than $\pm 50\%$.

By a bark chipping process, oleoresin or "gum" is extracted from slash pine for use in the naval stores industry. Gum yield is highly heritable, and a doubling of gum yield by mass and family selection seems possible (Squillace, 1965). Gum yield is also correlated with diameter growth to the extent that a doubling of gum yield would result from a 6% increase in diameter growth.

Monterey Pine

As of 1973 New Zealand had 238 hectares of clonal seed orchards of this species, all established with mass-selected clones from the country's extensive plantations. Most orchards are grafted, but because of graft incompatibility future orchards will probably be established with rooted cuttings. The orchards generally start to produce commercial quantities of seed by age 10. Many of the clones are being progeny tested (Figs. 11.1–11.3).

The mass selection was very effective. According to a 7-year-old progeny test reported by Shelbourne (1974), full-sib families resulting from selected × selected matings surpassed commercial controls by 46, 36, and 38% in volume production per tree, stem straightness, and branch distribution (a quadrinodal branching habit being considered more desirable than a uninodal one), respectively. These exceptionally high gains may be due in part to the fact that some commercial plantations in New Zealand are the product of introgression between Monterey and knobcone pines and are therefore exceptionally variable. With such high heritabilities as found by Shelbourne, mass selection deserves more emphasis than family selection.

Scotch Pine

Scotch pine has a large range in Europe and Asia, where it is an important timber tree. Many improvement projects are underway in its native range as well as in America where it is planted extensively. Many of these involve mass selection and the establishment of clonal or seedling seed orchards. In Finland alone, 140 orchards totaling 1700 hectares had been established by 1970. Extensive progeny testing of the selected clones is underway, but relatively few published data were available by 1974.

Those older experiments which are replicated indicate much less intrastand variability and less gain from mass selection than found in slash and loblolly pine. Nilsson (1968) measured a series of 10 half-sib progeny tests established 1938–1952 in Sweden to test parents included

FIG. 11.1. A plus tree of Monterey pine in New Zealand. The very large differences between the selected tree in the center and the surrounding ones in this plantation make it reasonably certain that this tree will prove to possess high general combining ability for growth rate and stem form. This proved to be the case. (Photograph courtesy of G. B. Sweet of the Forest Research Institute at Rotorua, New Zealand.)

in Swedish orchards. His data indicate almost no gain from the original phenotypic selection, as only 2 of 45 parent–progeny correlations that he calculated were significant, and 20 were negative. There were some differences among the offspring of different stands and among the offspring of different parents within a stand. His heritability and gain calculations indicate that gains of 0.5–1% are possible in traits such as number of branches, live crown length, branch diameter, and height–diameter ratio.

FIG. 11.2. Cross between two selected Monterey pines (left) compared with seedlings grown from commercial seed (right). One of the parents of the full-sib family on the left is the plus tree shown in Fig. 11.1. Monterey pine is one of the world's most productive and commonly planted trees. With genetic improvement, such as demonstrated here in New Zealand, it promises to be even more productive. (Photograph courtesy of G. B. Sweet of the Forest Research Institute at Rotorua, New Zealand.)

From 1953 to 1965 a total of 63 clonal seed orchards were established in Great Britain, using phenotypically selected parents from plantations in that country. Starting in 1957, open-pollinated half-sib progeny tests were established to test the genetic quality of the selected clones. Data

FIG. 11.3. Incompatible graft union in Monterey pine in New Zealand. This tree grew to a diameter of 40 cm and died. Because of the high frequency of incompatible grafts, rooted cuttings rather than grafts are being used in many of the new clonal seed orchards. (Photograph courtesy of G. B. Sweet of the Forest Research Institute at Rotorua, New Zealand.)

from the first progeny test, measured at age 15, indicate that the mass selection for growth rate had been ineffective, as only 2 of the 60 clones which were progeny tested produced offspring taller than the control seedlots used for comparison (Johnstone, 1973). An additional 405 clones are being progeny tested, in order to obtain 40 with high general combining ability.

In Michigan nine half-sib progeny tests were established in 1959, using open-pollinated seed collected from 10 or 20 parents in each of nine European stands located in southern Norway, East Germany, and Belgium. When measured at age 11 by Howe (1971), offspring of the different stands differed markedly in several respects, but differences among offspring of different parents within the same stand were small. In five of the nine stands, there were not statistically significant differences among half-sib families in height, winter foliage color, number of cones per tree, length of needle retention, resistance to Zimmerman moth, or number of branches per whorl. In three of the other four stands there were significant differences in 2 of the 6 traits, and in only one stand (which happened to be the one producing the slowest growing offspring) were there differences among half-sib families in as many as 5 of the 6 traits. In no case was there a significant correlation between parent growing in Europe and offspring growing in Michigan.

Such results are contrary to what might be expected on the basis of the known geographic variation pattern. Scandinavia was covered with glaciers until 10,000 years ago. In the 50 to 100 tree generations since the glaciers retreated a cline has developed with end points differing in growth rate by a 2:1 ratio and with correspondingly large differences in several other traits. Those differences presumably arose as the result of natural selection operating on the intrastand genetic variation.

Jack Pine

This species is a very common component of the forests of northeastern America and is transcontinental across Canada. The first inheritance study (Rudolph et al., 1959), conducted in Minnesota, showed that 52% of the offspring of trees producing nonserotinous cones (cones opening soon after ripening) were nonserotinous, whereas only 13% of the offspring of trees producing serotinous cones (cones remaining closed many years until opened by fire) were nonserotinous. The very high heritability of this trait explains why the species is changing rapidly by natural selection from predominantly serotinous to predominantly nonserotinous in areas protected from fire during the last generation. As already cited, Polk (1972) found that the tendency toward extremely acute branch angles (rarely found in nature) was highly heritable.

However, mass selection for growth rate, stem straightness, and branch size has not been effective. Canavera (1974) selected 400+ parents from several stands in Michigan for excellence or inferiority in those traits. At age 3 and again at age 8 the offspring of the selected plus trees were little different than the offspring of the selected minus trees. The latest measurements show very small differences among half-sib families from the same stand, or among the offspring of different stands within a radius of 100 miles.

Eastern White Pine

This is a native of northeastern United States and southeastern Canada. It is an important timber species but is susceptible to two important pests—white pine blister rust (causing death of young trees) and white pine weevil (causing stem crooks.)

Riker et al. (1943) started breeding for rust resistance with the selection of 143 rust-free trees in northern Wisconsin. The selections were made in areas having light to moderate rust infestations. They propagated each tree clonally by grafting and also conducted a half-sib progeny test. All the clonal propagules and seedlings were exposed to heavy rust infection conditions in the nursery. Judged by results of the clonal test, the mass selection was effective, as 40 of the 143 clones remained rust-free. Judged by the half-sib progeny test the mass selection was ineffective, as the offspring of selected trees were as susceptible as the offspring of average trees. There were differences among half-sib families, however. Recent developments in this project were summarized by Patton and Riker (1962).

In another study Kuo *et al.* (1972) tested the open-pollinated off-spring of 122 trees located in various parts of Michigan. When measured at age 9 the largest differences in growth rate were associated with region of origin—trees from the southern part of the state grew 8–32% faster than those from the northern part of the state whether tested in the southern or northern parts. There were also differences among different stands within region, but no statistically significant differences among the offspring of different parents within a stand. That is in sharp contrast to the results of Kriebel *et al.* (1972) who found differences in growth rate among full-sib families derived from the same Ohio stand.

In the experiment of Kuo *et al.* one particular type of leaf damage proved to be under a very high degree of genetic control. This was a tendency for certain trees to have brown current year needle tips; cause of the damage is unknown. Half the damage was confined to 9 of the 122 half-sib families. There were also statistically significant (but probably too small to be of practical importance) differences among families in resistance to white pine weevil; in studies of geographic variation, differences in weevil resistance have been small (see Chapter 15).

Of the various pests, the white pine weevil has caused the most damage. Improvement in weevil resistance is urgently needed, but the results do not hold much promise for either mass or family selection. Improvement in rust resistance is also desirable, but that would involve either a very great amount of work or a sacrifice in growth rate. Because of climatic factors the rust is serious only in cool moist northern areas inhabited by slow growing genotypes, and only in such areas would mass selection for resistance in native stands be feasible. The alternative procedure, requiring additional work, would be to establish large half-sib progeny tests of unselected southern trees in the northern areas where heavy rust infections could be induced.

Japanese Red Pine

This is one of two common hard pines in Japan. It is a commercially important species planted in modest amounts. Iwakawa *et al.* (1967) collected open-pollinated seed from 25 plantation trees in northern Honshu. The mother trees were 56 to 60 years old, 23 to 30 m in height, 29 to 44 cm in diameter, and 9 to 19 m in clear length. They collected seeds from each parent in each of 3 years and planted each half-sib family in each of 3 years.

At age 12 the test plantations averaged 8 m tall, with the tallest family exceeding that average by 8%. The half-sib family heritability was 0.13, and selection of the tallest 1 in 25 families could result in a 1% gain in growth rate. The parent–progeny regression for height was not significant statistically, indicating that mass selection for growth rate was ineffective.

Wood density averaged 0.40 in the parents and 0.33 in the offspring. As measured in one of the three replications, the parent–offspring correlation was $r = 0.67$ and the parent–offspring regression (single tree heritability) was 0.46. These were statistically significant and indicate that mass selection of the five parents with most dense wood could increase wood density in the offspring from 0.33 to 0.343.

Ponderosa Pine

This is a common tree of western United States and Canada. There have been several provenance studies, one of them 60 years old. Work on selective breeding is of much more recent date and is concentrated in Idaho.

Wang and Pattee (1975) collected open-pollinated seed from 271 trees located in various stands situated in central and southern Idaho. They practiced mild parental selection. They grew the offspring and planted them in four replicated permanent test plantations. When measured at age 5 there were large and statistically significant differences in growth rate among the offspring of different stands. Differences among offspring of different trees in the same stand were relatively small and unimportant. The half-sib progeny tests will be thinned, by removal of the poorest families. After thinning they will become seed orchards. Cone production is not expected for 15 to 20 years. In the interim, data from the half-sib progeny tests will be used to designate the genetically superior natural stands. These will be thinned and used as natural seed orchards.

Western White Pine

This valuable timber tree is distributed from British Columbia to California along the West Coast and also in the interior. Breeding work has been concentrated in northern Idaho, where the species is of greatest commercial importance. The excellent progress in resistance to white pine blister rust has already been mentioned (see Chapter 9). Less work has been done on other traits.

Steinhoff and Hoff (1971) measured periodic annual growth of 19 parents selected for high general combining ability with regard to rust resistance, crossed those parents, and measured the offspring at age 8. The 10 fastest growing parents grew 2.5 ft/year, 17% above average; their offspring grew 1.0 ft/year, 3% above average. Due to the small number of parents tested, the parent–offspring regression was not statistically significant, however.

Black Cherry

This is a valuable cabinet wood of eastern United States and southeastern Canada. Although common in natural stands, it has been planted little and breeding work is of recent origin. In one study, trees were selected for superiority (and inferiority) in form and growth rate from several stands in two parts of Pennsylvania and West Virginia. When the open-pollinated offspring were measured at age 2, Pennsylvania trees were 15% taller than West Virginia trees, but there were no significant differences among the offspring of different parents within the same stand (Pitcher, 1971).

In another experiment, started by Kingsley Taft of Tennessee Valley Authority, open-pollinated seeds were collected from 33 trees located in 7 stands scattered from North Carolina to Michigan. When one test plantation in southern Michigan was measured at age 10, the offspring of different stands varied markedly in height, flower production, crown width, and amount of forking; offspring of different trees in the same stand did not differ in those traits. In contrast, there were 6 to 8 day interfamily differences in time of leafing out, but no geographic variation in that trait.

Black Wattle

This extremely fast growing acacia is native to Australia, but is cultivated most extensively in the Republic of South Africa where several hundred thousand acres have been planted, mainly for the production of tannin from the bark. Wattle breeding for increased tannin production, started in 1950, is concentrated at the Wattle Research Institute, Pietermaritzburg, Republic of South Africa. There have been extensive half-sib and full-sib progeny tests, summarized in papers by Moffett and Nixon (1966), Zeijlemaker and Mackenzie (1966), and other papers published in the annual report of the Wattle Research Institute (Figs. 11.4 and 11.5).

Considerable work has been done to determine the validity of early measurements. Correlations between data taken at age 5 and data taken at age 10 are about $r = 0.9$ for bark thickness and bark tannin content and 0.8 for diameter, indicating that age 5 results are applicable to the normal 10 year commercial rotations.

Mass selection is effective. Single tree heritability, as measured by parent–offspring regression, is about 0.2 for diameter and bark thickness and 0.6 for bark tannin content. Half-sib family selection has been even more effective and has resulted in a 12% increase in 3 year height (i.e., from 25.5 to 29 ft) and an 8% increase in bark tannin content (i.e., from 38 to 41%). Most of the increase has been due to genes with additive effects.

Some of the half-sib and full-sib progeny tests have been thinned to leave only the best trees in the best families and thus become commercial seed orchards. Due to the fact that wattle grows rapidly and flowers quickly, a 5 year generation is possible. By 1974 third generation seed orchards had been planted, and the second generation orchards were supplanting the first generation ones as a source of commercial seed.

Eastern Cottonwood

This is a fast growing bottomland species of eastern United States. It and/or hybrids with other poplars are planted extensively. Most of the genetic work has dealt with hybrids, and studies of variation within the species are of recent origin.

Jokela (1966) collected open-pollinated seed and cuttings from 92 trees located in three parts of Illinois and grew the trees 4 years in central Illinois. He measured resistance to *Melampsora* rust, which causes leaf lesions and premature leaf drop. Damage varied from slight to heavy. As measured by the half-sib progeny test and the clonal test the heritability (as measured by parent-progeny regression) was 0.5 and 0.8, respectively, indicating that mass selection for resistance could be very effective.

Farmer (1970) collected open-pollinated seed from 81 selected and

FIG. 11.4. Differences in stem straightness among half-sib families of black wattle (*Acacia mearnsii*) growing in the Republic of South Africa. The 9-year-old family on the left is very crooked, that on the right is much straighter but not outstanding. (Photograph courtesy of Kathleen Nixon of the Wattle Research Institute at Pietermaritzburg, Republic of South Africa.)

FIG. 11.5. Two-year-old seed orchard of black wattle growing in South Africa. The seedlings shown here were grown from seed collected from the best individuals in 11 full-sib families resulting from crosses among the best individuals in the best half-sib families. There was family and mass selection in the first generation (half-sib

TABLE 11.2

Half-Sib Family Heritabilities and Gains Possible from Family Selection in Eastern Cottonwood from Mississippi

Trait	Half-sib family heritability h^2	Gain possible by selecting the 1 best family ΔG
Height	0.06	0.6%
Diameter	0.13	2%
Melampsora rust resistance	0.51	Nearly complete resistance
Date of leafing out	0.90	10 days advance
Wood specific gravity	0.20	1%
Fiber length	0.03	0.2%

[a] From Farmer, 1970.

average Mississippi trees and grew the half-sib families three years. His data, summarized in Table 11.2, indicate that selection can be very effective for rust resistance and date of leafing out, but not for four other traits.

Ying (1974) reported on a 7-year-old experiment conducted in Nebraska (using stock supplied by Jokela of Illinois) which included 498 clones belonging to 116 open-pollinated half-sib families from many stands scattered from Mississippi to Minnesota. He measured several traits and computed the variances due to region of origin, family within region, and clone within family (Table 11.3).

From Table 11.3 it is evident that there is appreciable genetic variation in eastern cottonwood at every level—among clones, among families, and among regions of origin. In growth rate, for example, it was possible to increase 7 year height from 39 to 42 ft by using Missouri rather than native trees, and to increase that still farther to 46 ft by selecting the best families and then the best clones.

Sakhalin Fir

Sakhalin fir is native to northern Japan and Sakhalin. It is one of the most important timber species of the northern Japanese island of

---

progeny test) and intrafamily selection in the second generation (full-sib progeny test). In this, the third generation, there is intrafamily selection; ⅓ of the trees were thinned at age 1 and ⅓ will be thinned at age 3 to leave a final spacing of 314 trees per hectare (125 trees per acre). The second generation orchard, now age 7, is the present source of most commercial seed. (Photograph courtesy of Kathleen Nixon of the Wattle Research Institute at Pietermaritzburg, Republic of South Africa.)

TABLE 11.3

Portions of Total Variance Due to Region of Origin, Half-Sib Family within Region, and Clone with Family for 7-Year-Old Eastern Cottonwood[a]

Trait	Portion (%) of total variance due to		
	Region of origin	Family within region	Clone within family
Height	10	8	12
Date of flushing	26	31	30
Amount of stem canker	9	8	14
Stem straightness	10	6	14
Branch habit	56	6	5
Bark thickness	30	7	24
Bark roughness	38	11	23
Leaf size	59	4	25

[a] From Ying, 1974.

Hokkaido and is one of the three most commonly planted conifers there. Okada and Mukaide (1969) collected open-pollinated seed from 117 trees in 4 native Hokkaido stands and grew the 117 half-sib families for 3 years. At that time offspring of one stand averaged 18.1 cm tall and offspring of the next best stand averaged only 15.7 cm tall. Differences among offspring of different mother trees in the same stand were statistically significant but much smaller than the differences between stands. The frequency of Lammas shoots followed a similar pattern, with the greatest differences being among the offspring of different stands.

Maximowicz Birch

This is also native to Hokkaido and is one of the largest and fastest growing of all birches. It has a white bark but is morphologically more similar to some of the dark-barked species than to other white-barked birches. An improvement program started with an open-pollinated progeny test of 55 parental trees growing in 12 scattered Hokkaido stands (Hatakeyama and Adachi, 1968). The experiment was replicated several times and the trees were measured after 7 years. The fastest growing trees (13 ft tall, 1.3 in. diam at age 7) were from an area about 50 miles across in central Hokkaido and grew 11% faster than a race 50 miles to the west. The area occupied by the fast growing race had favorable growing conditions (as measured by length of growing season, precipitation, and average temperature), although not so favorable as occupied by some slower growing genotypes.

Hatakeyama and Adachi calculated the variances in growth rate due to region of origin, stand of origin, and parent tree within stand; their calculations are shown in the following tabulation.

	Portion (%) of genetic variance due to		
Trait	Region of origin	Stand within region	Parent tree within stand
Height	61	39	0
Diameter	81	0	19

This tabulation shows that the majority of genetic improvement was obtained by testing the offspring of trees growing in different parts of the island. Additional improvement in rate of height growth can be obtained by testing the offspring of different stands within that region. Intrastand family selection can give additional improvement in rate of diameter growth but not in rate of height growth. The single tree heritabilities were too small to justify mass selection.

Seed Orchard Management in Loblolly and Slash Pines

General

Southeastern United States is a region with very large areas of forest land and with growth rates so fast that a 2 to 4 year wait for natural regeneration means a 10–25% increase in rotation length. Hence, clearcutting followed by planting has become accepted practice. Seed orchard development started in 1955. As of 1972, a total of 3500 acres (1580 hectares) of loblolly pine and 2600 acres (1200 hectares) of slash pine seed orchards had been established (U.S. Forest Service, 1974). Nearly all are clonal orchards established by grafting. The loblolly pine orchards are located along the Atlantic and Gulf Coasts from Maryland to Texas and inland to Tennessee and Arkansas. The slash pine orchards are along the southern coasts from the Carolinas to Texas (Figs. 11.6 and 11.7).

The following description of seed orchard programs in these species is taken largely from papers by Kellison (1969), Van Buijtenen et al. (1971), Zobel (1971), Zobel et al. (1973), and from annual reports of the

Fig. 11.6. Aerial view of the Weyerhauser Company's seed orchard complex at Washington, North Carolina. The orchards cover 250 acres (110 hectares) and include loblolly, pond, longleaf and slash pines, yellow poplar, and sweetgum. (Photograph courtesy of Bruce J. Zobel and Leroy C. Saylor of North Carolina State University at Raleigh.)

Texas Forest Service and the schools of forestry of the North Carolina State University at Raleigh and the University of Florida.

Organization of the Work

Seed orchards have been established by all those agencies owning large amounts of land—by wood using (mostly pulp) industries, states, and the United States Forest Service. To a certain extent, each company or public agency has an autonomous program, performing its own selection, grafting, seed orchard establishment, seed collection, and nursery propagation. Thus each has a corps of specialists, including people skilled in grafting, controlled pollination, progeny testing, etc. However, most of the industries are organized in cooperative groups and are associated with a university or state organization. In this way the programs for each region are coordinated, and there is cooperation involving exchange of clonal materials, pollen research results, etc.

Fɪɢ. 11.7. Hydraulically operated platform at work in a loblolly pine seed orchard. Such platforms can be used to pick cones, make controlled pollinations, etc. Newer versions have duplicate operating controls in the bucket and on the bed of the truck. (Photograph courtesy of Bruce J. Zobel and Leroy C. Saylor of North Carolina State University at Raleigh.)

Selection of Parental Clones

Initially the selection standards were very high so that each selected tree was the best of many thousands with respect to 15 to 20 traits. Selection standards have been relaxed slightly to permit the inclusion of more clones in each orchard and thus provide a broader genetic base for breeders in future generations. The Texas Forest Service, for example, included 134 loblolly pine clones in its seed orchards in 1974, and several hundred clones were included in slash pine clone banks or orchards in Florida in that year. Also, there has been a gradual reduction in the number of traits considered, and in recent years primary emphasis has been on selection for bole straightness (especially in loblolly pine), growth rate, and resistance to fusiform rust. A few small special purpose orchards have been established with clones selected for high specific gravity, drought resistance, high oleoresin yield, etc.

Most preliminary selections are made by field foresters in the course of their regular work. Nearly all are then checked by a specialist before being certified for inclusion in a breeding program or seed orchard. The selection work is usually done in native stands or plantations close to the area in which the seed is to be used. A company, for example, normally confines its selection work to its own lands. Generally speaking there was no preliminary provenance research to guide in the selection of the genetically best stands. With increased availability of data from combined progeny–provenance tests, there is now a greater tendency to concentrate selection work in areas known to have superior germ plasm.

Site Selection

Most orchards are located close to a nursery so that people are available for grafting work, seed collection, etc. As the pine growing regions are generally level to moderately hilly, most orchards are located on nearly level ground. Very low spots subject to frost damage are avoided. Soil fertility is not a major consideration, as the orchards are intensively managed and nutrient deficiencies are easily corrected by application of fertilizer. Good moisture retention capacity is important, and sandy loam soils overlaying permeable sandy clays are preferred.

Isolation from surrounding stands of the same species is necessary, otherwise the selected clones are pollinated by unselected outside trees and seed quality suffers. Width of the isolation strip is governed by size of the orchard. The smaller the orchard, the less pollen it produces and the greater the contamination by surrounding trees; therefore, the smaller the orchard the greater the need for a wide isolation strip. There are also practical considerations. Isolation by 1 mile or more is desirable, but tracts isolated by that much are often difficult to find in a region with millions of acres of pine. Hence, some orchards are isolated by only 500 ft.

Scion Collection and Grafting

Scions are collected from the upper branches of selected trees by climbing, shooting, or occasionally by the use of long-handled pruners. Sometimes only a few scions are collected per tree, with the idea that they will be placed in a clone bank, which is a future source of scion wood in quantity. Or, enough scions may be collected to perform all necessary grafting work in 1 year. After collection, scions are kept moist and cool and are usually used within a few days.

Some grafting work has been done during the dormant season, in December and January, using scions that consist of the last year's growth.

In Texas, however, success has been greater with grafts made during March or April, after the start of the growing season. In that case, the scions are a few inches long and consist of current year's growth.

Rootstocks are usually 1- or 2-year-old seedlings growing outdoors in soil and of the same species as the scions. Veneer grafts are most common. Some grafting is done in a nursery, with the grafted trees being moved to their permanent locations 1 year later as balled nursery stock. Whenever possible, though, the rootstocks are planted in the seed orchards prior to grafting, and grafting is done in the field. That necessitates some regrafting or shifting of spare grafts from nursery to seed orchard.

Newly made grafts are bound with grafting tape and protected from desiccation by enclosure in a plastic bag, which in turn is usually covered by a heavy paper bag to prevent overheating. Aluminum foil can also be used to protect the scions. As soon as a graft union heals (usually within a few weeks), the protective covering is removed. The grafting work proceeds at the rate of 100 or more grafts per man per day. With skilled workers and favorable weather conditions success rates of 99.9% have been achieved.

Orchard Design

A few specialty orchards are as small as 2 acres in size. Most general purpose orchards are 20 acres or larger, with a few covering 400 acres. The medium and large-sized orchards are more economical to operate with respect to fire protection, installation of an irrigation system, etc.

A commonly recommended spacing is 30 × 30 ft, equivalent to about 50 trees per acre. A spacing that wide enhances cone production and permits the operation of heavy equipment, such as trucks and cone pickers. Some orchards are originally planted to that final spacing. In such orchards, graft failures are quickly replaced. After a few years clones which do not flower (there are many) or which suffer scion–rootstock incompatibility (prevalent in loblolly pine) are replaced as the need arises. First generation clones may be replaced by second generation clones. Thus, over a period of years, an orchard established at the ultimate spacing tends to contain clones of several different ages. In orchards up to 15 years of age competition between trees of various ages has not been a serious problem. Whether it will be a serious problem as the orchards grow older is not known.

At the other extreme, some loblolly and slash pine orchards have been established with as small as a 10 × 10 ft spacing, equivalent to about 430 trees per acre. That is done with the thought that up to half the trees may be removed because of inability to produce cones or scion–rootstock incompatibility, and that up to three-fourths of the remaining clones may be removed because of low general combining ability as progeny test data become available.

The clones are normally planted in randomly located one-tree plots, in order to minimize self-pollination. Occasionally, special precautions are taken so that related clones will not be planted adjacent to each other, to reduce the likelihood of inbreeding still further.

Orchard Care

Southern pine seed orchards are usually disked the first year or two to conserve moisture and to promote good early growth. Then they are commonly reseeded to grass, as a sod cover minimizes erosion and makes it easier to operate heavy equipment.

Irrigation practices vary. Some orchards are not irrigated. Some are watered for the first 2 to 3 years by means of temporary sprinkler systems. In some there are permanent irrigation systems operated whenever rainfall is scanty. Generally speaking, supplemental watering has helped in the establishment and early growth of the seed orchards, but has not increased seed yields markedly in later years.

Except on soils with severe deficiencies, fertilizer application has not resulted in marked increases in cone and seed yield. In some loblolly pine experiments, however, application of high phosphorus fertilizer has caused dramatic increases in pollen production.

Cone insects are often a serious problem and must be controlled by appropriate sprays if present in epidemic amounts. As the insects are of various species and may attack during either the first or second year, it is important to determine the correct time and correct insecticide for each insect. Protection against fire is also important.

Efforts to increase cone production by girdling, banding, etc., have met with no success, nor have efforts to keep the trees low in stature by shearing the ends of the branches or leaders. Pine female flowers are borne on new growth, and removal of this new growth may lower cone production dramatically.

Cone and Seed Collection

Ripe slash pine cones develop an abscission layer and can be loosened by shaking. For that species tractor-mounted vibrators or shakers work well. These are devices which are placed against the boles of trees and vibrated rapidly so that the branches whip back and forth and loosen the cones. The cones drop onto canvas or the ground, where they are picked.

Shakers do not work well on loblolly pine, from which most cones are obtained by hand picking. In some places large vacuum cleaners are used. With this method, the cones are allowed to open naturally and shed their seed, which is then sucked up from the ground. Vacuum cleaners work best if the soil is covered by a closely mown grass sod.

Progeny Testing

By 1974 progeny testing was an integral part of the management of most southern pine seed orchards. Nearly all the progeny tests are well replicated; however, details vary. The principal types of progeny tests used are the following.

Open-pollinated, using seed collected from the selected parents at time of grafting (selected × average)

Open-pollinated, using seed collected from seleoted clones in seed orchards (selected × selected)

Half-sib, using control-pollinated seed produced by crossing selected females with a pollen mixture

Full-sib, using tester matings with each female crossed to the same 3 to 4 males

Full-sib, with each female crossed to different males

There has frequently been a lag of several years between a clone's incorporation into a seed orchard and the progeny testing of that clone; therefore, progeny test data were available for relatively few orchards as of 1974. Many of the progeny tests do not include offspring of average trees in the stands containing the selected trees; in such cases estimates of gain are made by comparing the offspring of an orchard with a commercial check.

Most of the progeny tests will provide data on general combining ability. In a few orchards these data are used to provide thinning recommendations and thus improve the quality of seed produced by those orchards. More commonly the data are used to guide the second generation breeding program or to select parental clones for incorporation into second-cycle first generation orchards providing higher gain than the original orchards.

Thinning, Replacement of Clones, and Second-Cycle Orchards

For those orchards established on narrow (i.e., 10 × 10 ft) spacings, it is possible to remove clones that suffer graft incompatibility, produce no cones, or have low general combining ability. This practice raises the genetic gain from the level possible with mass selection alone to a level close to that possible with full-sib family selection. However, the possible genetic gain is limited by the need to retain 25–50% of the clones for adequate seed production; in other words selection differential among the clones must be relatively low. For that reason it is common practice with slash pine in Florida to leave the original orchards intact and to plant second-cycle (but first generation) orchards containing a very few clones with the highest demonstrated general combining ability. With that practice gain during the first generation is approximately equal to that possible with full-sib family selection.

Most seed orchards in Texas have been established with a 30 × 30 ft spacing. In these orchards, clones are replaced one by one as they fail or as they are found to have low general combining ability.

Second Generation Orchards

When the 6000 acres of first generation orchards now in existence come into full production, they will satisfy most seed needs of the region. Hence, since 1970 several breeders have turned their attention to the additional gains possible through second generation breeding. A few second generation seed orchards and progeny tests were underway by 1974.

Generally speaking, the plan is to start with selections made in a first generation full-sib progeny test. Emphasis is generally on the same traits (growth rate, form, etc.) as in the first generation. Most of the selection work is done in progeny tests which are 8 to 10 years old. Design and physical management promise to be about the same for the second generation as for the first generation orchards.

Most breeders are looking beyond to the third and fourth generations, by which time inbreeding could be serious if not forestalled. Thus, several breeders are attempting to broaden their first generation base by crossing clones from different seed orchards. Also, most are limiting the number of clones selected from any one first generation family for possible inclusion in a second generation orchard.

Some gain in first generation orchards was achieved by selecting trees from stands which had a high average genetic potential. To the extent that this is true, no additional gain can be expected in the second generation. But some first generation gain was achieved by selecting from the variability within stands. Because of increased data about heritability and of advances in progeny testing, the second generation gain from individual tree variability can be expected to equal or exceed that obtained in the first generation.

Quality and Quantity of Seed Produced

Most progeny tests show that there has been considerable gain in oleoresin production and in growth rate in slash pine and in stem form of loblolly pine. The bulk of evidence also indicates that selection for growth rate and special wood properties of loblolly pine was generally effective. Selection for fusiform rust resistance was often ineffective, especially if the selections were made in stands with a low frequency of rust infection.

It is difficult, however, to make overall statements concerning the amount of gain achieved. Published estimates of gain in volume production run from a few percent to as high as 40%, but some of the higher estimates include advances due to better plantation management as well as to genetic improvement. Nevertheless, there is enough consistency to indicate that the overall gains are more than enough to justify the costs.

Yield data are only now forthcoming. Quantity production of female flowers starts at age 6 to 7 in loblolly pine and 8 to 9 in slash pine. Pollen production is usually delayed another 2 years. Hence the first cone crops are usually small, and the number of filled seeds per cone is usually low because of inadequate cross pollination. Yields of cones and seeds improve considerably by age 10.

The area devoted to seed orchards was planned on the assumption that 1 acre of grafted trees would produce about 500,000 loblolly pine or 200,000 slash pine seeds annually. The actual yield data shown in the following tabulation were taken from Texas Forest Service seed orchards for 1973 and are somewhat high, since 1973 was the best seed year in over two decades in east Texas. Nevertheless, they indicate that the goals can be achieved.

| Species | Age (years) | Yield per acre | | |
		Bushels of cones	Pounds of seeds	Seeds
Loblolly pine	7	12	10	180,000
Loblolly pine	10–16	57	65	1,170,000
Slash pine	8	0.7	0.3	4,000
Slash pine	9–16	44	35	470,000

The yields shown in the tabulation are appreciably higher than the yields obtained in natural stands. For example, the yield from a 43-year-old slash pine stand thinned for use as a seed production area was 17 bushels per acre, only 40% as great as from the older grafted seed orchards.

Management of Seedling Seed Orchards of Red Pine

General

This species, native to northeastern United States and southeastern Canada, was chosen as an example because it presents a marked contrast to the southern pines. Several million acres of plantations, mostly on public land, were established from 1930 to 1960, but the planting rate since then has slackened because of the unavailability of open land. As the plantations mature and are harvested, the planting rate is expected to increase.

Occasionally seedling plantations have fruited heavily by their seventh year, if that year happened to coincide with a generally good seed year. Grafting may reduce fruiting time slightly, but the results of grafting experiments are not clear-cut. Establishment of seed production areas by thinning of existing plantations has met with little success; the thinned trees required several years to develop heavy enough crown to produce even moderate cone crops. Canadian experiments show that cone production is much the heaviest on trees spaced widely enough apart to have full crowns to the ground and to be exposed all around to sunlight.

Sizable improvement programs are underway in Wisconsin and Michigan. The programs in the two states are autonomous but similar. In each state the "seed orchards" are half-sib progeny tests that are part of long-term improvement programs and also managed for seed production. For an account of progress in Michigan, see the paper by Yao *et al.* (1971). For simplicity, I will confine my discussion to the program in that state.

Description of the Progeny Test

The Michigan progeny test includes the open-pollinated offspring of 272 wild parents located in all parts of the state. The parents were average, as the wild parental stands were generally so small that they provide no opportunity for intensive mass selection. The progeny test includes four plantations established in 1964, each containing 5 to 10 replications. Two plantations are in the northern and two in the central part of the state. Initial spacing was 8×8 ft and four-tree plots were used. The total progeny test covers an area of 36 acres.

The planting sites were level, open, of moderate fertility, well isolated, and far from nurseries. There are no plans for irrigation, but preliminary studies indicate that high nitrogen fertilizer applied annually or biennially will increase cone production.

As of 1974, prior to any thinning, survival was generally high. Nearly all trees were straight-trunked and single-stemmed (a characteristic of red pine, in general). In that year the plantations averaged 12 to 15 ft tall and were growing at the rate of about 2 ft per year. Pollen and cone production were light. The tree crowns were separated by 1 to 2 ft, and crown closure was expected in another 2 to 3 years. Measurements made that year indicated the presence of interfamily differences in height but not in other traits.

Plans for Thinning

The first thinning is scheduled for 1975–1976, hopefully anticipating a good seed year by a season or two. At the time, the plantations will be 12 years old (15 from seed) but not quite closed so that the remaining trees will have adequate space for full crown development. A little pulpwood can be sold but most of the thinning debris must be disposed of in another manner.

The first thinning will remove approximately 136 (50%) families having the poorest average growth rate in all four plantations. Additionally, one tree (generally the shortest but some allowance must be made for evenness of spacing) will be removed from each remaining four-tree plot. Thus average growing space will be increased from approximately 64 to 171 ft² per tree, enough to permit full crown development for another 8 to 10 years.

The second thinning will occur about 1985 and will remove another 68 (25% of the original) of the slowest growing families. The third and final thinning will occur about 1995 and will remove another 34 (12.5% of the original) of the slowest growing families. During the second and third thinnings, the number of trees per plot will be reduced still further to the two tallest. Thus the ultimate final growing space will be about 1024 ft² per tree, equivalent to about 43 trees per acre.

Since the families were originally randomly located in each replication, thinning recommendations based solely on average family height will cause some inequalities in final spacing. To correct the worst of these, occasional plots of tall families will be removed, and occasional plots of shorter families will be retained.

The 1974 measurements indicate a general tendency for certain seed-lots to excel or do poorly in all parts of the state. This is the reason for basing the first thinning recommendations on average performance in all four test plantations rather than on the average in each plantation considered separately. A decision on this point must be made anew prior to each subsequent thinning.

Cone Harvesting

Red pine cones ripen in mid-September and shed their seeds quickly. For the present, hand-picking is contemplated. In the past, squirrel caches have been a primary source of seed for commercial planting, and squirrel harvesting may be possible. Mechanical shakers may also be used.

Expected Gain

Assuming a high correlation between performance as of 1974 and at maturity, thinning to the 36 tallest families will result in a genetic gain of about 2% in the rate of height growth or about 4% in the rate of volume growth. This is in comparison with seed collected from the best wild stands now known in the state. If the comparison is with seed collected from wild stands in the northern part of the state (which were a large source of commerical seed in the past) the expected gains are about 10% in the rate of height growth and about 20% in the rate of volume growth.

Calculations of individual tree heritabilities made from the 1974 progeny-test data indicate that an additional 0.1 or 0.2% gain might have been achieved if the original seed collections had been made from phenotypically selected trees rather than from average trees.

Plans for the F_2 Generation

Present plans call for the start of an F_2 generation at the time of the next good flowering year for red pine. The F_2 generation will consist of a full-sib progeny test resulting from mating of 100 to 300 of the tallest (female) trees in the 25 to 30 tallest families with 2 to 3 males each; the total number of males used will be large, however. Plantation design and total size will be similar to that used in the F_1 generation. Theoretically, the full-sib family selection will result in an additional gain of about 4% in rate of height growth (or double that achieved by intrastand selection in the first generation).

Special Cases

A few cases present problems that are so unique that they are worthy of mention.

Eucalypts

Those eucalypts planted for commercial purposes grow very rapidly (10 ft or more per year). They do not fruit until a few years old, i.e., when 60 to 70 ft tall, and rapidly lose their lower branches. Any breeding scheme which involves large-scale seed collection from living trees appears impractical. One feasible scheme is to establish a series of half-sib progeny tests as part of a commercial planting operation, planting one or more progeny test each year. Then, at commercial maturity (often 8 to 10 years), fell a progeny test and harvest seed from the best trees in the best families. As seed production per tree is heavy, a few such trees can satisfy the needs for the planting of a few hundred acres per year. Either store seed or fell another progeny test for the next year's seed needs.

Eastern Cottonwood

Vegetative propagation per tree is very simple; seed production per tree is very large; the species is dioecious; and YS and YG are nearly equal.

Due to dioeciousness, some full-sib progeny testing is necessary if a clonal seed orchard is to give more gain than a half-sib progeny test; otherwise the male clones included in the orchard will not contribute to gain. A simultaneous clonal orchard, consisting of one ramet per parent, would satisfy the seed needs of a moderate-sized planting program, so would a full-sib progeny test thinned to 40 to 50 trees (1 to 2 trees in each of 20 to 50 families). The two schemes would be about equal with regard to cost, gain, and time of seed production.

Taiwania

Taiwania is a valuable, fast-growing, rare, monotypic conifer genus endemic to Taiwan and little planted elsewhere. Conserving the species is of primary importance; improving it is secondary. Grafted trees 8 years old have not yet flowered; there has been little seed production on small plantations 20+ years old. Hence most seed must come from the few remaining forest giants.

One way to save the species and possibly to improve it is to use available seed for a very large half-sib progeny test in which the *Taiwania* trees would be planted at wide spacings and interspersed with cryptomeria or another less valuable species. In that way the *Taiwania* progeny tests could cover several hundred (possibly a few thousand) acres. They would, after commercial thinning of the interspersed trees at age 15 to 20, be permitted to grow to age 40 to 50. At that time they would be a valuable timber resource and would hopefully be a source of seed and of genetic information. Meantime, the small grafted seed orchard now in existence might also be a source of seed.

Metasequoia

Metasequoia is a fast-growing, rare, monotypic conifer genus endemic to a small area in central China but widely planted in other temperate zone countries. It can be rooted from cuttings, but the costs are too great to make large-scale forest planting feasible. It was introduced to the western world in the spring of 1948, and there have been sporadic reports of flowering on individual trees from 1952 to 1972. However, nearly all planted trees are single specimens not subject to cross pollination, and as of 1974 there is no reliable source of even small quantities of fertile seed.

A seed orchard of any type is needed. One approach is to root cuttings from a very large number of trees to ensure having many different clones. (Clonal pedigrees have not been maintained, and trees growing in Seattle and Philadelphia may be of the same clone.) This should be followed by the establishment of many multiclone plantations in various localities, with the hope that climatic conditions in one or more plantations would promote flowering and seed production.

White Spruce

Provenance tests have shown that white spruce from one part of Ontario grows best at many locations in the United States and Canada. Open-pollinated seed from that part of Ontario would be satisfactory but is not commonly available. A grafted seed orchard established with many clones of that seedlot, using scions collected from the 17 to 18-year-old provenance tests, was started in 1972 by the United States Forest Service and other cooperating agencies. The seed orchard will give some genetic improvement as a result of intraprovenance selection, but will serve principally as a reliable source of guaranteed high quality.

Study Questions

1. Compare the effectiveness of mass selection for growth rate in loblolly pine, slash pine, Scotch pine, black wattle, black cherry, and eastern cottonwood.

2. Do the same for family selection.

3. Why did mass selection for rust resistance have such different results in slash pine, western white pine, and eastern white pine?

4. In eastern cottonwood, how much gain in growth rate can be expected from family selection for *Melampsora* rust resistance and earliness of leafing out?

5. Describe the simplest type of seed orchard needed to provide seed with the highest genetic potential for rapid growth in each of the following: loblolly pine, slash pine, Scotch pine, eastern white pine, ponderosa pine, black cherry, Sakhalin fir.

6. Assuming equal genetic gains in the second and the first generation, what is the expected tannin content of the bark of black wattle trees raised from the seed orchard that will start bearing commercial seed in 1973–1974? Do the same for 3-year height.

7. List three traits and the species in which they occur which are controlled by dominant–recessive gene pairs.

8. How does the presence of a mass-selection gain help when practicing half-sib or full-sib family selection?

9. Describe the parental selection, method of grafting, spacing, care, time of cone production, and seed yields for a clonal seed orchard of slash or loblolly pine. How are the results of progeny tests used in the management of clonal orchards in these species?

10. Describe the use of a half-sib progeny test as a seed orchard in red pine.

11. Why are second generation seed orchards being planned, and in what respects do they differ from first generation orchards?

12

Estimation of Heritability and Genetic Gain

Experimental Procedures and Applicability of Estimates

Heritability and genetic gain can be estimated in either of two ways. The most direct estimates are derived from the relation between parents and offspring, obtained by measuring the parents, growing the offspring, and measuring the offspring. The other way is to establish a half-sib or full-sib progeny test, calculate the mean squares and variances, and compute heritability as a function of the variances.

Actual calculations require a knowledge of statistics, and formulas given in this chapter are intended primarily as handy references for persons with such knowledge. Even so, some of the formulas are complex. I have made no attempt to cover numerous possible variations that might result from irregularities in design.

Both heritability and gain estimates apply strictly only to the experiments from which they were obtained. They may be and frequently are very different when obtained from slightly different experiments. Therefore, when quoting them it is desirable to include pertinent details of experimental design and calculation procedures.

Also, it is good practice to state the statistical reliability of each heritability estimate because many such estimates of 0.2 are only reliable within ±0.3 or 0.4. Formulas for calculating reliability are, therefore, included in this chapter.

Most formulas given in this chapter are taken from Lerner (1958), Falconer (1960), Becker (1964), and Namkoong et al. (1966). Further details and references are given in those works.

Parent–Progeny Regression

Experimental Data Needed[1]

To calculate heritability and gain, establish an experiment as follows. (1) Select parents for superiority in a certain trait. (2) Choose average trees from the same stands. (3) Measure each selected and average tree and express the performance of each tree in the trait under consideration as X_a to X_z for parents A through Z. (4) Establish a half-sib or full-sib progeny test, preferably replicated, including the offspring of all superior and average parents. (5) Measure the trait in the offspring and calculate the mean for each family. (6) For half-sib progeny tests express the means of the offspring of parents A through Z as Y_a to Y_z. (7) If a full-sib progeny test, express the family means as Y_{ab}, Y_{ac}, Y_{ez}, etc., for families resulting from matings A $\times$ B, A $\times$ C, E $\times$ Z, etc.

Statistical Procedures

Calculate the regression coefficient b for the generalized regression equation

$$Y = a + bX$$

To calculate b

$$b = \frac{\Sigma\, XY - (\Sigma\, X)(\Sigma\, Y)/N}{\Sigma\, X^2 - (\Sigma\, X)^2/N}$$

where X and Y are the means for parents and families, respectively, and N is the number of families.

For a half-sib progeny test, X_a, X_b, X_c, etc., are the measurements made on parents A, B, C, etc. For a full-sib progeny test, X_{ez} is the mid-parent mean which is $\frac{1}{2}$ the sum of the measurements of parents E and Z.

This regression coefficient (b) is an estimate of the amount of increase (or decrease) in performance of the offspring per unit increase in performance of parents.

---

[1] For a true heritability estimate it is desirable to select parents randomly. However, the experimental procedures are so expensive that one is usually content to obtain a heritability estimate applicable to a particular applied problem.

Heritability

Size of the regression coefficient b depends on the measurement units used for parents and offspring. To counteract this problem, it is best to calculate a "standardized regression coefficient" b', which effectively states a given percentage increase in parents as a percentage increase in offspring. The calculation is as follows:

$$b' = b \left(\frac{\Sigma X^2 - (\Sigma X)^2/N}{\Sigma Y - (\Sigma Y)^2/N} \right)^{1/2}$$

where X, Y, and N are defined as above.

Single tree heritability applicable to the parental population is then as follows.

$$h^2 = 2b' \qquad \text{for a half-sib progeny test}$$
$$h^2 = b' \qquad \text{for a full-sib progeny test}$$

Gain

The general formula for gain is

Gain = selection differential $\times$ heritability

In this case the selection differential is the difference between selected and average parents. The proper multiplier depends on the use to be made of the selected trees. If seed is to be collected from them as they occur in a wild stand or plantation and are pollinated by average trees

Gain = selection differential $\times$ $(h^2/2)$

If the selected trees are to be control-pollinated among themselves or grafted and placed in a clonal seed orchard

Gain = selection differential $\times$ h^2

Reliability of Heritability Estimates[2]

The formula for the standard error of b is as follows:

$$\text{Standard error of } b = \left(\frac{\Sigma Y^2 - \dfrac{(\Sigma Y)^2}{N} - \dfrac{(\Sigma XY)^2 - (\Sigma X)^2(\Sigma Y)^2/N}{\Sigma X^2 - (\Sigma X)^2/N}}{(N-2)[\Sigma X^2 - (\Sigma X)^2/N]} \right)^{1/2}$$

[2] As pointed out by Gene Namkoong, the standard errors of regression coefficients and heritability estimates, as calculated from the formulas given here, are strictly true only when the parents are selected randomly and the offspring are randomly planted in one-tree plots. For actual estimates, which tend to be higher than would be calculated from the formulas given here, one needs to consider plot size and experimental design.

This is the standard error of the heritability estimate derived from a full-sib progeny test. It should be multiplied by 2 to apply to a heritability estimate derived from a half-sib progeny test.

Approximately, the "true" heritability is within ±1 standard error 67% of the time and within ±2 standard errors 95% of the time. For more accurate estimates of the confidence intervals, consult a statistics text.

Variance Analysis—Half-Sib Progeny Test

Experimental Data Needed

Establish a replicated progeny test consisting of open-pollinated offspring of F females, replicated B (for block) times at each of S sites, using N-tree plots. Measure a trait, such as height, and calculate the analysis of variance, using degrees of freedom as shown in Table 12.1.

Calculation of Variance Components

In ordinary statistical work the mean squares are divided by each other in various manners to obtain F values, which are then used to test significance. The mean squares themselves, however, are complex,

TABLE 12.1

Degrees of Freedom and Variance Components of Mean Squares for a Multiplantation Half-Sib Progeny Test

Source of variation	Degrees of freedom	Variance components of mean squares
Family	$F - 1$	$V_e + NV_{fb} + NBV_{fs} + NBSV_f$
Site	$S - 1$	$V_e + NV_{fb} + NBV_{fs} + NFV_b + NFBV_s$
Block-within-site	$S(B - 1)$	$V_e + NV_{fb} + NFV_b$
Female × Site	$(F - 1)(S - 1)$	$V_e + NV_{fb} + NBV_{fs}$
Female × Block-within-site	$S(F - 1)(B - 1)$	$V_e + NV_{fb}$
Tree-within-plot	$NSF(B - 1)$	V_e

[a] F, B, S, and N are numbers of females, blocks per site, sites, and trees per plot, respectively. V_e, V_{fb}, V_{fs}, V_f, V_b, and V_s are variances due to tree-within-plot, female × block-within-site, female × site, female, block-within-site, and site, respectively.

most of them containing variability due to several factors. To eliminate this difficulty, mean squares are apportioned into variance components according to the equivalents shown in Table 12.1.

In that table the symbols used are as follows:

F, B, S, and N are, respectively, the number of females, the number of blocks per site, the number of sites, and the number of trees per plot.

V_e, V_{fb}, V_{fs}, V_f, V_b, and V_s are the variances due to tree-within-plot, female $\times$ block-within-site, female $\times$ site, female, block-within-site, and site, respectively.

Having calculated the mean squares, set each equal to its variance components as shown in Table 12.1. Start at the bottom of the table and obtain each successive variance by a process of subtraction and division. That is, subtract within-plot mean square (V_e) from family $\times$ block mean square ($V_e + NSV_{fb}$) to obtained NSV_{fb}; then divide by NS to obtain V_{fb}. Proceed in a similar manner up the table.

Family Heritability

Having calculated the variances, calculate heritability of the half-sib family averages as follows.

$$\text{Family heritability} = \frac{V_f}{V_e/NBS + V_{fb}/BS + V_{fs}/S + V_f}$$

The family averages are more reliable than are the averages for any single plot or tree, and selection will be based upon family averages. That is the reason for division of V_e, V_{fb}, and V_{fs} by NBS, BS, and S, respectively, in the denominator of the heritability formula.

Single Tree Heritability

In a half-sib progeny test, differences among families account for only $\frac{1}{4}$ of the additive genetic variance; the remainder is accounted for by variation within the families. For that reason, V_f is multiplied by 4 when calculating single tree heritabilities. Also, since selection is based upon single trees, all variances are inserted *in toto* in the denominator. Therefore the formula for single-tree heritability is

$$\text{Single tree heritability} = \frac{4V_f}{V_e + V_{fb} + V_{fs} + V_f}$$

Shortcuts and Their Consequences

Assume the families to be tested in only one test plantation. Testing and calculating procedure are much simplified. Total degrees of freedom are $NFB - 1$; site and female $\times$ site mean squares and variances are eliminated from Table 12.1.

In this situation, families are measured at one site only. They might grow very differently at other sites. The calculated V_f is in reality a combination of V_f and V_{fs}. Therefore, heritability calculated on the basis of data from one plantation only is overestimated.

Recording and analysis of single tree data are the most laborious parts of measurement and calculation procedures, often accounting for 75% of the total effort. Estimates of V_{fb}, V_{fs}, and V_f are not changed if data are analyzed in terms of plot means rather than individual tree means, but V_e cannot be determined. The term V_e/NBS is often so small that it is inconsequential in the estimation of family heritability. However, single tree heritability is slightly overestimated if V_e is omitted.

Even more time can be saved by dealing solely with the means of families at different sites, i.e., calculating V_{fs} and V_f only. Elimination of the V_{fb}/BS term ordinarily causes a slight overestimate of family heritability. Elimination of the V_{fb} term may cause a greater overestimate of the single tree heritability.

Standard Deviation of Heritability Estimates[3]

The standard deviation (σ) of a single tree heritability estimate is

$$\sigma_{h^2} = \frac{(1 - h^2/4)[1 + (NBS - 1)h^2/4]}{[(NBS/2)(NBS - 1)(F - 1)]^{1/2}} \cong \frac{(1 - h^2/4)[1 + (NBS)h^2/4]}{NBS[(F - 1)/2]^{1/2}}$$

if NBS is the number of trees per family and is large.

The standard error of a sample of NBS trees is approximately $\sigma/(NBS - 1)^{1/2} = \sigma/(NBS)^{1/2}$ approximately. Therefore the standard error of a family heritability is approximately

$$\sigma_{h^2(\text{half-sib family})} \cong \frac{(1 - t)(1 + NBSt)}{[(NBS)(F - 1)/2]^{1/2}}$$

[3] See footnote 2, on p. 241.

where t is the intraclass correlation, which equals one-fourth of the single tree heritability.

The above formulas are correct if $V_e = V_{fb} = V_{fs}$. However, if one of these is much larger than the others, the term NBS should be reduced accordingly. If, for example, V_{fs} is much larger than V_{fb} or V_e, S might be substituted for NBS.

Heritability Estimates for Selection Based upon Family Means in a Single Plantation or on Plot Means

The above-calculated family heritability estimate is strictly applicable only if those families with the best overall performance in all plantations are selected. A breeder may select those families which are superior in one plantation only. In that case, family heritability is calculated as above except that V_{fs} is substituted for V_{fs}/S in the denominator.

If a breeder wishes to select on the basis of plot means only, family heritability is calculated as shown above, except that V_{fs} and V_{fb} are substituted for V_{fs}/S and V_{fb}/BS, respectively, in the denominator.

Use of Heritability Estimates

To calculate actual gain from a half-sib progeny test, use the formula

Genetic gain from family selection
= selection differential × family heritability
= (selected families − average families)
× family heritability

To calculate expected gain from mass selection in such a progeny test, use the formula

Expected mass selection gain
= selection differential × single-tree heritability
= (selected trees − average trees)
× single tree heritability

Note that the selection differentials and heritabilities must apply to the same things—individual trees or families. The same is true if selecting on the basis of plot means or of family means within a plantation.

Variance Analysis—Full-Sib Progeny Test

Experimental Data Needed

Establish a full-sib progeny test consisting of FM families resulting from matings between F females and M males. Replicate each family B (for block) times at each of S sites, using N-tree plots. Measure a trait, such as height, and calculate the analysis of variance, using degrees of freedom as shown in Table 12.2. If the experiment follows

TABLE 12.2

Degrees of Freedom and Variance Components of Mean Squares for a Multiple-Plantation Full-Sib Progeny Test[a]

Source of variation	Degrees of freedom	Variance components of mean squares
Family		
Female	$F - 1$	$V_e + NV_{fmb} + NMV_{fb}$ $+ NBV_{fms} + NBMV_{fs}$ $+ NBSV_{fm} + NBSMV_f$
Male	$M - 1$	$V_e + NV_{fmb} + NFV_{mb}$ $+ NBV_{fms} + NBFV_{ms}$ $+ NBSV_{fm} + NBSFV_m$
Female $\times$ Male	$(F - 1)(M - 1)$	$V_e + NV_{fmb} + NBV_{fms}$ $+ NBSV_{fm}$
Site	$S - 1$	$V_e + NV_{fmb} + NBV_{fms}$ $+ NFV_{mb} + NMV_{fb}$ $+ NBFV_{ms} + NBMV_{fs}$ $+ NFMV_b + NFMBV_s$
Block-within-site	$S(B - 1)$	$V_e + NV_{fmb} + NFV_{mb}$ $+ NMV_{fb} + NFMV_b$
Female $\times$ Site	$(F - 1)(S - 1)$	$V_e + NV_{fmb} + NMV_{fb}$ $+ NBV_{fms} + NBMV_{fs}$
Male $\times$ Site	$(M - 1)(S - 1)$	$V_e + NV_{fmb} + NFV_{mb}$ $+ NBV_{fms} + NBFV_{ms}$
Female $\times$ Male $\times$ Site	$(F - 1)(M - 1)(S - 1)$	$V_e + NV_{fmb} + NBV_{fms}$
Female $\times$ Block	$S(F - 1)(B - 1)$	$V_e + NV_{fmb} + NMV_{fb}$
Male $\times$ Block	$S(M - 1)(B - 1)$	$V_e + NV_{fmb} + NFV_{mb}$
Female $\times$ Male $\times$ Block	$S(F - 1)(M - 1)(B - 1)$	$V_e + NV_{fmb}$
Tree-within-plot	$FMSB(N - 1)$	V_e

[a] F, M, B, S, and N are numbers of females, males, blocks per site, sites, and trees per plot, respectively. V_e, V_{fmb}, V_{mb}, V_{fb}, V_{fms}, V_{ms}, V_{fs}, V_b, V_s, V_{fm}, V_m, and V_f are, respectively, variances due to tree-within-plot, female $\times$ male $\times$ block-within-site, male $\times$ block-within-site, female $\times$ block-within-site, female $\times$ male $\times$ site, male $\times$ site, female $\times$ site, block-within-site, site, female $\times$ male, male, and female.

a factorial design (slight changes in coefficients are necessary otherwise), the variance components of the mean squares are as shown in Table 12.2, and their calculation proceeds by a process of subtraction and division in the same way as discussed for half-sib progeny tests.

Family Heritability

Having calculated the variances, calculate heritability of full-sib family averages as follows:

$$\text{Full-sib family heritability} = \frac{V_f + V_m + V_{fm}}{V_e/NBS + V_{fmb}/BS + V_{fms}/S + V_{ms}/S + V_{fs}/S + V_{fm}/FM + V_f + V_m}$$

The above formula is correct if it is assumed that the selected families were to be reproduced exactly, as would happen if the female and male parents with the best combining ability were planted in a clonal seed orchard.

If, however, the selected families are themselves used as seed producers, as in a seedling seed orchard or as in a proceeding from the F_1 to the F_2 generation, nonadditive genetic variance (as measured by V_{fm}) will not contribute to superiority of the offspring. In that case, V_{fm} should be eliminated from the numerator of the heritability formula.

Single Tree Heritability

In a full-sib progeny test, differences among families account for one-half of the additive genetic variance, the remainder being accounted for by variation among trees within families. One-half the additive genetic variance equals $V_f + V_m$. Therefore the numerator of the single tree heritability formula is $2(V_f + V_m)$. As already explained under half-sib progeny tests, variances are included *in toto* in the denominator. The formula is

Single tree heritability

$$= \frac{2(V_f + V_m)}{V_e + V_{fmb} + V_{fms} + V_{ms} + V_{fs} + V_{fm} + V_f + V_m}$$

Assumptions

Implicit in the above formulas are the assumptions that the experiment is perfectly regular (i.e., equal numbers of plots per family, equal numbers of males crossed with each female, etc.) and that all effects are random.

Minor violations of these assumptions are numerous and usually have little consequence. The heritability estimates can be greatly in error, however, if the female and male parents are closely related. Papers by Namkoong and co-workers (1966; Kriebel *et al.*, 1972) discuss some of the problems. Advanced statistical texts discuss these assumptions and show the mathematically correct ways to correct for their violation.

Shortcuts and Their Consequences

As with half-sib progeny tests, full-sib families are often planted in one plantation only. In such a case the family variance is a mixture of unknown amounts of family × site variance, and its use leads to an overestimate, by an unknown amount, of heritability.

Even with a computer, calculation of a complete analysis of variance exactly following the model shown in Table 12.2 is tedious and rarely done. Consequences are almost negligible if the *FB* and *MB* interactions are not calculated separately but included in the *FMB* interaction. The same is true of *FS* and *MS* interactions, which can be included as part of the *FMS* interaction.

As is the case with a half-sib progeny test, 75% of the measurement and calculation time may be devoted to dealing with individual tree values instead of plot means, solely for the purpose of calculating V_e. Elimination of the V_e/NBS term from the denominator of the family heritability formula has almost negligible consequences; elimination of the V_e term from the denominator of the single tree heritability formula leads to a slight overestimate of heritability.

With most computer programs, one run only is needed to calculate family × site or family × block-in-all-sites interaction, but two runs are usually needed to calculate family × block-within-site interaction. In calculation of family heritability, it usually makes little difference whether one or both are calculated.

The V_f and V_m terms are the most important, and the V_{fm} term is sometimes ignored after it is calculated. However, it is so easy to derive that its calculation is always warranted.

Standard Deviation of Heritability Estimates[4]

The standard deviation of a single-tree heritability estimate is

$$\sigma_{h^2 \text{(individual tree)}} \cong \frac{(1 - h^2/2)[1 + (NBS)h^2/2]}{NBS[(F - 1)/2]^{1/2}}$$

[4] See footnote 2, p. 241.

The standard error of a full-sib family heritability is

$$\sigma_{h^2\text{(full-sib family)}} \cong \frac{(1 - t)(1 + NBSt)}{[(NBS)(F - 1)/2]^{1/2}}$$

where t is the intraclass correlation and equals one-half the single-tree heritability.

Note that these formulas are nearly the same as for half-sib progeny tests.

Use of Heritability Estimates

The family heritability estimates are used to calculate actual gain from a full-sib progeny test. They must be used with selection differentials based upon family means.

The single tree heritabilities are used to estimate probable gain from mass selection within the full-sib progeny test. They must be used with selection differentials based upon single tree measurements.

In the case of combined family + within-family selection, calculate the actual gain from family selection by using family heritability and family selection differentials. Then calculate probable gain from within-family selection by using single tree heritability and a selection differential based on variation within the selected families.

As in a half-sib progeny test, a breeder may select on the basis of family means in all plantations, family means in one plantation, or plot means. The family heritabilities are used in such cases, but should be adjusted as in the manner discussed under half-sib family heritabilities.

Conversion of Family to Single Tree Heritabilities

It is often easiest to calculate family heritabilities and to calculate single tree heritabilities from them. The general formulas for conversion (from Falconer, 1960) are as follows.[5]

$$h_{\text{hsf}}^2 = \frac{1 + 0.25(NBS - 1)}{1 + 0.25h_i^2(NBS - 1)} h_i^2$$

$$h_{\text{fsf}}^2 = \frac{1 + 0.50(NBS - 1)}{1 + 0.50h_i^2(NBS - 1)} h_i^2$$

[5] Falconer's formulas are strictly true only if offspring are grown in randomly located one-tree plots. Unfortunately there is no convenient way to make exact conversions when multiple-tree plots are used, as is most apt to be the case.

where

h_i^2 = individual-tree heritability
h_{hsf}^2 = heritability of half-sib family averages
h_{fsf}^2 = heritability of full-sib family averages
N, B, and S = numbers of trees per plot, blocks per plantation, and planting sites, respectively

Unfortunately, the above formulas are meant to be solved for family heritability in terms of single tree heritability. Hence it is necessary to perform a series of calculations with assumed values of h_i^2 to find the value of h_i^2 corresponding to the calculated value of family heritability.

That was done for values of h_i^2 varying from 0.01 to 0.30 and half-sib progeny tests containing 50- or 100-tree families; those comparisons are presented in Table 9.2 in Chapter 9. Further comparisons are made for half-sib and full-sib family heritabilities in Table 12.3.

Variance Analysis—Clonal Test

Experimental Data Needed and Calculation of Variance

Establish a half-sib progeny test consisting of the offspring of F females. At age 1 or 2, make cuttings of C trees per family, and establish a replicated clonal test consisting of C clones of each of F families, replicated B times, with N-ramet plots. For that experiment, degrees of freedom and variance components of the mean squares are as given in Table 12.4.

The above design was chosen for simplicity. In actual practice it would be better to establish S replicated plantations and to start with a full-sib rather than a half-sib progeny test.

TABLE 12.3

Relationship between Heritabilities of Family Averages and of Individual Trees in Half-Sib and Full-Sib Progeny Tests

Type of progeny test	Family size (No. of trees) (NBS)	Heritability of family averages if heritability of individuals is					
		0.10	0.20	0.30	0.40	0.50	0.60
Half-sib	50	0.60	0.76	0.86	0.91	0.94	0.96
Full-sib	10	0.38	0.58	0.70	0.78	0.85	0.89
Full-sib	25	0.59	0.77	0.85	0.90	0.93	0.95
Full-sib	50	0.74	0.87	0.92	0.94	0.96	0.97

TABLE 12.4

Degrees of Freedom and Variance Components of Mean Squares for a Combination Clonal-Half-Sib Family Test[a]

Source of variation	Degrees of freedom	Variance components of mean squares
Clone-in-family	$F(C-1)$	$V_e + NCV_{fb} + NV_{cb} + NBCV_f + NBV_c$
Family	$F-1$	$V_e + NCV_{fb} + NV_{cb} + NBCV_f$
Block	$B-1$	$V_e + NCV_{fb} + NV_{cb} + NCFV_b$
Clone(F) × Block	$F(C-1)(B-1)$	$V_e + NV_{cb}$
Family × Block	$(F-1)(B-1)$	$V_e + NCV_{fb}$
Tree-in-plot	$CFB(N-1)$	V_e

[a] C, F, B, and N are the numbers of clones per family, families, blocks, and trees per plot, respectively. V_e, V_{fb}, V_{cb}, V_b, V_f, and V_c are variances due to tree-within-plot, female × block, clone-within-family × block, block, female, and clone-within-family, respectively.

Family and Clonal Heritability

For the above experiment the formulas for family and clone-within-family heritability are as follows:

$$\text{Family heritability} = \frac{V_f}{V_e/NB + V_{fb}/B + V_f}$$

Clone-within-family heritability

$$= \frac{V_c}{V_e/NB + V_{fb}/B + V_{cb}/B + V_f + V_c}$$

Use of Clonal Heritability Estimates

Clone-within-family heritability estimates are used (in conjunction with selection differentials based on differences among clones within the same family) to estimate gain possible from clonal selection within families. Such gain estimates are then compared with estimates of gain from family selection. From a comparison between the two, the advisability of clonal selection can be determined.

Several estimates of clonal heritability have been made for poplars and willows. In most such estimates, between-family and within-family variances have not been separated, with the result that clonal heritability has been overestimated.

Genotype–Environment Interaction—Effect on Breeding Plans

Genotype–environment interaction exists if seedlot A grows faster than B at one place but slower than B at another. Its statistical significance is measured by relative sizes of the female × site, clone × site, etc., mean squares. If the interactions are nonsignificant or do not involve appreciable differences in rank among the best families or clones, they may be ignored, in which case selections should be based upon a genotype's average performance at all test sites.

If the interactions are large and can be interpreted sufficiently to permit forecasts of where particular genotypes will excel or grow poorly, they cannot be ignored. To determine this, group the data from several plantations according to the plantations' site characteristics (i.e., northern versus southern, dry versus moist, sterile versus fertile). Determine the amount of interaction within and between such groups. If a large portion of the interaction can be explained by the grouping, make separate selections for the sites typical of each plantation group. Then the correct statistical procedure is to make a separate analysis of variance (as outlined in Tables 12.1, 12.2, or 12.4) and heritability estimate for each plantation group within which the interactions are too small or too uninterpretable to be of practical importance.

Study Questions

1. What data are needed to calculate heritability from parent–offspring regression from an experiment designed to yield data on full-sib family heritability?

2. How can a half-sib family heritability be calculated from a full-sib progeny test? Is the reverse possible?

3. Why is it desirable to determine the statistical reliability of a heritability estimate?

4. How is heritability affected by establishing one instead of several plantations? By working with plot means rather than single tree values? By neglecting to calculate family × block interactions?

5. Describe how to determine the proper selection differential to be used with each heritability described in this chapter. Why is it necessary to match selection differential and heritability?

6. Describe the uses of heritabilities of half-sib family averages; of full-sib family averages, of single trees derived from parent–offspring regression, of single trees derived from progeny test data.

7. Why, in a clonal test, is it desirable to separate between-family and within-family variances?

8. Given half-sib families containing 60 trees and single tree heritabilities of 0.01, 0.02, 0.05, and 0.10, calculate the corresponding half-sib family heritabilities.

13

Provenance Testing

Role of Provenance Tests in Tree Improvement

"Provenance" is a synonym for "origin" or "source." The word has been used commonly by tree breeders to mean "ultimate natural origin." A "provenance test" is an experiment in which seeds are collected from a number of widely scattered stands (usually natural), and the seedlings are grown under similar conditions.

Provenance tests have been conducted in more than 50 temperate zone tree species. In those species for which there are also data on individual tree variability, genetic differences associated with place of origin have often been several times as great as those among individual trees in the same stand. That is the reason for provenance testing. Plus trees chosen for selective breeding without regard to provenance performance might be of an inferior race. If so, one to several generations of selective breeding might be needed to produce a strain equal to some already available natural race.

It is particularly necessary to do provenance testing prior to more intensive breeding work when dealing with an exotic. In some species there are 4:1 differences in growth rate and correspondingly large differences in other traits between different races of the same species, and a breeder should be sure he has the best race before starting crossing work and seed orchard establishment. However, provenance testing is also desirable in native species. Natural selection has tended to produce natural populations that are well adapted to the conditions in which they evolved, but the adaptation has not been perfect. In many experiments, native trees have not grown as well as trees from 50 to 500 miles away.

Some provenance testing is done for theoretical reasons, to uncover evolutionary trends and relate them to factors of the environments in which the trees evolved. In some respects a large provenance test is ideal for this purpose. There are very real limits, however, to what can be learned in this manner. Always there is a certain amount of residual unexplained variation, and usually one race will have a great many traits in common so that it is impossible to decide which ones arose in direct response to natural selection.

Provenance testing is also done for very practical reasons, to screen the naturally available genetic variation and to choose the best available types for reforestation or further breeding work.

Classic provenance experiments, starting in 1820, tended to include trees grown from seed collected in many widely scattered localities throughout a species' range; in them seeds were kept separate by stand but not by single tree. An example is the NC-99 experiment on Scotch pine (Wright *et al.*, 1966); trees grown from seed collected in 170 parts of Europe and Asia were grown at several locations in north central United States. More recently, there has been a tendency to combine provenance and half-sib progeny testing in the same experiment. An example is the study of loblolly pine in Mississippi by Wells and Switzer (1971); they grew trees from seed collected from 5 individual trees in each of 115 stands. That tendency may be carried even farther to include experiments in which primary emphasis is on plus tree selection and breeding, with the geographic variation being mainly a matter of technical importance.

Factors Influencing Amount of Geographic Variability

The size of a species' range is the principal factor influencing the amount of geographic variability. Species with very large natural ranges, such as Scotch pine (native to most parts of Europe and northern Asia) or ponderosa pine (native to the western half of the United States and parts of Canada and Mexico), contain much more genetic diversity than do species with very limited ranges, such as Monterey pine (native to six small areas in California and Baja California) or Japanese larch.

Another important factor is the amount of environmental diversity encountered within the species' natural range. Slash pine (native to the flat coastal plain of southeastern United States) and red pine (native to flat or somewhat hilly country in northeastern United States and southeastern Canada) do not encounter great climatic extremes in their native habitats and consequently do not contain a great amount of genetic variability associated with place of origin. There is much greater geographic variability in species inhabiting both level and mountainous land; Scotch pine and ponderosa pine are examples.

The third major factor is the extent of range discontinuities. In a species with a continuous range, there is opportunity for abundant gene interchange. Selection pressures may differ from one place to another, but if the two places constantly exchange pollen and seeds the two segments of the population may not become very different genetically. However, in a species inhabiting a mountainous region, different parts of the population may have been geographically isolated for hundreds of generations. There are several examples in which neighboring races separated by 50 to 100 miles are totally distinct genetically.

Fourth, unknown factors should be mentioned. Red pine, northern white cedar, and yellow birch have relatively large natural ranges, but relatively little geographic diversity. Taiwan red pine has a relatively small range, but a relatively great amount of diversity. The reasons for these exceptions are not known.

General Geographic Trends

North–South or Cold–Warm Trends

Most provenance testing has been done with species of the north temperate zone. Certain trends are common to many of these species. The most pronounced trends vary in a north–south direction and are enumerated below. As compared with northern seedlots, southern seedlots of the same species usually (Fig. 13.1):

Grow faster
Leaf out later in the spring
Are less subject to late spring frosts
Grow later and retain their leaves later in the fall
Have less intense autumnal coloration
Are less resistant to extreme winter cold

Actually, these trends probably reflect adaptation to cold and warm conditions more than to northern and southern conditions per se. Thus, similar trends can be observed in western United States in proceeding from the relatively cold interior to the relatively warm Pacific Coast.

FIG. 13.1. Geographic variation in white spruce, as shown in a 9-year-old (from planting) plantation in northern Minnesota. The four trees marked with placards were grown from seed collected in (left to right) eastern Quebec (latitude 47° N), eastern Ontario (latitude 46° N), Labrador (latitude 53° N), and Maine (latitude 45° N). This species exhibits many of the general geographic trends mentioned in the text. The same seedlots grew best at many test locations. (Photograph courtesy of Hans Nienstaedt of the United States Forest Service at Rhinelander, Wisconsin.)

Demonstration of such trends depends very much on test conditions. The faster growth of central European than of Scandinavian Scotch pine can be demonstrated in north central United States or Germany but not in central Sweden, where the central European trees suffer repeated winter damage. In frost-free parts of Argentina, south Florida slash pine grows faster in height and diameter than does north Florida slash pine; in northern Florida, where winters can be cold, south Florida slash pine grows faster in diameter but not in height than north Florida slash pine.

There are many interesting and important exceptions to such general trends. Under a variety of test conditions, eastern white pine from parts of Virginia grows more slowly than do more southern or some more northern trees. In tamarack, which is native all the way from Maryland to the shores of the Arctic Sea, southern trees grow fastest when tested under southern conditions. However, even though there are large differences in date of growth initiation in the spring, some trees from the far north leaf out at nearly the same time as do trees from Maryland. In Douglas-fir from interior United States, the trend for growth initiation in the spring is just the opposite of that found in most species—southern trees leaf out earliest.

Such exceptions are one good reason for gathering empirical data from actual tests rather than relying on general trends already observed.

Dry–Moist Trends

In progressing from a dry region to a moist region other trends can be demonstrated. As compared with trees from a dry region, trees from a moist region generally:

Grow faster
Have smaller seeds
Are less deeply rooted
Have greener foliage

Some of these trends are evident in comparisons among species. For example, species from eastern United States generally have smaller seeds than their close relatives in the more arid western part of the country. Species from southwestern United States, which has a high evaporation–precipitation ratio, generally have bluer foliage than their close relatives elsewhere in the world.

Elevational Trends

A 3000 ft elevational difference is often accompanied by as great climatic differences as may occur in a few hundred miles of level country. Therefore selection may operate in such a manner as to cause high- and low-elevation trees to become different. However, the rate of gene exchange is much higher between different elevations on the same mountain than between populations a few hundred miles apart. Such continuous gene interchange tends to keep the high- and low-elevation populations from diverging genetically.

Elevational clines are present in Scotch pine in Sweden and in ponderosa pine in California. In both cases the elevations change gradually. There is also some geographic isolation, and the genetic differentiation takes place over a distance of 50 to 75 miles.

In experiments that sampled regions of more abrupt topography, such as the Rocky Mountains and the mountains of Spain and Czechoslovakia, consistent elevational trends are generally lacking. In nearly all such experiments, trees from high and low elevations in the same area do not differ according to any set pattern.

Random Geographic Variation

The general moist–dry and north–south trends are usually evident only when comparing trees from regions 100 to a few hundred miles apart, regions which differ considerably in climate.

Over smaller areas, especially areas in which there is not a great amount of climatic variation, small geographic differences can often be demonstrated. Trees from some stands may grow 10 to 15% faster than trees from other stands, differ modestly in needle length or pest resistance, etc. Generally in such situations it has not been possible to demonstrate a relation between the genetic variation and the environment. Japanese larch is an example. Its total natural range is about 140 miles north–south and east–west. Trees from two mountains tend to grow fastest, and there are also differences in time of leaf fall, fruiting, etc. However, correlations between those traits and climate at place of origin are weak. North Florida slash pine is another example. Provenance tests covering the region from North Carolina to Mississippi have shown the presence of geographic variation in growth rate and other traits, but genetic differences between stands located a few miles apart are almost as large as those between stands at the opposite ends of the range.

Such randomly distributed variation has been demonstrated in nearly every intensively studied tree species. There is always a point at which breeders can no longer formulate general rules to forecast the genetic potentialities of a stand.

Among the possible explanations for random variation, the following three are worthy of mention. (1) The randomness may be more apparent than real. Investigators have usually attempted to correlate genetic variation with winter temperature, length of growing season, summer temperature, precipitation, soil type, etc. The differences may represent adaptation to environmental factors which have not been considered. (2) Climates and consequently selection pressure have changed. At any one locality there have probably been several cool–warm or dry–moist fluctuations in climate over the past 10,000 years. Some differences may represent adaptations to climatic conditions which prevailed some time in the past. (3) Genetic drift is a possible explanation in a species such as tamarack. In the southern part of its range, this species has a discontinuous distribution in swamps, and some of the isolated stands are very small.

Continuous versus Discontinuous Variation

In species with continuous distribution, the genetic variation is most apt to be continuous. If there are range gaps, especially range gaps associated with large differences in climate, the genetic variation is apt to be discontinuous, with neighboring races intergrading little if at all.

Scotch pine has a continuous range in Sweden. For convenience purposes, Swedish Scotch pine is often described as composed of three races—northern, central, and southern. These races differ very much from each other on the average, but intergrade so much that any lines drawn to separate them must be 50 to 100 miles wide. In contrast, the Vosges Mountain (northeastern France) race of Scotch pine grows 40% faster and differs in several other important respects from the French Central Massif race 300 miles to the southwest. There is no native Scotch pine in the intervening area, and the two races are so distinct that single trees in a provenance test can be immediately recognized as belonging to one or the other.

In western United States, northern Arizona and northern New Mexico are relatively low elevation areas occupied by desert or grassland. In four different species, races from immediately north of this range gap are very distinct from those immediately to the south.

The term "cline" is often used to denote a population in which the genetic variation is "clinal" or continuous. The term "race" denotes a population which differs from another; it may be a portion of a cline or a genetically distinct population. The term "ecotype" is used to denote a race whose distinctive characters represent obvious adaptation to the environment in which it evolved. When describing a species' variation pattern, it is better to describe the presence or absence of discontinuities and the amount of adaptation to the environment than to rely on special connotations of such words as defined above.

Knowns and Unknowns about Natural Selection

Certain north–south trends can be recognized in many species. In general these can be related to the general coldness of northern habitats or the general warmness of southern habitats. Beyond this, it is often difficult to say whether the differences represent adaptation to daylength, winter temperature, length of the growing season, summer temperature, precipitation–evaporation ratio, etc. Most of those environmental factors vary in a north–south manner, and even with complex statistical analysis one may not be able to prove that any one factor was most important.

The most definitive information has come from experiments at many different places, where the same seedlots were tested in regions with cold winters and warmer winters, etc. Yet the results of such experiments are often confusing, as when southern seedlots continue to grow rapidly without winter kill at places much further north of their natural habitats. Also, definitive information has come from a few growth chamber experiments in which trees were grown under different daylength conditions or temperature conditions, with all other factors held constant.

Even more puzzling are the several cases in which the usual north–south trends are reversed. In Douglas-fir and black walnut, for example, southern trees start growth earliest in the spring, whereas the reverse is true in most other species. We do not yet know why this should be so.

Nor do we know why there should be so great differences among species in response to given differences in selection pressure. Within the same geographic area, genetic differences are very much smaller in some species than in others. Efforts to relate this to a species' tolerance or soil preference have failed; that is, the slight amount of genetic variation in northern white cedar is apparently not due to the fact that it is an intolerant swamp species, because another intolerant swamp species (tamarack) from the same region is quite variable.

In a region without great climatic extremes, such as southeastern United States, it has been difficult to demonstrate that natural selection has indeed operated to cause geographic variation. There are geographic differences, but except in some extreme cases there is very little correlation between any growth trait and any factor of the environment. It almost seems that random gene fixation (genetic drift) had been more important than natural selection in causing the modern genetic variation patterns.

We know a great deal about the empirical genetic variation patterns, less about how they arose through natural selection, and much less about the internal physiological mechanisms controlling the development of trees that differ genetically in growth rate, leaf size, pest resistance, etc. In a few cases it has been possible to show that differences in growth rate are associated with differences in length of growing season, but in other cases this is not true.

Gaps in theoretical information about natural selection processes create difficulties in forecasting the amount or pattern of geographic variation within a given species or in forecasting how particular races or portions of clines will behave in certain circumstances. Therefore there is still a need for a considerable amount of empirical testing, even in well-known species.

Design of Provenance Tests

Range-Wide Tests

It is desirable that the first experiment in a species sample the entire range. This provides insurance against the possibility that some overlooked region is capable of providing the best trees. General east–west or north–south trends can be demonstrated with trees grown from seed collected in 10 to 15 widely scattered localities. However, it is usually desirable to learn much more, e.g., exceptions to general trends, whether the variation is continuous or discontinuous, and the approximate limits or the region(s) capable of supplying the best seed. Thus for a species with a comparatively small range it is common practice to test trees from 20 to 30 localities, and for a species with a large range it is common to test trees from 50 to 200 localities.

Seeds are usually obtained by correspondence, at least in part. This is satisfactory for species native to countries having a large network of forestry stations. For recent experiments with subtropical American pines, it has been necessary to send special collecting expeditions to Mexico, Central America, and the Caribbean islands.

Seeds should be collected from several trees per locality. It is best theoretically that the trees be average, but from the practical standpoint some additional improvement might result if the parents are selected. This is rarely practical, however. The seeds from a locality are usually bulked.

Data on exact geographic location and elevation of parents, collector's name, and collection data and method are easy to obtain and generally useful. Data on age and height of parents, branching habit, are also easy to obtain but have not generally been useful in analysis of the results. Information on soil type, aspect, possibilities of selfing, etc., might be important but are generally difficult to obtain as cooperators vary in their ability to assess such things.

The organization of a good provenance test takes time—sometimes 3 to 4 man months. It also requires much effort on the part of cooperators who furnish the seed. Therefore it has become common practice to undertake provenance testing work cooperatively, with one person starting the experiment and sharing seed or seedlings with people in other states or countries. Results of such experiments, with the same seedlots planted at many different places are usually more valuable than the results of several smaller, uncoordinated experiments. Commonly, 10 to 40 different plantations of the same experiment are established.

In most provenance tests established prior to 1955 large plots containing as many as 200 trees per seedlot were used, and each plantation contained 2 to 4 replications. More recently, small (1 to 10 tree) plots have become common, with 10 to 25 replications per plantation. The data tend to be most useful and most easily analyzed if all seedlots are represented in all plantations, if enough planting stock and space are available. It is not necessary that all plantations of the same experiment follow exactly the same design as regards plot size, spacing, number of replications, etc., but it is desirable that each plantation be established well to have low mortality and good growth.

Range-wide provenance tests are useful as producers of information, for demonstration purposes, and as breeding arboreta. All those things should be considered in their location, care, spacing, thinning, etc. They do not lend themselves to conversion to seed orchards, however. They often contain a vast array of genetic material which differs in flowering time. Thus thinning to leave only the genetically best material might leave too few trees per acre (some of which might not cross) to be a useful seed orchard. If local seed production is needed, some variations in design should be considered at the start. Nanson (1972) has discussed the general use of provenance tests as seed orchards.

Limited-Range and Combined Progeny–Provenance Tests

Limited-range provenance tests are often used as follow-ups to range-wide experiments to sample intensively the region(s) proved to give the best seed in general. Examples are Wells and Switzer's experiments with loblolly pine from Mississippi and Louisiana, the United States Forest Service experiment with ponderosa pine from the Rocky Mountains, and the NC-99 experiment with eastern white pine from the southern Appalachian mountains.

The approximate amount of genetic variability and most major trends will ordinarily be known as the result of range-wide experiments. Therefore the primary objectives are the detection of minor trends and the location of specific stands or specific small areas having the best germplasm. Additionally, many such experiments are designed to provide information on the comparative amounts of geographic and individual tree variability and to function ultimately as seed orchards.

The total amount of genetic variability is ordinarily much less than encountered in a range-wide experiment; geographic trends are often not pronounced, and much of the variability may be random. Therefore, intensive sampling is needed to satisfy the primary objectives. Often seeds are collected from 80+ different stands.

The smaller the range sampled, the more nearly does the amount of intrastand genetic variation approach the amount of interstand genetic variation. Therefore it is often desirable to combine the functions of an open-pollinated half-sib progeny test and a provenance test in the same experiment. This involves maintaining the identity of all seedlings by parent and by stand of origin. Several experiments of this type were cited in Chapter 11. There are two ways in which this can be done: (1) Completely randomize all families within each block, or (2) use what is known as a "compact family design" and group families from the same stand and randomize these family-groups within each block. The second method is easiest from the standpoint of measurement and analysis. (Fig. 13.2).

Limited-range provenance tests, since they include less range in genetic variation than range-wide provenance tests, are apt to include a greater proportion of valuable genotypes which flower at similar times. Therefore many can be converted to seed orchards by thinning.

Fig. 13.2. Geographic variation in 3-year-old Sakhalin fir growing in Hokkaido, Japan. This is a limited-range provenance test consisting of 117 half-sib families originating in several different Hokkaido stands. Families from the same stand were planted in adjacent plots. There were large differences in growth rate as well as in branchiness and ability to intercept snow.

Practical Use of Provenance Data

Planning Breeding Work

A range-wide provenance test usually indicates the total range of genetic variation within a species, and thus it gives a clue as to the amount of improvement which may be expected from more intensive breeding work. It also provides theoretical information about rates of evolution and responses to varying intensities of natural selection. Such data can be used to indicate probable responses to artificial selection. Also, the best trees in the best seedlots are frequently useful as breeding stock.

Generalized Planting Recommendations

Some or all of the general trends enumerated earlier can be demonstrated in most range-wide provenance tests. These trends can be restated as generalized planting recommendations, which permit a grower to order seed from a recommended area and grow trees which are 10 to 100% better than those he formerly planted, with little additional cost.

Recommendations for Seed from Specific Stands

There are usually exceptions to the general trends and a certain amount of unexplained variation among the offspring of different stands within the same region. Thus a grower can often obtain 5 to 15% additional gain by looking very closely at the provenance data and ordering seeds from exactly those stands that produced the best offspring. While many growers have used generalized recommendations to advantage when ordering seed, relatively few nurserymen or seed dealers have gone to the extra trouble of ordering seed from specific stands as yet.

Provenance Tests as Seed Orchards

Earlier it was pointed out that range-wide provenance tests are not as amenable to conversion into seed orchards as are limited-range provenance tests or open-pollinated progeny tests. However, there are situations in which such conversion is necessary if the data are to be put to practical use.

In the late 1960's the Imperial Forestry Bureau at Oxford University sent a collecting team to Central America to gather seeds from many natural stands of Caribbean pine. With this seed, provenance tests were established in many tropical countries. The rate at which Central American forests are being cut introduces a serious problem to the future survival of any particular stand that might prove to produce superior offspring.

Thus it is desirable to make some provision for future seed supplies. This could be done by (1) establishing isolated, widely spaced, unreplicated plantations of each seedlot at the time of experiment establishment, with the idea that some such isolated plantations can serve as future seed orchards; (2) sacrificing the data-producing aspects of an entire provenance test at an early age by removing all but the best seedlots; (3) establishing clonal seed orchards with grafts from the best trees in the best seedlots as soon as flowering starts; or (4) establishing seedling seed orchards by crossing the best trees in the best seedlots as soon as flowering starts.

Study Questions

1. Why is a provenance test a good way to start an improvement program?

2. Name three factors important in determining amount of geographic genetic diversity within a species.

3. Name three ways in which northern and southern trees of the same species may differ genetically; in which trees from dry and moist regions may differ.

4. Why are elevational trends less pronounced than latitudinal trends?

5. How do practical recommendations differ according to whether the geographic variation is random or follows a definite, predictable pattern?

6. Describe instances of continuous and discontinuous variation patterns. Under what conditions are these patterns expected?

7. Define cline, race, and ecotype.

8. Why is it difficult to learn the exact environmental characteristic responsible for the evolutionary development of a particular genetic characteristic?

9. Describe a range-wide provenance test for a widely distributed species with reference to numbers of seedlots, useful data, selection of parental trees, number of parents per locality, seed collection method, and number and design of test plantations.

10. How do range-wide and limited-range provenance tests differ with regard to intensity of sampling, study of general trends, formulation of general versus specific planting recommendations, conversion to seed orchards, and maintenance of separate identity for the offspring of each parent?

11. Under what circumstances can data from provenance tests be put to practical use through generalized rules? Through recommendations of specific stands as seed sources? Through conversion of the provenance test to a seed orchard?

14

Geographic Variation in Scotch Pine

Introduction

Scotch pine is Eurasia's most common and important timber tree. Its natural range extends from latitude 40° N (Madrid) to north of the Arctic Circle and from Spain and Scotland eastward to Turkey and eastern Siberia. Its range is continuous in northern Europe and Siberia. It is a less important tree in southern Europe, where it is confined to isolated mountains. Scotch pine has been planted on a large scale for centuries in countries such as France, Germany, and Czechosovakia. In modern times it has also been planted extensively in Canada and northeastern United States, primarily for ornamental and Christmas tree purposes.

Finnish foresters started small provenance tests of Scotch pine in the 18th century. The oldest well-documented study was started by a Frenchman, de Vilmorin, in 1821. He obtained seed from several parts of northern Europe and grew the trees on his estate near Paris. He reported his findings in 1857 and showed that much of the geographic variation, which had hitherto been attributed to environmental influences, was genetic in nature. His findings prompted researchers in Austria, Germany, and Sweden to install other small provenance tests prior to 1900. In 1908 the International Union of Forest Research Organizations (IUFRO) undertook the first replicated experiment, which consisted of 13 seedlots, collected from widely scattered localities in central Europe. Each seedlot was planted at 11 locations.

A second IUFRO experiment was started in 1938. This later one included 55 seedlots from widely scattered localities in northen and central Europe; there were replicated test plantations in several parts of Europe and at two locations in the United States.

From 1929–1936 Langlet (1936) of Sweden undertook an exhaustive study of geographic variation in Swedish Scotch pine. He assembled seed from 582 localities and grew the seedlings 2 to 3 years in a Swedish nursery. He concentrated on characteristics associated with winter hardiness, i.e., percent dry weight of the foliage, changes in moisture content with season, sugar content of foliage, etc.

The most recent large provenance test is the NC-99 experiment undertaken in north central United States. This experiment sampled 170 natural stands in all parts of the natural range, and included more than 40 test plantations. Many types of morphological, growth, and chemical characters have been measured in these plantations.

In quoting the experimental results it is often desirable to include a brief description of the experiment from which they were obtained. The major experiments listed above, as well as several smaller ones, differ in design and precision, so that the results are not always of equal value. Also, the results have varied with test conditions. For example, central European trees have consistently been winter-killed when tested in central Sweden, so that data on central and southern European races must be derived from experiments undertaken elsewhere.

Many of the published data are from plantations 10 to 20 years old. Nanson (1968) studied Belgian provenance tests that were more than 50 years old and found that results obtained at ages 10 to 15 were generally consistent with the results obtained at age 50+. Therefore it can be assumed that most of the early results are trustworthy.

With this long history of provenance testing, and with so many experiments conducted in Europe and the United States, Scotch pine is the world's most intensively studied species from the provenance standpoint. For that reason, it was chosen as the species to be described in detail here.

Phenotypic versus Genetic Variation Patterns

Early travelers noticed that native Scotch pine stands differed considerably in growth rate, branchiness, stem form, etc. Until the time of de Vilmorin, most foresters assumed these differences to be due solely to differences in soil, spacing, culture, etc. After his experiments there was a tendency to consider the differences as more probably genetic in nature. Working on this assumption, taxonomists started to describe and name geographic varieties. This process has gone on for 140 years, and more than 60 different geographic varieties have been given Latin names (Ruby, 1964; Staszkiewicz, 1970).

Taxonomists, working with phenotypic data, have generally been successful in delimiting genetically different populations. This is especially true of the more recent taxonomic work, where the investigators had access to large amounts of data and used mathematical analysis techniques.

However, phenotypic studies of geographic variation have been much less successful in forecasting the genetic potentialities of particular varieties. For example, variety A might be considered to be faster growing and to have larger branches than variety B on the basis of phenotypic study, whereas variety B might turn out to be fastest growing when the two are tested on the same sites.

Thus, published taxonomic studies of Scotch pine provide reasonably good indications of which geographic populations are and are not genetically different. The published varietal descriptions are best regarded as descriptive accounts of the natural forests as influenced by their genotypes and a number of environmental factors, but not as accurate indicators of their genetic potential. Data on genetic differences have been derived almost entirely from experiments in which many different seedlots were grown under similar conditions.

Amount and Pattern of Genetic Diversity

Amount of Diversity in Entire Species

The natural range of Scotch pine extends more than 27° in latitude, and from Scotland in the west to eastern Siberia in the east. The range includes regions with very different climates. Natural selection has therefore resulted in a great amount of genetic diversity within the species.

As examples, of this diversity, there are 4:1 differences in growth rate between the fastest and slowest growing varieties. Some varieties stay green during the winter, whereas others turn almost lemon yellow. Varieties differ in percent dry weight of the summer foliage, some having 27% and others 40%. Some varieties suffer 0% and others 100% mortality from winter cold if planted in central Sweden. Some varieties suffer 10 to 15% mortality and others 60 to 70% mortality if planted in areas infested by the pine root weevil. In some varieties 1% and in other varieties 50% of the cortical oleoresin consists of 3-carene.

This list can be expanded almost indefinitely. Indeed, the amount of genetic diversity which can be demonstrated seems to be limited only by the number of characteristics studied (Fig. 14.1).

Comparative Amount of Genetic Diversity in Entire Species, within Varieties, and within Stands

Data from taxonomic and genetic experiments have been used to delimit 19 geographic varieties (Ruby, 1964; Staszkiewicz, 1970). (Perhaps a few more should be recognized from eastern Europe and Siberia, which have been studied less intensively than other parts of the range.) A variety thus delimited is the largest population within which no consistent geographic or elevational trends can be demonstrated. Most of the varieties occupy ranges of 1000 to several thousand square miles. The smallest variety is *scotica*, limited to a few natural stands in Scotland;

Fig. 14.1. Geographic variation in Scotch pine, as shown in a 4-year-old (7 years from seed) plantation in eastern Nebraska. The tree on the left (enclosed in black rectangle) is from latitude 61° N in northeastern Siberia; it is a slow-growing type which turns yellow during the winter and appears nearly the same color as the dead grass. The tree on the right is from latitude 49° N in West Germany and is much darker green during the winter. Scotch pine varieties differ as much in several other traits as in growth rate and winter color. (Photograph courtesy of Ralph Read of the United States Forest Service, Lincoln, Nebraska.)

one of the largest varieties is *hercynica*, distributed over most of West Germany, East Germany, and western Czechoslovakia.

According to provenance test data, differences among varieties account for 70 to 85% of the total genetic variation in growth rate, winter hardiness, winter foliage color, resistance to some insect pests, moisture content of the foliage during the summer, cone production at early ages, needle length, concentration of certain monoterpenes (α-pinene, β-pinene, 3-carene, β-phellandrene, terpinolene) in the cortical oleoresin, date of leafing out in the spring, and date of terminal bud formation on young seedlings (Langlet, 1936; Wright *et al.*, 1966; Tobolski *et al.*, 1971). In these traits much of the remaining 15 to 30% of the genetic variation is accounted for by differences among stands belonging to the same variety.

The results of Howe, Nilsson, and Johnstone (see Chapter 10) indicate that genetic differences among offspring of different trees in the same stand are relatively small when compared with those among offspring of different trees in the same region, and very small in comparison with the amount of variation in the entire species.

Comparative Amounts of Genetic and Environmental Control and Genotype–Environment Interaction

The following comments are based largely on American plantations scattered from latitude 37° N (Missouri) to 47° N (Minnesota), since evidence is lacking as to what might happen to different genotypes planted in the native stands in all parts of Europe and Asia.

Winter foliage color and monoterpene concentration are under a high degree of genetic control and are not affected greatly by external environmental conditions. Winter-yellowing varieties from northern Eurasia became almost the same shade of yellow whether planted in the north or south and whether fertilized or not fertilized.

Growth rate, needle length, and cone production at early ages seem to be under a high degree of genetic and environmental control, without much genotype–environment interaction. A fast-growing, long-needled, early fruiting seedlot of *haguenensis* or *carpatica* grew about 50% faster, had about 40% longer needles, and fruited much more heavily by age 10 than did a seedlot of a slower growing variety, such as *septentrionalis* or *iberica*. But growth rate and needle length of all varieties could be 100% greater on a fertile than on a sterile site. Early cone production was heavy only in plantations south of latitude 42° N.

Genotype–environment interaction was evident in resistance to extreme winter cold and to various insect and bird pests. Presumably the internal factors responsible for differences were present in all environments, but the differences could be expressed only in very cold habitats or in the presence of the pest. Thus, the *iberica* variety could be shown to be very susceptible to winter cold if grown where the average January temperature was 20°F or below, but not in warmer climates.

Nearly all Scotch pines tend to grow straight if they suffer no damage to the terminal bud or leader. However, they do suffer such damage as the result of cold, insects, and birds. Thus, stem straightness proved to be a function of a seedlot's resistance to a particular damaging agent and the presence of that damaging agent. Thus, north European varieties were generally very straight, but became very crooked if grown where there were large flocks of pine grosbeaks which ate the trees' buds.

General North–South Trends

Proceeding from northern Scandinavia to central Europe there is a trend toward increasing growth rate (if trees are tested in north central United States). Then the trend is reversed, and varieties from southern Europe grow more slowly than those from central Europe. Much the same can be said for leaf length, except that south European varieties have even shorter leaves than those from the far north.

Winter foliage color varies most consistently from north to south. Varieties from the far north turn yellow in the early autumn and by midwinter are very yellow. The farther south the range of the variety, the less this occurs, and varieties from the Mediterranean countries have almost the same foliage color winter and summer. An exception to this general trend is the *scotica* variety, which remains almost summer green in the winter, even though it is from the same latitude as southern Sweden and the Ural Mountains.

Langlet (1936) also found a consistent north–south trend in succulence of the foliage. In the Spanish variety *iberica*, the summer foliage had a dry weight content of 27%, versus 40% for the north Scandinavian variety *lapponica*. Varieties with intermediate ranges had intermediate dry weight contents.

The proportion of early fruiting trees is lowest in slow growing northern varieties, generally high in fast growing central European varieties, and generally low–moderate in the medium growth rate southern varieties. One very noticeable exception is the southern French variety *aquitana*. Although it grows at only moderate rates, it is consistently among the first to produce cones, and per tree cone yields are generally high.

In monoterpene content and resistance to some insect pests, the varieties can often be categorized as "northern," "central," and "southern." In some cases the Scottish variety *scotica* can be included with the southern group of varieties; in some cases it is more similar to the central European varieties.

Adaptational versus Nonadaptational Traits

The differences in growth rate seem to represent adaptations to the climates in the regions to which the varieties are native. The fastest growing varieties are from Belgium, northeastern France, and eastern Czechoslovakia, regions which have more favorable temperature and precipitation conditions for Scotch pine growth than other parts of Europe and Asia.

Differences in moisture content of the foliage probably represent adaptations for increased cold hardiness, or drought hardiness, or both. Northern varieties have the highest dry weight percent in their foliage, and an appreciable portion of that is sugar. Thus they should be (and are) the most cold hardy. Trees from Spain are much more succulent; the extra moisture in their foliage reduces cold hardiness but provides a reservoir to protect the plants against desiccation in the dry Spanish climate.

There seems to be no particular selective advantage in the tendency of northern varieties to turn yellow during the winter. However, it is known that pigment changes are often connected with changes in sugars. The color changes may be mere byproducts of some more fundamental changes concerned with the development of increased cold hardiness.

Seed size increases almost threefold from northern to southern Europe. Undoubtedly this represents an adaptation to the generally drier conditions found in southern Europe. The large seeds provide greater food reserves, which in turn result in more rapid first-year root growth of southern than of central or of northern varieties (Brown, 1970).

Nothing is known of the adaptive significance of the differences in monoterpene concentration. The same could be said of the differences in resistance to insect pests. Most data on insect resistance were obtained from American tests and apply to insects not known or not important in the natural forests of Eurasia.

Continuous versus Discontinuous Genetic Variation

In Scandinavia and parts of central Europe, both the geographic ranges and the genetic variation patterns are more or less continuous.

Neighboring varieties intergrade with each other and are often given different names more for practical than for theoretical reasons.

Between the central and southern European varieties, however, there are large areas without Scotch pine, and there is almost no genetic intergradation. Most of the southern European varieties are genetically distinct from each other in at least one or two traits.

Description of Varieties

Tables 14.1–14.4 provide a tabular description of 19 Scotch pine varieties based on published papers by Wright *et al.* (1966, 1967, 1975), Wright and Wilson (1972), and Tobolski *et al.* (1971). The descriptions used here are based on 170 seedlot provenance tests conducted in Michigan, but the results are essentially similar in eight other north central states.

European provenance tests cover portions of the species range. There are Swedish experiments including many Scandinavian and a few continental seedlots, German and Czechoslovakian experiments including many seedlots from central Europe and a few from northern Europe, etc. Generally speaking, there has been good correspondence between varietal distinctions based on the European and American experiments. That is, both European and American experiments have shown the presence of large differences between varieties *rigensis* and *hercynica;* a cline through varieties *lapponica, septentrionalis,* and *rigensis;* small differences between varieties *hercynica* and *carpatica.*

However, the varietal descriptions as recorded here would probably be much different if based on work in another continent. Eiche's (1966) work in central Sweden indicates that most continental Scotch pines would suffer pronounced winter injury and consequently grow slowly if planted in Sweden. If the provenance tests were to be conducted in a southern country, such as Greece or Spain, most of the central and northern origins would probably suffer from drought injury and thus grow quite abnormally.

Some varietal distinctions are large and based upon several traits; compare the variety *iberica* and its neighbors with respect to winter hardiness; resistance to pine grosbeak, and monoterpene content. Some are small; varieties *hercynica* and *carpatica* differ mainly in growth rate and even overlap to some extent in that trait. Some varieties form clines and are given different names only as a matter of convenience.

The varietal differences in resistance to pests are especially marked. The patterns of resistance are not quite the same for any two pests,

TABLE 14.1

Differences in Growth Traits among 19 Varieties of Scotch Pine, as Determined from 11- to 12-Year-Old Plantations in 8 Parts of Michigan

| Variety | Country of origin[a] | Relative height % of mean | | Needle length (mm) | Autumn color (0 = yellow 100 = green) | Cold-damaged trees, North Michigan (%) |
		South Michigan	North Michigan			
Varieties from northern Europe and Asia						
lapponica	FIN, SWE, SIB	51	53	57	3	0
septentrionalis	FIN, NOR SWE	81	86	56	13	0
rigensis	LAT, SWE	95	108	53	19	0
jakutensis	SIB	75	82	74	0	0
'Krasnoyarsk'	SIB	86	92	68	9	0
uralensis	URA	99	103	68	4	0
Varieties from central Europe						
polonica	POL	118	125	73	38	0
hercynica	W GER, E GER, CZE	117	124	71	56	0
carpatica	E CZE	128	126	71	49	0
haguenensis	BEL, FRA, W GER	127	125	74	72	0
pannonica	HUN	114	115	72	62	0
Varieties from western and southern Europe and Asia						
'E. Anglia'	ENG	112	118	63	80	0
scotica	SCO	98	94	49	70	2
iberica	SPA	87	66	51	100	91
aquitana	S FRA	105	92	50	96	8
subillyrica	ITA	100	98	69	65	0
illyrica	YUG	107	107	64	70	0
rhodopaea	GRE, BUL	102	99	58	81	1
armena	GEO, TUR	95	94	60	89	4

[a] BEL, Belgium; BUL, Bulgaria; CZE, Czechoslovakia; E GER, East Germany; ENG, England; FIN, Finland; FRA, France; GEO, Georgia SSR; GRE, Greece; ITA, Italy; LAT, Latvian SSR; NOR, Norway; POL, Poland; SCO, Scotland; SIB, Siberia; SPA, Spain; SWE, Sweden; TUR, Turkey; URA, Ural Mountains between Europe and Asia; W GER, West Germany; YUG, Yugoslavia.

TABLE 14.2

Differences in Susceptibility to Insect and Bird Pests among 19 Varieties of Scotch Pine[a]

		Proportion of trees attacked by				
Variety	Country of origin[b]	Pine root collar weevil (% dead)	Zimmerman moth (% dead)	European pine sawfly (% attacked)	White pine shoot borer (% attacked)	Pine grosbeak (% attacked)
Varieties from northern Europe and Asia						
lapponica	FIN, SWE, SIB	14	7	0	7	77
septentrionalis	FIN, NOR, SWE	38	18	2	21	64
rigensis	LAT, SWE	45	22	6	31	30
jakutensis	SIB	12	33	1	18	32
'Krasnoyarsk'	SIB	42	33	1	18	33
uralensis	URA	40	31	3	19	25
Varieties from central Europe						
polonica	POL	67	11	19	37	5
hercynica	W GER, E GER, CZE	69	22	21	34	10
carpatica	E CZE	43	26	20	42	7
haguenensis	BEL, FRA, W GER	65	35	26	38	7
pannonica	HUN	45	19	20	47	0
Varieties from western and southern Europe and Asia						
'E. Anglia'	ENG	55	37	27	36	28
scotica	SCO	31	26	7	41	46
iberica	SPA	17	3	11	58	7
aquitana	S FRA	12	7	10	49	11
subillyrica	ITA	11	3	12	44	21
illyrica	YUG	10	7	19	56	20
rhodopaea	GRE, BUL	19	6	9	53	21
armena	GEO, TUR	12	3	7	51	20

[a] As determined from 11- to 12-year-old plantations in 8 parts of Michigan. Heavy damage from a particular pest was usually sustained in one or two plantations only.

[b] See Table 14.1 for abbreviations of country names.

TABLE 14.3

Differences in Frequency of Crooked Trees among 19 Varieties of Scotch Pine[a]

		Trees (%) with stem crooks			
Variety	Country of origin[b]	Kellogg Forest	Rose Lake	Dunbar Forest	Lake Linden
Varieties from northern Europe and Asia					
lapponica	FIN, SWE, SIB	3	10	17	15
septentrionalis	FIN, NOR, SWE	5	7	14	19
rigensis	LAT, SWE	9	10	19	20
jakutensis	SIB	3	10	10	26
'Krasnoyarsk'	SIB	0	—	5	25
uralensis	URA	9	7	16	17
Varieties from central Europe					
polonica	POL	16	37	8	22
hercynica	W GER, E GER, CZE	18	26	25	33
carpatica	E CZE	20	22	19	23
haguenensis	BEL, FRA, W GER	29	41	28	48
pannonica	HUN	18	28	25	37
Varieties from western and southern Europe and Asia					
'E. Anglia'	ENG	18	—	12	—
scotica	SCO	9	0	11	15
iberica	SPA	12	15	70	35
aquitana	S FRA	16	28	22	—
subillyrica	ITA	16	40	17	31
illyrica	YUG	2	12	0	22
rhodopaea	GRE, BUL	6	3	17	21
armena	GEO, TUR	14	22	13	29

[a] As determined from four 11- to 12-year-old plantations in southern and northern Michigan.

[b] See Table 14.1 for abbreviations of country names.

indicating different internal resistance mechanisms for resistance to each pest.

The differences among varieties in stem form varied with the test site. Variety *haguenensis* tended to be among the most crooked at all places, but no one variety could be regarded as straightest at all places (Table 14.3). Actually, nearly all the crooks were due to some external factor. Thus the amount of crook suffered by a particular variety depended very much on the resistance of that variety to the particular pest which happened to be most abundant in a particular plantation.

Geographic variation is as pronounced in chemical traits as in growth triats (Table 14.4) The biological significance of the chemical differences is not yet known.

TABLE 14.4

Difference in Monoterpene Composition of the Cortical Oleoresin among 19 Varieties of Scotch Pine

Variety	Concentration (% of total monoterpenes)					
	α-Pinene	β-Pinene	3-Carene	Limo-nene	β-Phellan-drene	Terpin-olene
Varieties from northern Europe and Asia						
lapponica	9	17	41	9	11	3
septentrionalis	8	17	44	6	10	4
rigensis	7	14	45	6	20	4
jakutensis	15	29	9	16	16	1
'Krasnoyarsk'	11	17	26	14	17	4
uralensis	9	20	25	11	17	3
Varieties from central Europe						
polonica	10	22	39	8	6	3
hercynica	11	26	29	10	7	3
carpatica	8	15	40	7	12	4
haguenensis	10	20	42	5	5	3
pannonica	8	13	50	6	9	5
Varieties from western and southern Europe and Asia						
'E. Anglia'	17	32	26	4	3	2
scotica	11	22	40	5	4	3
iberica	46	15	1	7	15	0
aquitana	18	47	5	12	3	0
subillyrica	17	42	4	6	4	1
illyrica	12	24	20	13	8	2
rhodopaea	12	35	6	23	8	1
armena	19	43	4	10	3	1

Practical Use of Scotch Pine Provenance Data

In Europe, Scotch pine is grown mainly as a timber tree. For the most part it is planted in regions to which it is native. Provenance tests have shown the extent to which growth rate or form could be increased by using seeds from another region. In some parts of Sweden it is possible to increase growth rate by using seed collected 300 to 400 km south of the planting site without risk of serious winter injury (Eiche, 1966). On exceptionally cold microsites it is possible to lessen cold damage by using seed collected north of the planting sites. Fast-growing German trees can be planted successfully in limited portions of southern Sweden. As another example, growers in parts of West Germany and East Germany can take advantage of the inherently faster growth of var. *carpatica* from eastern Czechoslovakia and of var. *haguenensis* from countries immediately to the west.

In general, European gains in forest productivity as the result of provenance testing have been more than enough to justify the work, but nevertheless have been relatively modest. By transferring seed a few hundred kilometers there have been 5 to 10% increases in growth rate or similar advantages in form.

Scotch pine was introduced into the United States about 1650, and for the next 300 years was grown primarily as an ornamental, with occasional timber plantations. It started to be planted on a very large scale for Christmas tree production. At first the same seed—probably from central Europe or southern Sweden—was used as had been used formerly. Few plantations had good winter color. Many plantations were located on dry sites infested by the pine root collar weevil, and mortality in such places was often high.

As provenance data became available, American Christmas tree growers used it. Most growers switched to one of the southern varieties with a color rating of 80 or better (see Table 12.1) and high resistance to pine root collar weevil (see Table 12.2). Where the climate was suitable, many growers preferred the Spanish variety *iberica* as producing the highest quality tree. From 1950 to 1974 there were many instances of 50% gains in tree value or 50% reductions in work necessary for shearing, pest control, etc.

Most American provenance tests provide data beyond that shown in Tables 12.1–12.4. Here I refer to data about the genetic variation to be found within varieties. By using seed from progeny-tested stands, it is often possible to obtain an additional 5 to 10% gain in some trait. Such data have not yet been put to practical use. Most American nurseries dealing with Christmas tree species offer several of the preferred varieties. As of 1974 none offer seeds or planting stock from the best parental stands within those varieties. Hopefully, that will also be done.

The following specific example is quoted to show how the provenance tests are being used in long-term improvement programs. Variety *uralensis* has the hardiness, growth rate, resistance to European pine sawfly, and resistance to white pine shoot borer desired of a Christmas tree; it lacks the desired resistance to pine root collar weevil, resistance to Zimmerman moth, and autumn color. Variety *iberica* has the desired growth rate, autumn color, needle length, and resistance to pine root collar weevil and Zimmerman moth but lacks the desired winter hardiness and resistance to European pine sawfly. The amount of intravariety variation is such as to indicate that very many generations of intravariety selection would be needed to correct the deficiencies of either variety. Hence H. D. Gerhold of Pennsylvania State University and others started an intervariety hybridization program in 1966. They already have young F_1 hybrids, will raise F_2, F_3, etc., progeny, and will select in those later generations to produce strains with desired combination of characteristics. The total work involves many different intravariety combinations, with the objective of improved types for timber, ornamental, and Christmas tree production.

Typical and Atypical Aspects of Scotch Pine

I chose to describe Scotch pine in some detail because it is one of the world's most intensively studied tree species. Now I wish to discuss briefly the extent to which results obtained in it are or are not typical of those to be expected in other species.

Considering the size of its geographic range, the amount of genetic variability in Scotch pine is about the same as encountered in some large-ranging western American species and is far greater than encountered in other well-studied European species, such as Norway spruce and European larch. As to the exact list of geographically variable traits, Scotch pine is probably unique. At least in American tests, it varied considerably in growth rate, needle length, winter foliage color, resistance to certain pests, and monoterpene concentration. Yet it did not vary nearly so much in winter hardiness or phenology as some other species.

Scotch pine has been planted for centuries in East Germany, West Germany, Czechoslovakia, etc. There has been some moving around of seed so that the variation pattern as determined from a modern provenance study is not entirely natural. However, the amount of planting in northern and southern Europe has not been great, so that in those regions the variation pattern now detected is essentially natural.

While the amount of variation within Scotch pine is much greater than in most species, the amount and pattern of variation within a limited region, such as that occupied by var. *hercynica*, is similar to that found in many species with small to medium ranges. Within that region the geographic variation seems to occur at random. This may be due to the amount of planting within the region, but the lack of pattern is certainly typical of many other species occupying a similar range of site conditions.

Scotch pine is native to many different countries, and taxonomists from all those countries took part in the naming of the geographic varieties. This is partly responsible for the multiplicity of varietal names. However, this is not the whole story, as many of the varieties are distinguishable in the wild as well as in provenance tests, and there are probably many other large-ranging species in which taxonomic varieties could be recognized if there were a desire to do so.

Two of the geographic varieties that differ markedly in growth rate are also apt to differ markedly in foliage color, terpene chemistry, pest resistance, etc. The extent to which variation in one trait is paralleled by variation in another trait seems to be unique to Scotch pine.

Study Questions

1. To what extent are the phenotypic and genetic variation patterns in Scotch pine similar? Dissimilar?

2. What is the total amount of genetic variation in Scotch pine in growth rate? Winter foliage color? 3-Carene content? Resistance to pine root collar weevil? Resistance to pine grosbeak?

3. What proportion of the total genetic variation can be accounted for by differences among varieties, within varieties, and within stands in growth rate? Winter hardiness? Terpinolene content? Needle length?

4. What traits are under such genetic control as to vary little with environmental conditions? What traits are under strong genetic and environmental control? What traits are indirect genetic control so that performance is a complex relation between genotype and environment?

5. Under what conditions would it be possible to mistake a long-needled for a short-needled variety?

6. In what traits can Scotch pine varieties be categorized as northern, central, and/or southern? In what traits are there reversals in trends when processing from north to central and from central to south? In what traits is there a fairly steady progression in proceeding southward?

7. In what traits are the east-west trends similar in the northern, central and southern groups of varieties?

8. What traits are of the most obvious adaptive significance? Of the least obvious adaptive significance? What about autumn color?

9. What governs stem form?

10. How have provenance data been put to practical use in Europe? In the United States?

11. How would you produce a new strain having maximum possible growth rate, stem straightness, greenness of winter foliage, and resistance to Zimmerman moth?

15

Geographic Variation in American Trees

Although individual species from other parts of the world have been studied intensively, trees native to United States and Canada have received most attention with regard to provenance research as a group. Therefore they are used as examples of the diversity of results which may be expected in different species. The story is most coherent if the species are grouped by region.

The Pacific Coast, California to Alaska

Along the Pacific Coast, from latitude 37° N in central California to latitude 60° N at the northern limit of the Alaskan panhandle there is a dense coniferous forest—one of the most productive in the world. The climate is generally regarded as moist and favorable for tree growth. There are large north–south variations in temperature, length of the growing season, and photoperiod. The southern part of this forest is occupied by redwood, about which relatively little is known genetically. From Oregon northward, three of the most important species are Douglas-fir (Munger and Moris, 1936; Rowe and Ching, 1973), western hemlock (Lines and Mitchell, 1969), and Sitka spruce (Lines and Mitchell, 1965; Burley, 1966) (Figs. 15.1–15.3).

The two major Douglas-fir experiments included seedlots from California to southern British Columbia and test plantations in various parts of coastal Oregon and Washington. There was considerable genetic variation in growth rate, some seedlots growing 35% faster than others. There

FIG. 15.1. Provenance test of lodgepole pine (*Pinus contorta*) from western United States, growing in New Zealand. Each 50-tree plot represents a seedlot from a different part of the natural range of this widely distributed species. The differences in growth rate and mortality are so large as to need little comment. (Photograph courtesy of G. B. Sweet of the Forest Research Institute at Rotorua, New Zealand.)

was no noticeable tendency for trees of local origin to excel in growth rate, but there was a tendency for particular seedlots to grow well wherever planted. In one experiment, for example, trees from a particular stand in British Columbia grew fastest at nearly all test sites, and in the other experiment trees from Darrington and Granite Falls, Washington grew fastest at nearly all test sites. There were also large differences in phenology, some seedlots flushing 1 to 2 weeks earlier than others. Elevational and latitudinal trends were not pronounced, except that trees from extreme southern Oregon and northern California, where

FIG. 15.2. Differences among provenances of Douglas-fir from western United States growing in New Zealand. Most of these seedlots were grown from seed collected along the West Coast. Differences in growth rate are evident, as are differences in susceptibility to the fungus *Phaeocryphopus govemanii* which has nearly killed some trees at left center. (Photograph courtesy of G. B. Sweet of the Forest Research Institute at Rotorua, New Zealand.)

growth conditions are somewhat less mesic than farther north, grew relatively slowly.

The western hemlock experiments, including 17 seedlots, were conducted at 16 locations in Great Britain. At northern test sites (latitude 54° to 58° N in Scotland and northern England) trees from northern British Columbia and Alaska (also latitude 54° to 58° N) grew three times as fast as more southern trees. At southern test sites (latitude 50° to 53° N in Wales and southern England) trees from latitude 49° N in southern British Columbia grew fastest, surpassing those from farther south as well as farther north. The growth rate differences were much less pronounced at the southern that at the northern test sites. As with Douglas-fir, there was a tendency for a particular seedlot to surpass others at different places.

Geographic variation in dates of flushing and of growth cessation are especially marked in Sitka spruce. In British experiments seedlings

FIG. 15.3. Differences among interior Douglas-firs in a 10-year-old provenance test in eastern Nebraska. In (A) the tree is from Arizona, the one in (B) from northern Idaho. Squares are 1 ft on a side. In Nebraska as well as in other north central states these are the fastest and second fastest of the interior races, some of the others being very slow. (Photographs courtesy of Ralph Read of the United States Forest Service, Lincoln, Nebraska.)

from latitude 60° N flushed 15 days earlier and ceased growth 30 to 50 days earlier than seedlings from more southern parts of the range. The date of growth cessation seems to be under photoperiodic control and to coincide with the date at which summer daylength shortens to 15 to 16 hr. There were also important growth rate differences, and some seedlots from Oregon and Washington grew three to four times as fast (at least in southern England) as some from latitude 60° N.

Western United States—A Region of Great Genetic Diversity

Amount of Genetic Diversity

From the dendrological standpoint, western United States includes the area from the Great Plains to the Pacific Ocean. There are several mountain ranges with interspersed areas of grassland and desert. The native tree species encounter extremely varied environments, from the moist and relatively mild climates in the mountains facing the West

Coast to the severe continental climates found inland. Progressing from south to north, the same species generally grows at an increasingly lower elevation (with an important exception when progressing from Arizona–New Mexico to Utah–Colorado). This difference in elevation compensates only partly for the latitudinal differences; a species encounters large differences in daylength, temperature, and precipitation distribution from south north. Thus there are very great differences in selection pressure. Also, there are innumerable range gaps, so that isolation is also an effective differentiating mechanism.

Thus the wide-ranging trees native to western United States possess a very great amount of genetic diversity; certainly very much more than is found in species native to the eastern part of the country. The diversity is evident in a large number of traits (Figs. 15.4 and 15.5).

Western versus Interior Differences

Several tree species have large ranges along the Pacific Coast and also in the mountains of the interior. In all those that have been studied intensively there are very large differences between the western and interior populations. However, the differences are not quite the same in any two species. This is evident from the following listing of contrasts for four species.

In Douglas-fir the crest of the Cascade Mountains is the approximate dividing line between the two populations. As compared with the interior population, the western one is more homogeneous genetically, much faster growing in mild climates, much less resistant to severe winter cold, less blue-green, and has different north-south intrapopulation trends (Kung and Wright, 1972; Wright et al., 1971a).

In ponderosa pine the western population (var. ponderosa) includes the region from California to British Columbia and the mountains of northern Idaho and western Montana (classified as "interior" in Douglas fir). The interior population includes the Rocky and other outlying mountains to the east; it is called var. scopulorum. As compared with the interior population, the western one has greener and longer leaves (which are more palatable to jack rabbits in Nebraska), more nearly erect cone prickles, browner and more loosely appressed bud scales, higher foliar concentrations of N, K, P, Ca, and B, and more nearly cylindrical stem form (Wells, 1964; Wright et al., 1969; La Farge, 1971; Read, 1971).

In grand fir, which is nearly confined to the northwestern states, the western population occurs west of the Cascade Mountains, and the interior population occurs in northern Idaho and adjacent parts of western Montana and northeastern Washington; intermediate forms probably

FIG. 15.4. Geographic variation in a 9-year-old provenance test of ponderosa pine in southern Michigan. S. K. Hyun of Korea is pointing to tree from northern Washington and has his back to a dying (the light foliage was actually brown) tree from California. California trees do not thrive in the cold winters of Michigan but are the fastest growing in milder climates such as found in California and New Zealand.

FIG. 15.5. Geographic variation in southwestern white (tallest trees from Arizona and New Mexico) and limber pines (others, from more northern states) in a 10-year-old plantation in eastern Nebraska. From left to right the seed sources are Pine Bluffs, Wyoming, (left tallest) near Flagstaff, Arizona, (small tree center foreground) central Montana, Castle Rock, Colorado. One-tree plots were used in this experiment. Parental data indicate that the trees in this test are growing as fast or faster than in their native habitats. Commercial use of southwestern white pine increased rapidly after such provenance test data became available. (Photograph courtesy of Ralph Read of the United States Forest Service, Lincoln, Nebraska.)

occur in the intermediate areas. As compared with the interior population, the western one when tested along the West Coast grows slower, flushes earlier, is less disease resistant, and has yellower foliage (Douglass, 1974). This appears to be an unusual species in that West Coast Christmas tree growers can profitably switch to interior seedlots.

In white fir the western population is confined to California and western Oregon, where it hybridizes extensively with the more northern grand fir. Grand fir occupies a lower elevational zone (up to 4000 ft) than the white fir (up to 10,000 ft) and is consequently the least winter hardly of the two. Because of introgression, there is a north–south trend in California white fir toward increased winter hardiness, increased blueness, and increased needle length in the south and increased growth rate in the north. The growth rate and winter hardiness trends are the opposite of those found in the interior population, which is confined to the southern Rocky Mountains (Conkle *et al.,* 1967; Hamrick and Libby, 1972; Wright *et al.,* 1971b).

Because of the introgression, one must make two different contrasts when comparing the western and interior populations. In the north the western population (i.e., from northern California and southern Oregon) grows much faster, has greener needles, has more nearly two-ranked needles, and is much more variable than the interior (i.e., Colorado and Utah) population. In the south there is not a great contrast between the western (i.e., southern California) and interior (i.e., Arizona and New Mexico) population.

The Discontinuity in the Southern Rocky Mountains

To describe the western versus interior contrasts is to describe only a small portion of the variation in most species. Some of the genetic variations found along the West Coast have already been described. Even greater differences are to be found in the interior. One contrast found in the southern Rocky Mountains is especially interesting.

Arid grasslands cover large areas of northern Arizona and New Mexico, creating large gaps in the distribution of tree species between Arizona–New Mexico and Utah–Colorado. Peculiarly, some tree species grow nearly 1000 ft lower in elevation in Arizona–New Mexico than in states immediately to the north. The Arizona–New Mexico and Utah–Colorado tree populations do not interbreed and experience great differences in selection pressure because of the differences in latitude and altitude, and have therefore developed into distinct races in some species (Kung and Wright, 1972; Steinhoff and Andresen, 1971). The following comparisons show racial differences in some traits. They are based on provenance tests conducted in Michigan, Nebraska, and Pennsylvania with ponderosa pine, white fir, Douglas-fir, and the limber–southwestern white pine complex. As compared with Utah–Colorado trees, Arizona–New Mexico trees:

> Grew 50 to 100% faster if not winter killed
> Were less winter hardy (not a serious problem on white fir or southwestern white pine)
> Had longer needles and cones
> Had bluer or grayer foliage (all except white fir)
> Had more glaucous twigs (ponderosa pine only)

These differences are of great importance to Christmas tree growers in eastern United States, for they can now grow some species on rotations only 50% as long as used formerly. There are no data on relative performance of these races in their native habitats.

Northern Utah and Colorado—A Center of Slow Growth Genes

Another interesting generalization arose from the provenance tests of the four species mentioned in the previous section. In all four, trees from northwestern Colorado and northern Utah grew more slowly than trees from almost any other part of the range, whether to the north or to the south. This center of slow growth is mountainous and has a continental climate with rigorous long winters. It is probable that resistance to extreme winter cold and slow growth are physiologically associated in such a manner that natural selection for one has resulted in changes in the other. If so, the slow growth of the northern Colorado–northern Utah trees is a necessary adaptation to the rigorous growing conditions found there, and the introduction of faster growing races would not be of great benefit. If that hypothesis is incorrect, the introduction of more southerly or more northerly races might be advantageous.

The Inland Empire—A Region of Rapid Growth

The Inland Empire includes parts of eastern Washington, northern Idaho, and adjacent parts of western Montana. The climate, while not so moist or warm as found along the West Coast, is less continental and more favorable for tree growth than in the regions farther to the east. Native to this region are several important species not found elsewhere in the interior.

The rapid growth of Inland Empire grand fir when planted in western Oregon and Washington has already been mentioned. Provenance tests of Douglas-fir and ponderosa pine have shown that in these species, too, Inland Empire seedlots generally grow faster than most others from the interior (if tested in northeastern United States), except possibly those from Arizona–New Mexico. The evolution of rapid growth in this region probably occurred at the expense of winter hardiness and is an adaptation to the relatively favorable growing conditions.

Elevational versus Geographic Differences

Callaham and Liddicoet (1961) found an elevational cline in ponderosa pine from northern California. Trees from medium elevations grew faster (at low and medium elevation test sites) than trees from high elevations. This cline exists along a transect from the Sacramento Valley to near the crest of the Sierra Nevada Mountains. Along this transect the rise in elevation is gradual, so that the end points of the cline are many miles apart. Thus there was some geographic separation. The differences between high and medium elevation trees were much less pronounced than between geographic races.

A similar cline can be demonstrated for Douglas-fir from northern Idaho. In the extreme north this species grows at elevations 1000 to 2000 ft less than those at which it grows 200 miles farther south. Trees from the northern counties grow 10 to 20% faster than trees from farther south.

In the southern interior, most tree species tend to have a 3000 to 4000 ft elevational range within a given area. The changes in topography are abrupt so that high- and low-elevation forests may occur within a few miles of each other. Thus there is opportunity for gene exchange up and down a mountain. In this part of the range, no consistent low- to high-elevation trends have been found.

Intraregion and Intrastand Variability

In all the species mentioned, differences among trees from different geographic regions are large and of great practical importance. Recent experiments that sample certain portions of the range intensively show the presence of a large amount of genetic variation within region.

The experiment by Wang and Pattee (1975) may be mentioned as an example. They performed a combination half-sib progeny test and provenance test in ponderosa pine, using open-pollinated seed collected from several trees in several stands in various parts of Idaho. As regards growth rate, there were only small intrastand differences (probably not enough to justify mass selection), but important differences of 10 to 15% among the offspring of different stands. Wells (1964) found much the same thing in parts of Colorado and the Inland Empire. According to his results some stands located only 20 miles apart differed in growth rate and some other traits by 10 to 15%. So far it has not been possible to relate much of this intraregion variation to local environmental factors.

Boreal America—A Region of Clines and Great Genetic Diversity

General

The boreal forest is transcontinental, extending from New England and Canada's maritime provinces northwestward across Canada to Alaska. It is a dense forest consisting of relatively few species, each of them common and having a continuous range. Only toward their southern limits do the species have discontinuous ranges. The forest has a north–south distribution of about 25° in latitude. The winters are generally cold, with average January temperatures ranging from +20°F in the south to −20°F in the north.

There are limited-range and range-wide provenance test data for three common species. All experiments have been conducted in the southern parts of the species' ranges, and most include several 8- to 20-year-old plantations per experiment. Those on jack pine have been summarized by King and Nienstaedt (1965), King (1971), and Canavera and Wright (1973). Those on white spruce have been summarized by Nienstaedt (1968). Those on tamarack were started by the late Scott S. Pauley and are being followed by Carl Mohn, both of the University of Minnesota. Additionally, unpublished data are quoted here.

Broad Geographic Trends

The variation is predominantly clinal and strongly correlated with latitude. There are fairly constant trends in progressing from northwestern Canada southeast to the Great Lake States. Those trends differ slightly from the trends observed in progressing from Labrador southwest to the Great Lake States, so that trees from the same latitude in the western and eastern portions of the range differ in some traits.

The greatest differentiation has been in traits of obvious adaptive value, such as growth rate, date of flushing, time of terminal bud formation. Variation in these traits is nearly as great as in western species. As compared with northern trees, southern trees grow three to four times as fast (in southern habitats), produce cones at earlier ages, have slightly longer needles, flush 1 to 2 weeks later, and set terminal buds earlier. Only in jack pine are there large differences in autumnal coloration; northern trees become yellow in the autumn, whereas southern trees remain green.

Although the differences in growth rate and phenology are large, the total amount of variability does not seem to be so great as in most widely distributed western American trees. It is not possible, for example, to identify the state or province of origin of a single branch of jack pine or white spruce by looking at the twig color, leaf twist, bud scale color, etc., as is possible with ponderosa pine.

Constancy of Behavior at Different Test Sites

The same 24 seedlots of white spruce (from such diverse places as Alaska and Michigan) were planted at each of 14 locations, scattered from North Dakota to New Brunswick. Even though there were large differences among test plantations, there was almost no genotype–environment interaction. The same seedlots grew exceptionally well or exceptionally poorly wherever they were planted.

This was not true in jack pine. In one experiment, started in 1952, trees were grown from seed collected in 29 Great Lake States localities and field planted at 17 localities in the Great Lake States. Trees grown from seed collected in southern localities grew fastest in southern but not in northern plantations, where they suffered some winter injury.

The Upper Michigan–Lower Michigan Discontinuity

An exception to the general clinal variation pattern occurs in Michigan, which is composed of two peninsulas. From tree line to tree line, they are separated by about 4 miles, which is enough to prevent gene exchange. The northern (upper) peninsula generally has colder winters and shorter growing seasons than the southern (lower) peninsula. Differences in selection pressure, plus the range gap, have resulted in the development of races that are more or less distinct (Wright, 1972). In four species for which data are available, trees from the southern peninsula grow 10 to 20% faster than trees from the northern peninsula, if tested in the southern peninsula; the growth rate differences are less pronounced if the tests are conducted in the northern peninsula.

There is also considerable climatic and soil variation within either peninsula, but because of the continuous nature of the forest there has been an opportunity for continuous gene exchange. Consequently, genetic differences within a peninsula are much less pronounced than differences between the peninsulas.

Other Anomalies

The natural range of white spruce overlaps that of Sitka spruce in Alaska and that of Engelmann spruce in the Canadian Rockies. In both cases introgression has resulted, which is evident at some distance from the zone of contact. This introgression, particularly with Engelmann spruce, which tends to be slower growing and have bluer foliage, has resulted in a reversal of some of the trends evident elsewhere in the species' range. Thus, white spruce from British Columbia and Montana tend to be much bluer and slower growing than would be expected from their latitude.

There is an interesting and unexplained anomaly in tamarack from northern localities in west central Canada (northern Alberta to be more exact). Trees from that region flush as late in the spring as trees from the most southern part of the range, 1 to 2 weeks later than trees from 200 to 300 miles farther south.

A third anomaly is of great practical importance. Throughout the southern part of the range of white spruce, the fastest growing trees are not from the most southern areas or from stands closest to the test sites. Rather, they are from an area in eastern Ontario and western Quebec roughly stretching from 45° to 46° N and from 75° to 80° W. They surpass others by 5 to 10% and are being used in seed orchards for many different states and provinces.

Slight to Moderate Variation in Northeastern Species

General

In the forests of northeastern United States and southeastern Canada, there are several species with predominantly northern ranges but with range extensions down the southern Appalachians. I shall describe the variation pattern in four of these.

Most of them have essentially continuous ranges. They experience climates that can generally be categorized as humid, with moderate to heavy amounts of precipitation during the growing season. The largest climatic differences are in minimum temperature during the winter and in length of the growing season. There is considerable variation in those respects, but not nearly so much as encountered in the western part of the countries.

Three Species with Little Geographic Variability

Three species with little geographic variability are northern white cedar (unpublished data from 8-year-old experiments started by the late Scott S. Pauley of the University of Minnesota), yellow birch (unpublished data from 10-year-old experiments started by Knud Clausen of the United States Forest Service), and red pine (data from four large provenance tests up to 40 years old summarized in Fowler and Lester, 1970; Wright et al., 1972).

The reasons for the small amount of variability are unknown. Each species has a large range extending from eastern Canada to the Lake States and south in the southern Appalachians, at least to West Virginia. Two are pioneer species, and one is a member of the climax forest. One grows naturally in swamps, one is mesic, and one occupies relatively dry sandy loam soils. Two are easily planted, but yellow birch is a difficult species to plant and might have been expected to develop quite different races in order to invade forests successfully as different as those of Maine and northern Georgia.

Northern white cedar is the least variable of the three geographically. Even in low-mortality plantations that are growing well, it has not been possible to detect statistically significant differences in any trait. This is especially surprising in view of the fact that horticulturists have propagated a number of clones differing in foliage or growth habit.

Yellow birch is a member of the climax forest. In the three most successful test plantations, differences in growth rate and most other traits have been slight and of no practical significance. However, there are surprisingly large differences in time of leafing out. Trees from the southern Appalachians and a portion of Wisconsin leaf out 1 to 2 weeks later than trees from other parts of the range. Apparently the differences in length of growing season are unrelated to total growth during the growing season.

Red pine, which has been planted very extensively, has also been studied extensively. Yet the only discernible variation is in growth rate. Trees from central Wisconsin and central Michigan grow 5 to 10% faster than trees from the northern parts of those states, and 15 to 20% faster than trees from the extreme northwestern part of the range, in test plantations located near the southern parts of the range. In more northern plantations the total range of variation is less.

Eastern White Pine

Eastern white pine has approximately the same natural range (eastern Canada to the Great Lake States and south along the Appalachians to northern Georgia) as the three species mentioned above, but presents a vastly different story. It is geographically variable in a number of traits and has been studied intensively.

The largest and oldest provenance test is one established in 1955 by the United States Forest Service. It includes trees grown from seed collected in 30 widely scattered localities covering the entire species range and tested in 35 localities, also widely scattered throughout the species range (Wright, 1970a; Garrett et al., 1973). The 10- to 12-year growth data from that experiment are summarized in Table 15.1, for which the plantations were grouped as southern or central (i.e., areas from Pennsylvania and southern Michigan southward having an average January temperature of 22° F or warmer) and northern (i.e., areas from New York and central Michigan northward having an average January temperature of 20°F or colder).

TABLE 15.1

Geographic Variation in Relative Growth Rate of Eastern White Pine[a,b]

Seedlot No. and origin[c]	Relative height (% of plantation mean) in	
	24 southern or central plantations	11 northern plantations
1 GA	129	90
2 NC	125	93
3 TE	125	95
30 VA	95	87
5 WV	105	92
6 PA	119	103
9 PA	115	106
12 NY	96	103
13 MA	106	107
14 ME	83	99
16 OH	113	105
32 S MI	95	111
29 N MI	89	92
19 MN	83	98
15 IA	84	99
24 S ONT	110	102
25 N ONT	92	100
20 NS	92	97
21 NB	89	93
23 QUE	76	91

[a] From Wright (1970a) and Garrett et al. (1973).

[b] The seedlots are arranged in approximate north–south order.

[c] GA, Georgia; NC, North Carolina; TE, Tennessee; VA, Virginia; WV, West Virginia; PA, Pennsylvania; NY, New York; MA, Massachusetts; ME, Maine; OH, Ohio; S MI, southern Michigan; N MI, northern Michigan; MN, Minnesota; IA, Iowa; S ONT, southern Ontario; N ONT, northern Ontario; NS, Nova Scotia; NB, New Brunswick; QUE, Quebec.

Within the southern group of plantations there was a general north–south trend in growth rate, with trees from the southern part of the range generally growing fastest. There was, however, an interesting and very important exception to that trend, as trees from a relatively warm part of Virginia grew considerably more slowly than those from farther south or from some distance to the north. This exceptionally slow growth of Virginia trees has been confirmed in Genys' (1968) range-wide provenance test and by unpublished data from the NC-99 experiment, which intensively sampled the southern Appalachian portion of the range.

Within the southern plantations there was relatively little genotype–environment interaction. The same seedlots from Georgia, North Carolina, and Tennessee were among the leaders wherever tested, even several hundred miles north of their native habitats. There was considerable interaction between the northern and southern groups of plantations. Southern Appalachian trees suffered winter injury and grew slower than others at northern test sites. At the northern sites, however, the best growth was usually by a seedlot originating somewhat south of the planting site, and there was a tendency for certain seedlots to grow well at a number of places.

The height differences were accompanied by diameter differences, but the correlation was far from perfect (La Farge, 1971). There were large differences in the height–diameter ratio, indicating that different genes may be involved in the control of these two components of growth. Differences in these two traits, as measured at age 11 in a southern Michigan plantation are as shown in the tabulation below.

Seedlot No. and origin[a]	Relative size (% of mean) at age 11		
	Height	Diameter	Height (%) / Diameter (%)
1 GA	103	107	96
3 TE	112	113	99
24 ONT	109	99	110
20 NS	91	84	108
19 MN	83	85	98

[a] See Table 14.1 for abbreviations of origins.

Among other measurements made on this plantation, those by Carl Lee on wood density are of special interest. He found a significant positive correlation between rate of height growth and wood density, the tallest seedlots having 1 to 2% more dense wood than the slowest growing ones.

By 1974, when the trees were 18 years old from seed, cone production had started in several of the centrally located plantations of the

United States Forest Service provenance test. Peculiarly, cone production was practically limited to the slowest-growing northern seedlots.

Fowler and Dwight (1964) found differences in stratification requirements of eastern white pine seed (see Chapter 4). Seeds from northern stands germinated well if sown fresh; seeds from southern areas germinated well only if stored moist at near-freezing temperatures for 60 to 90 days before sowing.

In any temperate zone tree there are seasonal variations in the ability of the leaves and twigs to withstand cold temperatures. Twigs sampled in the summer are damaged by much higher temperatures than are twigs sampled in midwinter. In eastern white pine there are also geographic differences in cold hardiness at any given time of the year (Maronek and Flint, 1974). In midwinter, for example, trees from northern areas growing in southern Michigan could withstand temperatures of $-50°$ to $-60°C$, whereas trees from the southern Appalachians growing in the same plantation suffered severe damage at temperatures of $-30°$ to $-40°C$. These lethal temperatures are several degrees colder than are usually encountered in the regions to which the trees are native. Thus, natural selection seems to have resulted in the development of more cold hardiness than is absolutely necessary. This helps to explain why some southern seedlots have grown so well for 18 years, without apparent winter damage, in areas several hundred miles north of their native range.

Small amounts of geographic variation have been demonstrated in several other traits, such as foliar concentration of N, Ca, Cu, and Zn, cortical concentration of certain monoterpenes, and needle length and width. Generally speaking, such variation has not followed any recognizable geographic pattern and has not been related to economically important growth traits. Disappointingly, there seems to be very little geographic (or local) variation in resistance to the white pine weevil, the species' most important pest.

A limited-range half-sib progeny and provenance test was started in 1965 cooperatively by several north central states. Preliminary data from the 1973–1974 measurements substantiate most of the broad trends found in the earlier range-wide experiments. The new experiment also shows an appreciable amount of genetic variation in growth rate among stands 50 to 100 miles apart, with some of the differences seemingly related to elevation of origin and some not. For example, offspring of different stands in North Carolina differ in growth rate by as much as 10%. In a few stands there were significant differences in growth rate among the offspring of different trees.

Geographic Variation in the Central Hardwoods

General

The central hardwoods region of the United States contains a great many valuable trees. Many have large geographic ranges extending from southern Canada nearly to the Gulf of Mexico. In most cases the ranges are more or less continuous.

Selection Differentials within the Region

The climate can generally be described as moist, with greater precipitation during the summer than during the winter. The most valuable hardwoods are generally confined to the more mesic sites. From south to north the total annual precipitation decreases, as does the average annual temperature. Differences in the precipitation–evaporation ratio are not great. Probably the largest climatic differences are to be found in average winter temperature; average January temperature varies from 42°F ($+6$°C) in northern Alabama to 20°F (-7°C) in southern Michigan.

The large provenance test of eastern white pine has been mentioned. It includes plantations in several parts of the central hardwoods region. There have been similar experiments with other species, in which the same seedlots were planted at many places from Ohio to Illinois and from southern Michigan to Missouri. At least through the first 10 to 15 years, genotype–environment interaction has been relatively small, and the same seedlots have been superior in growth rate at places several hundred miles apart (Wright, 1973). Only when the experiments include plantations in the extreme southern states or the extreme northern parts of the Great Lake States has interaction been considerable.

Such results indicate that differences in selection pressure are not great between points 200 to 300 miles apart. This is borne out to some extent by data from provenance tests of central hardwood species. Preliminary 15-year data indicate relatively little geographic variation in northern red oak. On the other hand, there is considerable geographic variation in some species, three of which are discussed in greater detail below.

Sugar and Black Maples

Sugar maple is a widely distributed and common hardwood native to most of eastern United States and southern Canada. Black maple is a northern relative of uncertain taxonomic and genetic status. Typical forms of the two are different and easily recognizable. However, where the two grow together they hybridize freely and have the same growth and use characters and site preferences; therefore, they are considered together here. Geographic variation has been described by Kriebel and Gabriel (1970).

The southern third of the range is occupied by var. *floridanum*. As compared with typical var. *saccharum*, this has smaller leaves, fruit, and flowers, a much branched stem, relatively slow growth, less resistance to winter cold, late growth start in the spring and late growth cessation in the autumn, and a more branched root system. The two varieties intergrade over a narrow transition zone.

The northern two-thirds of the range is occupied by the typical variety or by sugar–black maple mixtures. Within this region some traits vary clinally and others in a more complex manner. In Ohio growth tests trees from northern states started growth earliest in the spring, continued growth latest in the fall, developed most intense autumn color, were most sensitive to hot summer weather, and required a long period of winter chilling. There were some exceptions to these general trends. Trees grown from seed collected within a few hundred miles of the planting site grew faster than those grown from seed collected farther away to the north, south, east, or west. Trees grown from seed collected in Wisconsin started growth later than expected, whereas trees grown from seed collected in the Missouri Ozarks started growth earlier than expected from their southern origin.

Sugar maple is a very tolerant tree, which reproduces naturally in the shade of other trees. Perhaps because of this, it is sensitive to summer heat. In the nursery stages of the Ohio experiments, leaf scorch (or death of the leaf edges) was noticeable after hot dry spells. Damage was greatest on trees grown from seed collected in northern states and at high elevations in the southern Appalachian Mountains.

Black Walnut

This is a widespread component of the mixed hardwood forests of eastern United States and extreme southern Canada. It is absent along the Gulf Coast and in the coniferous zone of northern United States. The tree is valued highly for its wood and nuts. A paper by Bey (1973) summarizes 6 to 8 year results of a range-wide provenance test, including the offspring of 78 different stands and test plantations in many localities.

Contrary to the situation in most species, southern trees start growth 5 to 10 days earlier in the spring than do northern trees. The adaptive significance of this is not certain. Black walnut is extremely sensitive to damage from late spring frosts, and even in native stands most seedlings suffer dieback 1 year out of 3. Presumably natural selection should have favored late flushing types, which would not suffer such damage. Maybe that selection was counteracted by selection favoring early flushing types, which would have a growth advantage in years with no frost. This seems to have been the case, since there is considerable local variation in flushing date, and there are no consistent growth rate advantages over a several year period between early and late flushing types from the same locality (Fig. 15.6).

There is the usual north–south trend in growth rate, the fastest growing trees being southern. There is the usual north–south trend in winter hardiness, southern trees being least hardy. The differences in winter hardiness are so slight that trees from as far south as North Carolina grew 5 to 20% faster than native trees in southern Michigan; they were large enough, however, that southern seedlots grew less well than northern ones when tested in cold localities, such as Iowa and Minnesota. Presumably part of the faster growth and lower winter hardiness of southern trees is due to a lengthening of the growing season, by 5 to 10 days in the spring and almost a month in the fall. There is also considerable random local variation in growth rate.

One trait seems to vary in a discontinuous manner. Trees from the southeastern states produce new leaves that are reddish, whereas trees from all other parts of the range produce new leaves that are green. Two other traits—frequency of insect damage and leaf angle— vary in a random manner. Offspring of certain stands are exceptional, but such stands are scattered throughout the range.

Tree planters have raised many questions about the relative effects of seed source, fertilization, nursery treatment, etc., on the later growth in plantations. An experiment to determine these effects was started in 1971, when seeds from five different states were sown at different spacings in nurseries in different states and subsequently were transferred to field plantations in several states. The first report (Williams et al., 1974) on winter injury indicates that seed source was most important, but that the amount of damage suffered also varied with spacing and fertilizer treatment. The damage was least on northern trees and on trees fertilized with nitrogen.

Eastern Cottonwood

Eastern cottonwood is a fast growing bottomland species that is much valued for pulp. It is planted to a modest extent in south central

Fig. 15.6. Variation in date of flushing in 4-year-old black walnut growing in southern Illinois. In this species southern trees (from Tennessee on right) flush earlier and grow faster than northern trees (from Indiana on left). In many species southern trees flush latest. (Photograph courtesy of Calvin Bey of the United States Forest Service, Carbondale, Illinois.)

United States and on large scales in some other countries. As it has a linearly continuous range along major rivers, somewhat greater geographic differentiation is expected in this than in upland species with areally continuous ranges. This seems to be true.

A combination provenance–progeny–clonal test was started in 1965 by J. J. Jokela. He obtained seeds from many stands in the Mississippi River basin, all the way from Mississippi to Minnesota, and distributed the planting stock to cooperators in many states in that region. The first published report is that by Ying (1974) on 7-year data from a Nebraska plantation.

Some of the most important differences were phenological. There was considerable local variation in this respect but also large regional differences. Trees from the extreme north leafed out about April 21, those from the extreme south about April 25, and trees from some north central areas about April 28. There were much larger differences in time of growth cessation in the autumn. Trees from northern and central areas stopped growth early enough to become winter hardy, whereas trees from Mississippi grew until the onset of cold weather. As a consequence, 90% of the Mississippi trees died or suffered repeated winter injury.

As for growth rate, trees from 100 miles south of the planting site grew fastest, exceeding the slowest growing ones by 50%. There was east–west differentiation in some traits. For example, trees from the east (Indiana and Ohio) had rougher bark and larger branches than trees from farther west. There were also important regional differences in other traits, such as stem canker, leaf shape and size, and flower size. Variation in those traits did not easily follow described north–south or east–west patterns. The relative sizes of the geographic, family–intrastand, and clone–intrafamily components of variation in several traits are given in Chapter 11.

Southeastern United States—A Region of Moderate Genetic Diversity

Amount of Climatic and Genetic Diversity

The pine region of southeastern United States extends from Maryland south to Florida and west to Texas. Most of the pine forests are on the Coastal Plain or in the Piedmont zone, within a few hundred miles of the coast. Some of the hardier species grow northward in the interior to southern Illinois. The region as a whole is characterized by abundant precipitation, long growing seasons, and mild winters. However, even as far south as northern Florida, there are occasional short periods with temperatures near 0°F (−18°C), so that all species except those limited to southern Florida have evolved some resistance to winter cold. Most of the common species have more or less continuous ranges.

Because of the absence of climatic extremes and because of the continuous ranges, geographic variation is much less pronounced than in other parts of the country. Also, much of the variation appears to be random and not correlated with climate or site characteristics at the place of origin.

The two most important species are loblolly and slash pines. Each has been the subject of several small and one or two extensive provenance tests. Much of the provenance research on loblolly pine is summarized in the papers by Wells and Wakeley (1966) and Wells and Switzer (1971). Much of the provenance research on slash pine is summarized in the paper by Squillace (1966). An extensive provenance test of short-leaf pine has been summarized in papers by Wells (1973) and Wells and Wakeley (1970). A recent paper by Genys et al. (1974) describes results in Virginia pine.

Major Regional Differences

Slash pine grows from South Carolina south to southern Florida and west to Louisiana. The species, *Pinus elliottii*, is divided into varieties *densa* (southern Florida) and *elliottii* (remainder of range). These are recognizable in the field by cone and leaf characters. They are very different in important growth traits (Table 15.2) (Squillace, 1966). They also vary in other traits. As compared with south Florida slash pine, the typical variety has fewer seedlings with three-needled fascicles, fewer stomata per millimeter, lower concentration of β-pinene, and higher concentration of β-phellandrene in the cortical oleoresin.

Note in Table 15.2 that the height–growth trend is the opposite of that found in most species—northern trees grew tallest. This is probably due to the fact that the experiments were conducted in northern Florida, where winter temperatures may be too low for good growth of the south Florida variety. A similar experiment was conducted in a nearly frost-free part of northern Argentina, with opposite results (Barrett, 1969, 1970). Under warm growing conditions the south Florida variety grew fastest in both height and diameter.

Loblolly pine grows naturally from southern New Jersey south to central Florida and west to eastern Texas. There is one major range gap. The Mississippi River Valley, 25 to 120 miles wide, is essentially pineless and offers a barrier to east–west migration. Trees from Texas and Arkansas west of the river differ from those to the east in two respects; they are more resistant to fusiform rust and to drouth (Fig. 15.7). This east–west difference seems to be due in part to the range gap and in part to introgression from shortleaf pine in Texas and Arkan-

TABLE 15.2

Growth Characteristics of Two Varieties of Slash Pine and the Intermediate Population When Grown for 1 Year in Northern Florida[a]

Region of origin and variety	Characteristic[b]		
	Height (cm)	Diameter (cm)	Needle length (cm)
South Florida (var. *densa*)	9–19	0.82–1.02	16–20
Central Florida	10–28	0.66–0.74	16–19
North Florida and South Carolina to Louisiana (var. *elliottii*)	24–30	0.64–0.79	14–15
Source of variation	% of total variance		
	Height	Diameter	Needle length
Region	66	25	44
Stand–within–region	5	6	6
Mother tree–within–stand	4	8	5
Error	25	61	45

[a] From Squillace, 1966.
[b] Range in stand–progeny means.

sas. There are also large and important differences in growth rate (Fig. 15.8).

There is conflicting evidence regarding the amount of north-south differentiation in loblolly pine. In greenhouse experiments conducted in northern Florida, Maryland trees started growth 25 days later, ceased growth 85 days earlier, grew only 40% as tall, produced 2 instead of 6 to 7 branch whorls per year, and responded more to an increase in summer daylength than did Florida trees (Perry *et al.*, 1966). In some field experiments conducted in cool climates (e.g., Illinois, Maryland, central Japan) northern seedlots survived best, grew fastest, and suffered the least winter injury. However, Maryland and southern trees performed similarly in several southern test plantations.

The range of shortleaf pine is divided into western and eastern portions, just as is true of loblolly pine. This has resulted in differentiation with regard to phenology but not in other respects. Trees from west of the Mississippi leaf out earlier than almost all trees from east of the Mississippi.

Field tests of shortleaf pine revealed greater north–south differences

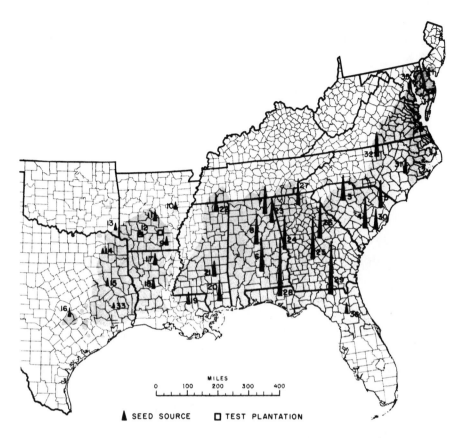

Fig. 15.7. Geographic variation in resistance of loblolly pine to fusiform rust, as determined from a provenance test plantation in Arkansas. The height of the cone indicates the relative incidence of attack and the numbers refer to seedlot numbers. The broad Mississippi River valley separates the more resistant western from the less resistant eastern seed sources. (From Grigsby, 1973.)

in growth rate in shortleaf than in loblolly pine. There was a strong correlation between height at age 9 and average annual temperature at the place of origin. At that age, when the trees were grown in southern Mississippi, the most northern trees were 13 ft tall and trees from the warmest areas were 20 ft tall.

Virginia pine has a more inland and more northern range than the three previously mentioned species. It grows naturally from central New Jersey southwestward to northern Mississippi and Alabama. In northern test plantations, seedlots from the extreme southern portion of the range grew more slowly, suffered more winter injury, and pro-

FIG. 15.8. Top: Geographic variation in loblolly pine, as shown in P. C. Wakeley's Bogulusa, Louisiana plantation established in 1926–1927. From left to right (tallest to shortest) the trees were grown from seed collected in Louisiana, Texas, Georgia, and Arkansas, respectively. In the winter of 1948–1949 (age 22) the plantation was thinned by removing every other tree in each plot to produce the wood piles shown at the bottom. The living trees were photographed when 35 years old. (Photograph courtesy of P. C. Wakeley and O. O. Wells of the United States Forest Service.)

duced fewer cones than seedlots from the central and northern portions of the range. In a southern test plantation, the reverse trends occurred. Disregarding the extreme southern seedlots, the geographic variations in growth rate were not large, the fastest and slowest growing seedlots differing by 10 to 15% at most test sites.

Intraregion Variation in Loblolly Pine

Wells and Switzer (1971) collected open-pollinated seed from each of 5 trees in each of 115 stands in Mississippi and southeastern Louisiana. In the resulting test plantation, trees grown from seed collected in one portion of southeastern Louisiana grew rapidly and were resistant to fusiform rust. Trees grown from seed collected in a part of northern Mississippi also grew rapidly but were susceptible to the rust. Trees from other places grew more slowly and were rust-susceptible. The variation was clinal, and the transition from an area of low rust resistance to one of high rust resistance occurred gradually over a distance of 50 to 75 miles.

Unexplained Geographic Variation

Squillace's (1966) provenance test of slash pine included trees grown from seed collected from 36 stands of the typical variety extending in a belt southwest from South Carolina to Louisiana. Those seedlots can be denoted by their relative growth rates (from 1 = slowest to 4 = fastest, 20% taller than 1) and listed in a northeast–southwest sequence. The sequence is: 22332 22222 13111 14333 23221 22222 21312 1. Note that there is no consistent trend, although there is a tendency for trees from certain areas to excel or do poorly. Nor was it possible to detect consistent trends of other types, such as north–south, east–west, high–low rainfall, high–low winter temperature. Other measured traits also varied in a more or less random manner throughout much of the range of the typical variety.

If we exclude the most northern and most southern seedlots, much the same can be said of the variation in loblolly and Virginia pines. Within a distance of 300 to 400 miles, 10 to 20% differences in growth rate may occur but seemingly not be correlated with any measured characteristic of the environment.

It is possible that differences in selection pressure in the southern pine region are small and that much of the variation is due to genetic drift rather than to selection. However, this is a matter of conjecture, and there are at least two other possible explanations. The differences may be adaptive, but not to aspects of the environment that have been measured. Or, the differences may represent adaptation to past rather than to modern environments.

Variation in Wood Density

Most southern pines are grown extensively for pulp, in which wood density is a very important consideration. There have been extensive studies of wood density (summarized in Wells, 1973). These studies have been of two types: phenotypic (sampling trees cut in native stands) and genetic (sampling trees grown in provenance tests). The phenotypic variation pattern has usually been one of increasing density from north-west to southeast. The genetic variation pattern has been different (random in shortleaf pine or decreasing from northwest to southeast in some species). Also, there is considerable local genetic variation in density.

In the shortleaf pine provenance test, the range in specific gravity of different seedlots was from 0.46 to 0.50, whereas the range in volume was from 318 to 972 ft³/acre. There was a slight positive correlation ($r = 0.18$) between specific gravity and growth rate, the fastest growing seedlots tending to be the densest. Thus, from the practical standpoint of producing the most usable wood per acre, one can concentrate almost solely on growth rate.

Practical Usefulness of the Provenance Data

A few broad trends have been found in the southern pines permitting one to make generalized recommendations about the best seed sources to use for specific areas. However, much of the geographic variation follows no definite pattern, so there may be little gain from such generalized recommendations unless they are based on intensive sampling. Considerable gain may result from a study such as that on Mississippi–Louisiana loblolly pine, which pinpointed a specific area from which to collect high quality seed.

The presence of geographic variation is very important in breeding programs involving the selection and crossing of individual trees, however. The studies by Squillace (1966), Wells and Switzer (1971), and La Farge (1974) indicate that differences among the offspring of different stands in the same general region are often as great or greater than the differences among the offspring of different trees within a stand. Thus the greatest gain can usually result if one searches for genetically good stands and then for genetically good trees within those stands.

Study Questions

1. Why is there more geographic variation and why are races more sharply delimited in western American than in southeastern species?

2. There are western versus interior contrasts in four western American species. To what extent do these contrasts differ among species?

3. Given optimal test conditions, what region supplies trees with the greatest potential for growth rate in ponderosa pine? In Douglas-fir? In white fir?

4. Suppose the above three species were to be grown in a cold part of northeastern United States. Why would the regions supplying the fastest growing race for a particular test site differ from those enumerated in question 3?

5. Why is there such a large genetic difference between Arizona–New Mexico and Utah–Colorado races of the same species?

6. How do the amounts and patterns of genetic diversity in adaptive traits vary between boreal American and western American species? In nonadaptive traits?

7. To what extent is the genetic variation continuous in boreal American species? To what extent discontinuous?

8. What eastern American species have the least amount of genetic diversity associated with place of origin of the seed? The greatest amount?

9. In what eastern American species do the geographic trends agree best with the general trends enumerated in Chapter 13? Agree the least?

10. Compare the amounts of genetic diversity in southeastern and western species. Why the difference?

11. To what extent does the geographic variation in southeastern species follow definite geographic trends that can lead to generalized planting recommendation? To what extent is the variation in these species unexplained?

12. What is meant by genotype–environment interaction in a provenance test? Give examples of its occurrence. Of its nonoccurrence.

13. In the southern pines, how do the phenotypic and genetic variation patterns for wood density compare? How much weight should be placed on specific gravity when selecting a seedlot for pulpwood production?

14. If geographic variation is seemingly random, can it be put to practical use? If so, how?

16

Species and Racial Hybridization

Use of Hybridization in Agriculture

Hybridization is the crossing of diverse types. The term may be used to include crossing organisms that differ by a single gene, but in forestry it usually refers to crossing different species or different races within a species.

Hybridization between different strains of the same species is a constantly used tool of the agricultural crop plant breeder. A wheat breeder wishing to produce a new variety with good milling characteristics and with good rust resistance usually starts by crossing a strain having the necessary milling qualities with another having the necessary resistance. Then he practices several generations of selection. Much the same procedure is followed to produce a nutritive and productive new corn variety. First a strain with high lysine (an amino acid essential in human nutrition) content is crossed with one having good yield characteristics. Then several generations of selection are practiced.

Corn breeders use hybridization in another way. They produce uniform and genetically different strains by selfing for several generations, then mass produce those combinations between selfed lines that exhibit a great deal of hybrid vigor. The success of this particular technique depends very much on the floral biology of the corn plant. The male flowers or "tassels" are borne at the top where they can be easily removed. Thus if two selfed lines are interplanted, one can be detasseled and produce only hybrid seed.

While hybridization within species is a common improvement

313

method, hybridization between distinct species may have played an even more important role in the distant past. Several important crop plants are believed to have arisen as the result of chance hybridization between totally different plants. Among these are bananas, sugarcane, some wheats, some cottons, apples, and tobacco. Is it possible to emulate nature's accidents and produce something as interesting as the sweet, seedless banana? Most banana relatives have seedy, starchy fruits. Some breeders think so. The success ratio is not high but some of the successes promise to be startling. The development of a wheat–rye hybrid, under way for a half-century, seems to be finally approaching fruition.

Goals of Hybridization in Trees

General

Most agricultural crops are bred for their seeds (wheat, corn, mustard, rape, rice), seed appendages (cotton), or fruit (apples, bananas, oranges). Hybrids between distinct species have sometimes had vigorous vegetative growth but lacked the proper seed or fruit characteristics to be commercially valuable. Also, many have been partly sterile and for that reason have not been commercially acceptable.

Forest trees are usually bred for their wood-producing ability or appearance. Vegetative vigor is important. Except in the poplars and willows, which are propagated vegetatively, complete sterility would be a barrier to commercial propagation, but partial sterility would often be acceptable. From that standpoint, species hybridization can probably play a larger role in modern tree breeding than in modern agricultural plant breeding.

The possibilities of species hybridization are relatively great in forest trees. In several genera, related species from different regions do not seem to have developed strong genetic barriers to crossing; they maintain their identities in nature mainly because of geographic isolation. When they are interplanted so that they can cross-pollinate naturally, or when they are crossed artificially, a great many hybrids are possible.

Much of the hybridization work in trees was done prior to 1950. At the time, breeders had access to collections of different species, but not to provenance tests or progeny tests which include a variety of genotypes within a species. Consequently, "hybridization" during that early period was usually interpreted as "species hybridization," and most of the crosses were between average parents. Many hybrid combinations are still known only as the result of a cross between one tree of one species and one tree of another species. Some that are presently regarded as unsuccessful might be more worthwhile if repeated by crossing several selected trees of one species with several selected trees of another species.

The possibilities of crossing different races or portions of clines of the same species have been appreciated for a long time, but only in the last decade has this type of hybridization received much emphasis. Even now it is in its infancy. As more becomes known of geographic variation, interracial crossing can be an important way of producing new trees having special combinations of desirable traits.

Capture of Hybrid Vigor in F_1 and F_2 Generations

Hybrid vigor has been the goal of much species hybridization work. Hopefully, such crossing could lead to F_1 combinations that grow faster than either parent and could be mass produced at a reasonable cost. Several examples of hybrid vigor have been discovered in forest trees, but in many cases the hybrid vigor was not useful because it occurred only on "hybrid habitats" or because the hybrids were surpassed by nonhybrids of another species. Consequently there is a modern tendency to confine this type of breeding to special cases.

The majority of breeding work for hybrid vigor has stopped at the F_1 generation. That is so for two reasons: lack of promise in many F_1 generation hybrids and belief that hybrid vigor is confined to the F_1 generation. However, of the four possible explanations of hybrid vigor only one (the overdominance hypothesis) precludes its fixation in later generations. The success with F_2 hybrids between pitch and loblolly pine in the Republic of Korea indicates that F_2 and F_3 breeding should receive more attention.

Combination of Traits by Selecting in Segregating Late Generations

This is the usual reason for hybridization in agricultural crop plant breeding. Often one variety has many desirable traits but is deficient in one or two major respects and does not contain within itself enough genetic variability to permit rapid progress by intravariety selection. Hence diverse varieties are crossed, and selection is then practiced for several generations.

This possibility has not been tempting to tree breeders because of the long generation length. Nevertheless, it appears to be the easiest way to accomplish certain objectives. The production of a tall, straight, fast-growing, blight-resistant, timber-type chestnut is an example. American chestnut has all the desired characteristics except one; it is extremely and uniformly susceptible to the fatal chestnut blight. Chinese and Japanese chestnuts are blight-resistant but are small trees. The F_1 generation hybrids are intermediate but that means only that they are semi-good forest trees and can live 8 to 10 instead of 4 to 5 years. The possibilities of success seem good if the work is continued for 4 to 5 generations.

This approach also has promise in Scotch pine. The Spanish variety has good winter color, high resistance to Zimmerman moth and pine root collar weevil, moderate growth rate, and low resistance to winter cold. The Ural Mountain variety has moderate growth rate and high resistance to European pine sawfly, but is deficient in the other respects. Intervarietal hybridization followed by several generations of selection, could produce trees excellent in all respects.

Information about Evolution

Presumably, the shorter the time since two species became differentiated, the smaller the genetic differences between them, and the greater the ease with which they can be crossed. Thus, species hybridization studies have contributed much to our knowledge of evolution in trees. They have been a valuable adjunct to studies of comparative morphology and have taught us more about recent evolutionary trends than has the fossil record, which is very incomplete for trees.

Information about evolution is mainly of theoretical interest but sometimes has practical applications. Data about evolutionary rates of change are very useful in estimating improvement rates in applied breeding programs. In the spruces, the evolutionary data seem to be of greater practical importance than any hybrids produced so far because there is a relationship between evolutionary history, ease with which natural reproduction can be obtained, and ease with which a species can be managed silviculturally.

Crossability Patterns

Barriers to Successful Crossing

1. Geographic isolation is one barrier to natural hybridization. In poplars it seems to be the principal barrier. In general there is only one cottonwood species native to any particular part of the world. Artificial crossing experiments have shown why. The different species cross so easily that if two or more occupied the same range, they would long ago have merged into one.

2. Differences in flowering time prevent species, such as red and silver maples, from crossing in nature. If pollen from the early blooming silver maple is stored and applied to red maple, hybrids can be produced rather easily. Under natural conditions this does not happen.

3. Genetic differences may be great enough to prevent pollen from one species from germinating and effecting fertilization when applied to the female flowers of another species. If, for example, pollen of a hard pine is placed on the female flower of a soft pine, nothing happens. It is just as if the hard pine pollen was so much dust. Presumably a female flower secretes enzymes and other substances which enable a pollen grain to germinate and grow. These seem to be specific enough to prevent most types of pollen from being effective except on a very few closely related species.

4. Genetic differences may be small enough to permit pollen to germinate and effect fertilization, but large enough to prevent the hybrid embryo from growing. Kriebel (1972) found this to be the case in the white pines. He studied embryo development in a number of species crosses that had been unsuccessful up to that time. In several, the fertilized embryos developed to the suspensor stage, then aborted. The stimulus of pollen germination or of fertilization was enough to stimulate seed coat formation, so the result was a large number of empty seeds. This happens to varying degrees. In the pines, there are many species crosses that regularly result in large numbers of empty seeds but rarely if ever result in filled seeds. There are other species crosses that regularly result in 1 to 10% as many filled seeds as can be obtained from crosses within species. Still other crosses between species show no or slight amounts of reduction in the set of filled seeds.

5. Genetic differences may be small enough to permit fertilization and normal embryo development, but large enough to prevent normal chromosome pairing in the hybrids which are produced. In such a case the hybrids can be produced but may not themselves produce seed. Such hybrid sterility seems rare in the pines and oaks and possibly in other tree genera.

Crossability within Species

Karrfalt *et al.* (1975) made an extensive series of hybridizations among races of Scotch pine in 1969 and 1970. The crossing work was done in a provenance test where trees of various races grew close to each other. Between the most northern and the most southern races there were differences in flowering time that were sufficient to prevent natural hybridization. However, there were no apparent genetic barriers to crossing, and it is probably safe to assume that almost any Scotch pine could be crossed with almost any other Scotch pine.

Crossability between Genera

The hybrid × *Cupressocyparis leylandii* is believed to have resulted from a cross between the genera *Cupressus* and *Chamaecyparis,* both members of the family Cupressaceae. Several intergeneric hybrids are also known in the subfamily Pomoideae of the Rosaceae. Among them are × *Crataegomespilus* (*Crataegus* × *Mespilus*), × *Pracomeles* (*Pyracantha* × *Osteomeles*), × *Sorbaronia*   (*Sorbus* × *Aronia*) × *Sorbopyrus* (*Sorbus* × *Pyrus*), and × *Pyronia* (*Pyrus* × *Cydonia*). The fact that these are considered intergeneric hybrids may be due in part to the definition of a genus. The genera that have hybridized are similar to each other. A supposed hybrid between *Tsuga* and *Picea* of the Pinaceae has been described. However, the evidence for hybridity is weak, resting almost entirely on a few aberrant pollen grains.

The practical possibilities of intergeneric hybridization are small. Possibly several additional genera could be crossed with each other. However, except in those cases where taxonomic distinctions are exceedingly fine, the chances of success are probably very low—perhaps 1 seed for each 10 years' effort.

Factors Affecting Success of Crosses between Species in the Same Genus

Closeness of taxonomic relationship is a good clue to the ease with which two species belonging to the same genus can be crossed. To obtain data on such taxonomic relationships it is desirable to consult a monograph such as Critchfield and Little's (1966) "Geographic Distribution of Pines of the World." When planning crossing work, it is desirable to be guided by the groupings in such a monograph. For example, Critchfield and Little subdivide the genus *Pinus* into 3 subgenera, 6 sections, and 15 subsections. Of the 60+ successful species combinations that have been produced, the vast majority are between species belonging to the same subsection. Before that fact was generally known, many hybridizers tried crosses between less closely related pine species, usually without success.

Many other examples could be quoted. The maple genus is a large one, which can be divided into 13 series according to Rehder (1940). Of approximately 30 species combinations that have been attempted, 6 have produced viable seeds as the result of controlled pollination.

The 6 successful crosses involved species belonging to the same series; the unsuccessful crosses involved species belonging to different series. In this particular genus it should be mentioned that some taxonomists of the past have considered some of these series as different genera. The eucalypt genus, native to Australasia, is comprised of 250 to 500 species according to various authors. It has been divided into subgenera and sections by Blakely (1955). Many natural hybrids have been reported (Pryor, 1957), most of them between species belonging to the same subgenus.

Geographic distribution also provides clues as to the ease with which species can be crossed. The rules for determining probable crossability are somewhat complex and can be summarized as follows.

1. Species occupying the same ecological niches in the same areas usually do not cross.

2. Species with neighboring ranges often cross rather easily.

3. Species occupying distant natural ranges usually do not cross readily.

4. Crossability is apt to be rather high in a genus or genus subdivision comprised of only one species per region.

The above rules have an empirical basis—studies of natural hybridization and controlled pollination work by many different investigators. They also have a theoretical basis. Presumably two species can cross more easily the more recently they became differentiated from each other geologically speaking. Two species inhabiting neighboring ranges probably became differentiated more recently than did two species inhabiting distant ranges. For two species occupying the same range it is usually necessary to develop some crossability barrier; otherwise they would merge naturally into a single species. In groups such as sections *Aegeiros* and *Tacamahaca* (cottonwoods and balsam poplars) of the genus *Populus* there is only one species in any particular region, a fact that indicates that geographic rather than genetic isolation has been responsible for species differentiation. If so, one might expect (as happens to be the case) that nearly every species can be crossed with nearly every other species.

Numerous examples can be quoted showing the operation of these rules. In spruce, morphologically similar species with neighboring or slightly overlapping ranges can be crossed with each other in most cases, whereas hybrids among species occupying the same range or very distant ranges are much more difficult to produce. In subsection *Strobi* of the pines, which contains one species in each of many regions of the northern hemisphere, most species can be crossed with each other. In the one case in which two species have greatly overlapping ranges (sugar and western white pines in Oregon and California), it has not proved possible to obtain hybrids, and in fact sugar pine can be crossed with almost nothing else. The eucalypt forests of Australia are usually composed of several species, but usually there is only one species of any particular subgenus or section in any particular area. Numerous hybrids are possible, but in nearly all cases these are between species from different regions. Subsection _Sylvestres_ of the pines contains many Eurasian and a few American species. Crossing experiments have dealt with one American species (red pine) and several Eurasian species. Nearly complete failure has resulted from crosses involving American red pine and other species; many hybrids are possible between the various Eurasian species, however.

There are exceptions, such as those involving the pine species native to southeastern United States. Loblolly, shortleaf, slash, longleaf, and pitch pines have overlapping ranges and overlap sufficiently in flowering time for natural hybridization to occur. There may be introgression between loblolly and shortleaf pines (Hare and Switzer, 1969), but for the most part these species remain distinct. Most of the hybrids that have been produced artificially have not grown as well as the pure species and have therefore been at a selective disadvantage.

Differences in chromosome number do not in themselves prevent species from crossing; hybrids between diploid and tetraploid or hexaploid trees are known in several genera and in some cases can be produced quite easily. Such hybrids are usually sterile, however.

Hybridization and Evolution

Natural Hybrid Swarms and Introgression in a Species

Hybrid swarms are numerous in the southern pines and oaks. Most are small and composed primarily of F_1 generation trees with a few F_2 or backcross generation trees; they usually persist for 2 to 3 generations only.

If the hybrids have a selective advantage, the hybrid swarms may become larger and persist indefinitely. In Texas there are several examples of oak species with very different soil requirements but with overlapping ranges. If the hybrids happen to grow on sites intermediate between those to which the parental species are adapted, moderately extensive hybrid swarms may develop (Muller, 1952). Very extensive hybrid swarms are found in northern California (white × grand fir), the Canadian Rockies (white × Engelmann spruce), and northern Scandinavia (Norway × Siberian spruce). In each of those cases there are large areas in which it is difficult to find trees typical of either parental species and even larger areas in which the trees are generally recognizable as typical of one species but have received genes and characteristics of the other through "introgression" or gene migration from one to the other. Such introgression affects the geographic variation patterns in species such as white spruce, white fir, loblolly pine, and shortleaf pine (see Chapter 15).

Large hybrid swarms and introgression are probably more numerous now than formerly. As man has exploited the forest, habitat conditions have changed, frequently in a manner to favor hybrids more than pure species. Thus on disturbed habitats in Ohio, hybrid swarms of buckeyes have been found, although the species had previously remained distinct. Similarly hybrid swarms of eucalyptus are found on some disturbed habitats in Australia. Also, man has moved species around the world and planted them together in such a manner that they could cross with one another in a manner that is impossible in nature. In the United States, Japanese walnut has been planted close to butternut with which it crosses easily. As a result, one is never quite certain that a "Japanese walnut" seen in the foreign habitat is the pure species or a hybrid.

Formation of New Species through Introgression

Bornmuller fir of northwestern Turkey is intermediate in a great many respects between Greek fir (of Greece) and Nordmann fir (of eastern Turkey). It may have arisen as the result of hybridization when the ranges of these two previously isolated species coalesced because of some large climatic change. Actually in this case it is difficult to tell because the Bornmuller fir could be an intermediate stage in the evolution of three species from one.

A new species does seem to be in the process of formation in southern Europe now. Eastern cottonwood was introduced into Europe from America two centures ago. It hybridizes freely with the European black poplar and the hybrids grow well. Man has selected mostly hybrid clones for cultivation, and it seems that even without man's further intervention a new hybrid species could replace the former native one.

A new hybrid species seems to be in the process of formation in the Republic of Korea. About two decades ago workers there produced F_1 hybrids between the American loblolly and pitch pines. The hybrids grew well and were produced in quantity. Then F_2 hybrids were produced, and they proved to be surprisingly fast growing and comparatively uniform. From them another generation will be grown. By the F_4 generation a new species will have been formed.

Formation of New Species through Remote Hybridization

The previous section was devoted to species so similar taxonomically that they cross readily once they are brought into contact. Taxonomic and/or cytological evidence indicates that many species have arisen as the result of remote hybridization, that is, by hybridization between species so little related that they bear little similarity to each other now and would if crossed yield filled seeds only once in several thousand or several million attempts. Thus the late Karl Sax postulated that the Pomoideae subfamily of the Rosaceae may have arisen as the result of ancient hybridization between a rose-like and a spiraea-like plant.

Such postulates are difficult to prove or disprove. To do so would require a very great amount of artificial hybridization work. But the morphological or cytological evidence, if not overwhelming, is often sufficient to indicate that a hybrid origin is the most likely explanation.

The possibilities of producing something entirely new by emulating nature's very wide crosses are enticing. Agricultural breeders have attempted to do so in synthesizing a new banana and in developing the wheat–rye hybrid. In both cases the work has progressed far enough that success seems likely. The work has not been easy, however. Much patience will be necessary to do the same for a forest tree.

Crossability Patterns by Genus

Larix

The larch genus contains about 10 species; 3 in America, 6 in Asia and 1 in Europe. All are allopatric, occupying separate ranges. They are more similar to each other morphologically than are most species of pine or maple, indicating a relatively slight amount of genetic differentiation. Where neighboring species have overlapping ranges, small hybrid swarms occur. Thus there are hybrid swarms involving Lyall's and western larch, Master's and Potanin's larch, and Korean and Siberian larch (Bobrov, 1973).

Much of the larch hybridization work has been summarized by Larsen (1956). Only one-quarter of the 45 possible species combinations have been attempted artificially. General indications are that almost any two species can be crossed with each other, but usually with reduced seed set. For example, only one tree was obtained by crossing European and western larches, and some other crosses yielded only a few trees.

However, two combinations can be produced with ease. Japanese × European larch hybrids have been known for about 75 years and are currently being mass produced at several places in Europe. As they are vigorous, they are under test at a great many places. Japanese × Korean larch hybrids, which are more resistant to voles than the Japanese larch, are being produced in quantity for commercial planting in northern Japan. Interestingly, in the Japanese × Korean larch combination, hybrid seed can be produced on Japanese larch even when the female flowers are pollinated with Japanese larch pollen and with Korean larch pollen applied a few days later.

Another combination, European × Siberian larch, has proved of value in Finland and is produced there on a moderate scale (Fig. 16.1).

Pinus

SUBGENUS *Strobus*

This subgenus, the soft pines, is divided into 2 sections and 5 subsections. Subsection *Strobi*, which contains 14 species, is by far most important and largest. Most of the crossing work has been concentrated in that subsection (Fig. 16.2). Among the 14 species, there is a group of 6 (eastern white, western white, Mexican white, Japanese white, Himalayan white, and Macedonian white) that can be crossed with each other in almost every possible combination, usually with seed sets 10 to 50% as great as can be obtained from crosses made within species. Another member of the subsection, southwestern white pine, can be crossed readily with Himalayan white pine but not with many other species.

Most members of subsection *Strobi* are allopatric, that is they occupy different geographic ranges. Two members—sugar and western white pines—are sympatric (occupy the same range) over a considerable part of Oregon and California. Apparently as a result of occupying the same range, genetic barriers to crossability have been formed, not only between sugar and western white pines but between sugar and nearly all other white pines (Duffield, 1952; Wright, 1959).

Attempts have been made to cross most of the 6 above-mentioned species with four other members of subsection *Strobi* (limber, southwestern white, Armand's, and sugar pines and with three members of subsection *Cembrae* (Korean, Swiss stone, and whitebark pines). In most such attempts, cone set and set of empty seeds have been high, indicating

FIG. 16.1. Three-year-old grafted trees of hybrids between European and Siberian larch (clone E 302) planted in southern Finland. In older plantations, 12-year-old trees of this clone grew 18 cm in diameter, very fast for northern conditions. (Photograph courtesy of Lauri Kärki of the Finnish Forest Tree Breeding Foundation.)

that fertilized embryos were formed but aborted after a few cell generations (Kriebel, 1972). The purely tropical species have not been included in crossing studies.

SUBGENUS *Pinus*, SUBSECTION *Sylvestres*

Subgenus *Pinus* contains the hard pines, mostly with 2 to 3 needles per cluster. The subgenus is divided into nine subsections. A relatively slight amount of work has been devoted to crosses between species in different subsections, nearly all of which have failed.

FIG. 16.2. Ten-year-old eastern white pine (left) and eastern white × Himalyan white pine hybrids (right) growing in Ohio. Several white pine species can be crossed with each other and produce vigorous F_1 offspring. (Photograph courtesy of H. B. Kriebel of the Ohio Agricultural Research and Development Center.)

Much more work has been devoted to crosses between species in the same subsection, particularly to five subsections important in North American forestry. The largest and most important in cool temperate regions is subsection *Sylvestres*. Crossing work in this group is described in papers by Duffield (1952), Duffield and Snyder (1958), Wright and Gabriel (1958), and Critchfield (1963a). The crossability pattern can be summarized as follows.

Most crosses can be made between the seven Asiatic species (Japanese red, Japanese black, Yunnan, Taiwan red, Masson's pines, *P. tabulaeformis* and *P. insularis*), and seed sets are usually high (Fig. 16.3).

Three of those Asiatic species can be crossed with Austrian pine from Europe and one can be crossed with Scotch pine from Eurasia, but seed sets are low except in the case of Austrian × Japanese red pine.

Three north and central European species (Scotch, mugo, and Austrian pines) can be crossed with each other, but seed sets are very low (Fig. 16.4).

Two south European species (*P. brutia* and *P. halepensis*) cross readily with each other, with high seed sets; a third south European species (maritime pine) has not been included in crossability studies.

One American species (red pine) has been crossed only with Austrian

FIG. 16.3. Rapid juvenile growth of F₁ hybrids between Japanese red and Japanese black pines growing in New Hampshire. Many of the Asiatic species of the subsection *Sylvestres* can be crossed with each other, with relatively high seed sets. Hybrid vigor is more evident in the "hybrid habitats" of northeastern United States than in Japan. (Photograph courtesy of Peter W. Garrett of the United States Forest Service, Durham, New Hampshire.)

FIG. 16.4. F₁ hybrid, 7 years old, between Austrian and Scotch pines growing near Zagreb, Yugoslavia. This hybrid is more vigorous than Austrian pine, but less vigorous than the male parent. This particular combination is very difficult to produce and it is probable that backcross or later generation hybrids will be much more useful than F₁ generations. (Photograph courtesy of Mirko Vidakovic of the University of Zagreb.)

pine (four hybrid trees) and with *P. tropicalis* (one hybrid tree) in spite of repeated attempts.

This crossability pattern illustrates a principle listed earlier, namely, that species with different but neighboring ranges tend to cross more readily than species with distant ranges. It also illustrates another principle, namely, that morphologically similar species cross most easily.

Commercially, Scotch and red pines are the most important species of this subsection. Unfortunately from the practical standpoint, hybrids between them and other species are the most difficult to produce, so that species hybridization has a limited role in their improvement.

Austrian (♀) × Japanese red (♂) pines are among the easiest of all tree hybrids to produce, with approximately 22 filled seeds per cone after open pollination in a mixed plantation (Wright *et al.*, 1969b). The hybrids also grow rapidly in some habitats. The probable explanation for the high seed set rests on the fact that pines are polyembryonic, with 4 to 5 egg cells in the female gametophyte which will develop into a seed after fertilization. Even if all the eggs are fertilized, only one normally develops into a mature embryo. Apparently in this particular combination, if one fertilized egg cell is hybrid, it can express hybrid vigor during the embryonic stage and suppress the nonhybrid embryos in the same developing seed. Thus, even if only 25 to 35% of the pollen landing on an Austrian pine tree may be from Japanese red pine, 85% of the seed produced may be hybrid.

SUBGENUS *Pinus*, SUBSECTION *Australes*

The *Australes* (from austral = southern) includes the commercially important timber trees of southeastern United States and the Caribbean area. Crossability data are nearly limited to the United States species, most of which can be crossed with each other (Table 16.1). In spite of repeated attempts, nearly all crosses with species in other subsections have failed; seed sets were very low in the one success. Seed sets are very high for crosses between loblolly and shortleaf pines (*P. taeda* and *P. rigida*); these hybrids have been mass produced for commercial planting in the Republic of Korea.

Although these species do hybridize with each other occasionally in their native habitats, the hybrids sometimes grow more slowly than their parents and thus are at a selective disadvantage. Limited introgression may have occurred between shortleaf and loblolly pines west of the Mississippi River (Hare and Switzer, 1969), but for the most part the species have maintained their individuality in spite of the occasional crossing. That has not been true, however, for the two subspecies of

TABLE 16.1

Summary of Species Crosses in Subsections *Australes* and *Ponderosae* of the Hard Pines[a,b]

Subsection and species	Subsection and species									
	Australes eastern America					*Ponderosae* western America				
Australes (Eastern America)	elli-ottii	taeda	palus-tris	echi-nata	gla-bra					
elliottii										
taeda	H									
palustris	H	H								
echinata	H	H	F							
glabra	—	—	—	H						
rigida	—	H	—	H	—					
clausa[c]	H	—	—	—	—					
cariabea	H	—	—	—	—					
Ponderosae (Western America)						pon-derosa	arizo-nica	monte-zumae	engel-mannii	jeff-reyi
ponderosa	F	F	F	F	—					
arizonica	—	—	—	—	—	H				
montezumae	—	F	—	—	—	H	H			
engelmannii	—	—	—	—	—	H	—			
jeffreyi	F	F	—	F	F	H	—	H	—	
coulteri[d]	—	—	—	—	—	F	F	—	F	H

[a] After Duffield (1952), Duffield and Righter (1953), Hyun (1956), Righter (1955), Critchfield (1963b), and Saylor and Koenig (1967).

[b] H, hybrids; F, failure.

[c] *P. clausa* is classified in subsection *Contortae*.

[d] *P. coulteri* is classified in subsection *Sabinianae*.

pitch pine (*P. rigida* subsp. *rigida* and subsp. *serotina*). Whenever they grow together, they hybridize and lose their separate identities (Smouse and Saylor, 1973).

Schmitt (1968) has summarized the possibilities of southern pine hybrids tested in southern Mississippi. In general the hybrids have been intermediate in growth rate. Parental selection seems to be important, as the offspring of some particular tree × tree combinations grow better than the offspring of other combinations. Transfer of useful genes from one species to another over a period of several generations presents more promise than the production of heterotic F_1 generations.

SUBGENUS *Pinus*, SUBSECTIONS *Ponderosae, Contortae,*
AND *Oocarpae*

Subsection *Ponderosae* includes pondersa pine and 12 of its relatives
from western United States, Mexico, and Central America. Most crossing
work has been done in California using species native to the United
States. Four United States species are taxonomically similar and have
neighboring but nonoverlapping ranges. They can be crossed readily
with each other (Table 16.1 and Fig. 16.3). Two species (Jeffrey and
ponderosa pine) are very similar to each other (having at one time
been considered to be varieties of the same species) but are sympatric.
Probably because of this, genetic barriers to crossing have arisen; they
can be crossed but seed sets are very low.

Jeffrey and Coulter pines are exceptions to the general rule that
taxonomically similar species cross most easily. They are classified in
different subsections and are easily identifiable yet occasionally cross
with each other in nature as well as artificially. The F_1 hybrids are
fully fertile.

Subsection *Contortae* contains jack pine from eastern United States
and Canada and lodgepole pine from western United States and Canada.
They are morphologically and ecologically similar and are neighboring
species in the sense that their ranges overlap in the Canadian Rockies.
There are extensive hybrid swarms in the zone of overlap and many
artificial hybrids have been reported. The subsection contains two more
species—Virginia and sand pines—from southeastern United States.
Limited crossing work indicates that they cross slightly with each other
but not with jack and lodgepole pines.

Subsection *Oocarpae* includes seven species from western United
States and Mexico (*P. radiata, P. attenuata, P. muricata, P. patula, P.
greggii, P. oocarpa,* and *P. pringlei*), among which are some of the
most commonly planted trees in subtropical southern hemisphere countries. *P. patula* has been crossed with *P. greggii* and *P. oocarpa,* both
of the same subsection; and with *P. teocote* and *P. lawsonii* of subsection
Ponderosae. Many other crosses between subsections have failed.

The most intensive hybridization work in subsection *Oocarpae* has
been carried on at the Institute of Forest Genetics in Placerville, California. Results of that work are summarized in Table 16.2 (Critchfield,
1967). As shown in the table, Critchfield detected probable differences
(significant at 10% level) in crossability among varieties of Monterey
pine (*P. radiata*) and Bishop pine (*P. muricata*) as great as those among
some species.

According to Stebbins (1950) Monterey pine hybridizes slightly

TABLE 16.2

Crossability Pattern in the Closed-Cone Pine Complex of California, Based upon Placerville Experience since 1928[a]

	Male parent[b]				
		radiata		*muricata*	
Female parent	*attenuata*	var. *radiata*	var. *binata*	Channel Islands	Baja California
attenuata (California)	HYB	HYB	HYB	HYB	—
muricata (California, Mexico)					
Mendocino County	FAIL	FAIL	FAIL	PROB	FAIL
Monterey County	PROB	FAIL	FAIL	PROB	PROB
Channel Islands	HYB	HYB	PROB	HYB	HYB
Baja California	HYB	FAIL	PROB	—	HYB
patula (Mexico)	—	PROB	—	—	—

[a] From Critchfield, 1967.
[b] HYB, hybrids certain; PROB, probable hybrids with low seed sets; FAIL, failure.

with Bishop pine and knobcone pine (*P. attenuata*) at two places in California. This may account for some of the great variability encountered by southern hemisphere foresters who have planted Monterey pine on a large scale (Bannister, 1958).

Picea

The spruce genus contains about 40 species native to cool parts of the northern hemisphere. The species are much less distinct than in the pines and there is no subdivision into subgenera or sections. In general the greatest morphological similarities are among species occupying neighboring ranges and the greatest differences are among species occupying exactly the same range or widely separated ranges.

Crossing work in this genus is summarized in Table 16.3. Altogether 76 crosses have been attempted. The results showed a definite relationship between geographic distribution, morphological similarity, and ease with which a cross can be made. The ratio of success–failure ratio was high (19:3) for crosses between species that are morphologically similar or have neighboring ranges. The success–failure ratio was very low (8:42) for species that are morphologically distinct and have widely separated ranges. Some hybrids are shown in Fig. 16.5.

TABLE 16.3

Relationship between Morphology, Geographic Distribution, and Species Crossability in the Spruces, *Picea*[a]

	Number of species crosses	
Morphological similarity and type of natural range	Successful	Unsuccessful
Similar, with neighboring ranges	10	1
Similar, ranges widely separated but connected by an intermediate species	5	2
Distinct, ranges sympatric	2	2
Distinct, with neighboring ranges	5	0
Distinct, with widely separated ranges	8	41

[a] After Wright (1955) and Fowler (1966), with personal communications from J. W. Hanover and Gilbert Fechner.

S 31   S 33   S 34   S 35

FIG. 16.5. Hybrid and nonhybrid spruce growing near Zagreb, Yugoslavia. From left to right the trees are (S 31) *Picea omorika* × *P. sitchensis*, (S 33) *P. omorika* × *P. abies*, (S 34) and (S 35) *P. omorika* × *P. omorika*. (Photograph courtesy of Mirko Vidakovic of the University of Zagreb.)

The forests of Mount Yatsuga on the Japanese island of Honshu contain five spruce species, all morphologically distinct from each other and easily identifiable. They probably do not cross with each other, but this has not been tested experimentally. There are four other regions of the world with sympatric pairs of dissimilar species. In two cases rare natural hybrids have been reported; in the other two cases hybrids have not been reported.

Crossability data in spruce help explain some of the ecological and silvicultural problems encountered by some species (Wright, 1955). To explain this, it is desirable to categorize spruce species as northern (growing in Canada, Siberia, northern China, Scandinavia, etc.), intermediate, and southern (growing in the southern Appalachians, Mexico, Taiwan, southern Asia, southern Europe, etc.). When so defined, northern species have the following general characteristics as compared with southern species.

Have much larger geographic ranges

Are usually in contact with another species with which they can cross

Reproduce naturally in the open rather than under an overstory

Are biologically more successful, not being in danger of extinction as are some of the southern species with very restricted ranges

Attain less maximum height than their southern counterparts on the same continent

Show no obvious morphological adaptations to northern growth conditions.

Geologically speaking, the spruces are an ancient group and have presumably evolved relatively slowly in the past several million years. Southern species, initially adapted to reproduce under an overstory, found their habitats constantly changing because they were growing in company with more rapidly evolving angiosperms. To maintain their competitive status they should have evolved rapidly, but they could not because of inherent conservatism, limited population size, and lack of opportunity to exchange genes with neighbors. Northern species, reproducing in the open, do not encounter greatly changed seedbed conditions from those encountered far in the past. They did not need to change as much as southern species to maintain their competitive status, but could because they have large ranges and large store of intraspecies variability as well as the opportunity to exchange genes with neighboring species. Hence the northern species are now biologically most successful and easiest to manage silviculturally.

Absolute proof that hybridization played a major role in the modern success of such northern species as white, black, Sitka, Siberian, and Norway spruces is of course almost impossible to obtain. However, there is abundant evidence that introgression is occurring now (between white and Engelmann spruces in western Canada and between Siberian and Norway spruces in northern Scandinavia). In each case the hybrids are more successful on certain sites than the pure parental species, so there is an opportunity for the parental species to change if site conditions should become different.

Acer

The maple genus contains about 115 species widely distributed in temperate parts of the northern hemisphere. Most species are shrubs, but many are large trees and among the best timber species in their respective regions.

Maples are insect pollinated and as a result have become well differentiated as regards floral characters as well as foliage and growth traits. In comparison with the oaks and spruces, the species are very distinct and easily identified. Moreover there are very great differences in the degree of similarity among species, so that they may be easily grouped by sections. Rehder (1940) recognized 14 such sections. Most of the sections are confined either to America or to Eurasia, but a few contain species in both continents.

Of the 16 natural and artificial hybrid combinations which have been reported (Rehder, 1940; Wright, 1953), 11 involve species belonging to the same section. Most of those were easily made, with seed sets 25 to 50% as great as for crosses made within the same species. Each of the five intersectional crosses was reported as a single natural hybrid that occurred in an arboretum where different species were grown together. Only 1 of 24 attempted artificial intersectional crosses was even partially successful; the result in this case was a few incompletely developed seeds which did not germinate. Thus the crossability pattern in this genus illustrates principles mentioned earlier, namely, that morphologically similar species with neighboring ranges tend to cross most easily.

Two American species belonging to section *Saccharina* (black and sugar maples) hybridize frequently in nature, and extensive hybrid swarms occur in northern United States and southern Canada. In fact the hybridization is so frequent as to cast doubt on the validity of the species. Otherwise, however, natural hybrid swarms are rare in spite of the ease with which some species can be crossed artificially. This is due in large part to the fact that closely related species often have very different flowering times. It was necessary to store pollen for 2 to 3 weeks in order to make most of the artificial hybrids. Such differences in flowering time tend to eliminate any improvement program based on the mass production of F_1 hybrids in natural crossing gardens.

Aesculus

This genus, the buckeyes and horse chestnuts, includes about 25 species native to the warm temperate parts of Eurasia and North America. The species are grouped into five series. Some are shrubby, others form large trees. Most have large, showy flowers and are useful as ornamentals, but they are rarely managed for timber. Genetic work has been confined to the study of natural hybrids (Hardin, 1956; Li, 1956).

One natural hybrid ($A. \times carnea = A.$ hippocastanum $\times A.$ pavia) is between species belonging to different series. This particular hybrid is a true-breeding allotetraploid, having double the usual chromosome number. It presumably arose as the result of pollination of an unreduced egg cell by an unreduced pollen grain, a very rare occurrence in trees.

The other reported hybrids involve four American species having neighboring and slightly overlapping ranges and belonging to the same series. The four species could be combined in six different ways, and five of these combinations have been reported. In only one combination ($A.$ pavia $\times A.$ sylvatica) has there been extensive introgression. In the other cases hybrids were rare and reported only from arboreta, or hybrid swarms were small and confined to the immediate zone of overlap or to disturbed habitats. Apparently more extensive introgression has been prevented by (1) sterility of the hybrids and (2) selection against the hybrids on anything but an intermediate habitat.

Castanea

The chestnut genus contains 13 tree and shrub species native to United States, Europe, and Asia. American chestnut, a very desirable timber tree, and European chestnut, a very desirable nut and timber tree, are susceptible to the chestnut blight, a disease introduced from Asia.

American chestnut is almost extinct, being kept alive mainly because of root suckers. The European chestnut is also in danger of extinction. In neither species is there evidence of enough genetic variation in disease resistance to warrant much hope for the development of a resistant variety by selecting races or individuals within the species. The Asiatic species, of less desirable timber form and growth rate, are resistant to the blight, however. Because of this, species hybridization has played a large role in chestnut improvement. Species hybridization work started in the 1880's with crosses made by Luther Burbank. Sporadic crossing work was done from then until 1924 when the large-scale crossing program of the United States Department of Agriculture at Beltsville, Maryland was started. Another large program was started in 1929 at the Connecticut Agricultural Experiment Station. Altogether in these two programs about 38,000 hybrid nuts were produced. Similar work has been conducted in Spain, Portugal, and Italy. Among the numerous publications dealing with progress in chestnut hybridization are those by Brevigliere (1951), Jaynes and Graves (1963), and Jaynes (1964).

The 13 species of chestnut are divided into three sections. The species, even those belonging to different sections, are less distinct from each other and less easily identified by a layman than are most species of pine or maple. Of the 49 attempted interspecific combinations (including reciprocals), only one was a definite failure, although the hybridity of several other combinations has not yet been proved. Hybridization between species belonging to different sections has been as successful as hybridization between species belonging to the same section.

In general it appears that almost any two species of chestnut could be crossed with each other if the cross was repeated enough times. However, there are few cases in which crosses can be made as easily between species as within species, and many of the successful hybrid combinations produced weak seedlings or trees with reduced fertility.

At least in the United States, greatest interest centers around American, Chinese, and Japanese chestnuts. Unfortunately, the problems of sterility and weak growth have been especially marked in crosses among these species. There appears little possibility of mass producing F_1 or F_2 trees, but the prognosis for producing a late generation segregate variety is favorable.

Controlled pollination of chestnut is time-consuming, and seed sets per man hour of crossing work are low, even under the best circumstances. As a consequence, any long-term hybridization project can utilize mass and half-sib family selection but not full-sib family selection.

Eucalyptus

Eucalyptus is the principal genus of trees in Australia and Tasmania. The number of species is variously estimated at between 250 and 500, of which approximately 100 are believed to be of hybrid origin (Blakely, 1955; Pryor, 1957). The genus illustrates three basic principles governing the occurrence and distribution of hybrids in general: (1) The majority of hybrids involve closely related species. (2) Most hybrids are between species occurring in neighboring areas rather than between species occupying the same ranges. (3) Hybrids are most common on intermediate sites or on disturbed habitats.

Most of the species are grouped in one of four sections. The 140 species of the section *Renantherae* cross freely among themselves when given the opportunity, but have not yet been crossed with members of other sections. The same is true of the members of the section *Platyantherae* and of subdivisions of other sections. The natural eucalypt stands in Australia are commonly dominated by two or three species of different sections, and these do not cross. However, each may cross with a related species from another territory. Many hybrids originated before settlement, but the great spread of hybrid swarms has taken place on man-disturbed lands. Pryor described one such example in the vicinity of Lees Springs, Australian Capital Territory. The forest was once a pure stand of *E. dalrympleana*, but extensive fires occurred, and the forest on the disturbed sites is now composed mostly of hybrids between *E. pauciflora* and *E. dives*.

Eucalypts are planted more commonly outside than inside their natural ranges. They are used extensively in other countries with Mediterranean-type climates. It is very likely that many sites in these other countries can be considered to be hybrid habitats, which are better adapted to hybrids than to pure species. In spite of the ease with which many species can be crossed, the F_1 generations will probably not be planted on a large scale because of the difficulties encountered in controlled pollination. Presumably the emphasis should be on the production of segregate varieties from F_2, F_3, and later generations.

Populus

This genus, the cottonwoods, balsam poplars, and aspens, contains about 30 species, widely distributed in the northern hemisphere from southern United States and southern Europe nearly to the limit of tree growth in the Arctic. Eastern cottonwood was introduced into Europe in the late 1700's and hybridized there naturally with the native European black poplar, producing the hybrid which was named *P.* × *canadensis* in 1789. Clones of it are widely planted in Europe. In 1912, the Englishman A. Henry undertook the first controlled pollination experiments, crossing the American species eastern cottonwood with northern black poplar (Larsen, 1956). In 1924 an extensive hybridization project was started by Stout and Schreiner (1933) at the Oxford Paper Company of Maine. Since then there have been several poplar hybridization projects in Europe, America, and Korea. Publications on them are numerous. Among the general papers are those of Piccarolo (1952) and Schreiner (1959) and annual reports of the Korean Institute of Forest Genetics.

The genus has two principal subdivisions. Species in different subdivisions can be crossed, but for practical reasons most breeding work is confined within a subdivision. One subdivision is called section *Leuce,* which includes the aspens and white poplars. They are northern species that grow on upland soils, which is frequently of poor quality. They have slow to moderate growth rates and spread by root suckers; often one clone can occupy nearly an acre. All seven species in the section are allopatric, except trembling and largetooth aspens of northeastern America. Those two can cross but usually do not in nature because of differences in flowering time. Of the 21 possible combinations involving the 7 species, 12 have been attempted and produced good seed; all species can probably be crossed (Fig. 16.6).

The other subdivision includes the cottonwoods and balsam poplars belonging to the sections *Aegeiros* and *Tacamahaca.* These are fast growing bottomland species that have been planted extensively and therefore have received the most attention from tree breeders. All are allopatric and maintain their identities in nature because of geographic isolation. Of approximately 100 species combinations that have been attempted, nearly all have succeeded, frequently producing almost as many seeds as could be obtained from crosses made within a species. It is likely that all can be crossed with each other. In fact, there has been so much artificial hybridization between species and the hybrids have been distributed so widely that it is difficult to locate a tree typical of a pure species in the arboretum and park collections of the United States and Europe. Most poplar planting in southern and central Europe, northeastern United States, and Argentina involves hybrids (Fig. 16.7). Many native stands of southern Europe are gradually being replaced with hybrids between European black poplar and American eastern cottonwood.

Quercus

Oak is a very large genus of warm temperate and subtropical hardwoods. The oaks are important timber and ornamental trees. Artificial hybridization work started in 1909 in Texas with the production of a few hybrids between live and overcup oaks. A second American hybridization project started in 1937 at the Northeastern Forest Experiment Station, but the work was dropped in 1948 because of low seed sets: one sizeable hybrid progeny (white × English oak) was obtained,

FIG. 16.6. Hybrids between European and trembling aspen (*Populus tremula* × *tremuloides*) growing in an industrial forest in Finland. This hybrid aspen stand is 17 years old and has produced over 200 m³ of wood per hectare. Hybrid aspen seed can be mass produced indoors. (Photograph courtesy of the Finnish Forest Tree Breeding Foundation.)

but the seeds were eaten by mice before germination. The largest oak hybridization project is at the Ukrainian Scientific Research Institute of Forestry and Agreforest Reclamation and the Kharkov Agricultural Institute in the Ukrainian S.S.R. (Piatnitsky, 1960). There, over a 30-year period, some 15,000 control-pollinated acorns (hybrid and nonhybrid) have been produced as the result of control-pollinating some 200,000 female flowers. The hybrids produced belong to 27 different species by species combinations, but the authenticity of some is in doubt because the "hybrids" resembled the female parent only.

Most of our knowledge about oak hybridization comes from the study of natural hybrids occurring in nature or as the result of chance crossing between trees growing in arboreta. Much of the voluminous

FIG. 16.7. Plantation of hybrid poplars planted by E. J. Schreiner at Williamstown, Massachusetts. Numerous species combinations between balsam poplars and cottonwoods are represented in the plantation. Each clone was planted in a square 16-tree plot and differences among the clones are clearly evident. Several of Dr. Schreiner's selected clones have been planted in various parts of the world. (Photograph courtesy of Peter W. Garrett of the United States Forest Service, Durham, New Hampshire.)

literature is summarized in Palmer's (1948) catalogue of United States hybrids, Rehder's (1940) manual, Muller's (1952) paper on hybrids in the American Southwest, and Tucker's (1952) paper on hybrids in California. An additional source of information about the frequency of natural hybrids is the account of the Michaux Quercetum study by Schramm and Schreiner (1954).

Most temperate zone oaks of Europe and United States belong to the subgenus *Erythrobalanus* (red or black oaks, confined to America) or to the subgenus *Lepidobalanus* (white oaks, Europe and America). These two subgenera differ in many respects. Only one successful cross between them has been reported, and its authenticity is doubtful. For all practical purposes, hybridization seems limited to species belonging to the same subgenus.

Two instances of extensive introgression have been reported. In southwestern United States, Muller found hybrid swarms of *Quercus gambellii* × *stellata* (= *Q.* × *margaretta*) which had been in existence for several generations and resulted in movement of genes from one species to the other away from the zone of contact. Johnsson (1952) found a high frequency of intermediates between *Q. petraea* and *Q. robur* growing in mixed stands in Sweden.

Otherwise, however, hybridization among the oaks seems to be a relatively uncommon phenomenon. Palmer listed a great many species by species combinations that had been authenticated, but noted that in most cases the hybrids were few in number and confined to one or two localities. The few hundred hybrid trees of which he knew were a very small percentage of the millions of trees growing in Europe and America. Even where two or three closely related species grow sympatrically, as happens frequently in the United States, hybrid swarms are small and endure for only a generation or two.

Such conclusions may seem in direct contrast to the experience of those who have attempted to identify oaks in the wild. Most species are very plastic. The flowers possess few diagnostic characters, so one must usually rely mainly upon leaf characters. Leaf shape changes with age, position in crown, growing conditions, etc., and in such a manner that the leaves of one species may seem similar to those of another species. Thus for positive identification it is desirable to know not only the characteristics of each species but the tremendous range of variation in those characteristics. This difficulty is not serious, however, if seedlings are grown in a nursery under moderately uniform conditions. This was done as part of the Michaux Quercetum study, which included growing open-pollinated offspring of several hundred trees of several different species. Nearly all the half-sib families were uniform and easily recognizable as belonging to one pure species; only a few hybrids were evident.

The concensus of controlled-pollination experiments is that artificial hybridization is too difficult to ever be considered as a mass production possibility. The concensus of natural hybridization studies is that many hybrid combinations are possible but that the actual number of hybrid trees is small; usually large numbers of hybrids exist only on hybrid habitats.

Study Questions

1. Name three agricultural crop plants that are believed to have arisen as the result of hybridization between very different species.

2. Contrast the roles of hybridization in the improvement of corn and wheat.

3. Why could a hybrid with relatively low seed-producing ability be of greater value in a forest tree than in many agricultural crop plants?

4. List the three goals of species and racial hybridization in trees and give one example showing how each goal has been achieved in specific cases.

5. Contrast the ease with which hybrids can be obtained between different races of the same species; between different species in the same section or subsection; between species in different sections or subgenera; between species in different genera.

6. Explain the relationship between degree of morphological similarity, amount of genetic differentiation, length of time since differentiation, and ease of crossability between two species.

7. Why do sympatric species cross less easily than allopatric species with neighboring ranges? Why do allopatric species with neighboring ranges cross more easily than those with distant ranges? Give examples illustrating these principles from three different genera.

8. In what genus do commercially important sympatric species cross with each other? How have these species maintained their distinctness?

9. Contrast the crossability patterns between subgenera *Strobus* and *Pinus* of the genus *Pinus*.

10. Describe the general features of the crossability pattern in the spruces. What is the evolutionary importance of hybridization in this genus?

11. Has genetic or geographic isolation been most important in the poplars? Why, with only one exception, is there only one species of aspen or of cottonwood in any given region?

12. Most oak forests contain several species of oaks. Judging from that fact alone, what do you infer as to the relative ease with which species of oak, white pine, and cottonwood can be crossed?

13. What are the possibilities of developing F_1 hybrid varieties of maple which can be mass produced by seed? What is the limiting factor?

14. In what tree genus are a great number of modern species believed to have hybrid origins?

15. Contrast the crossability patterns in the chestnuts and the oaks. In the pines and poplars. In larch and eucalyptus.

17

Species and Racial Hybridization—Practical Application

Historical Background

The mule, a hybrid between a male donkey and a female horse (the reciprocal hybrid is called a hinny), is a well-known example of an F_1 hybrid possessing heterosis. Mules are known for endurance, being capable of hard labor day after day, which would tire horses. The fact that mules are sterile and cannot reproduce themselves is relatively unimportant because enough can be produced by mating donkeys with horses.

Other heterotic F_1 hybrids are known, and many vegetatively propagated ones have been used in horticulture. It remained for corn breeders to make "hybrid" a household word. Corn is a special type of plant, with female inflorescences ("ears") along the stalk and male inflorescences ("tassels") at the top. The structure enables a corn grower to plant one row of variety A and several rows of variety B, to detassel the rows of variety B, and thus to obtain large quantities of hybrid $A \times B$ seed from the rows of variety B. The theoretical foundations for hybrid corn were laid in the 1880's, but the practical development of commercial hybrid varieties did not take place until the second two decades of the present century. American farmers started planting hybrid corn about 1930, and a decade later there were more acres planted to hybrid than to open-pollinated varieties. As of 1974, hybrid corn is very big business in the United States, requiring tens of thousands of acres just for production of hybrid seed.

The impact on American agriculture has been very great. The introduction of hybrid varieties resulted in 20 to 25% increases in per acre yields. Those increases stimulated other types of research on fertilization, irrigation, spacing, etc., and with each yield increase due to research, farmers were stimulated to practice more intensive culture. The annual increases in productivity were especially dramatic from 1930 to 1950, but have also been considerable since then. For example, there was a 20-year (1949–1969) increase in United States production of corn from 74,000,000 to 114,000,000 metric tons.

With such an example, it was natural for breeders of other crops to concentrate on hybridization as an improvement method. Thus, onion breeders introduced male-sterility genes into many different lines so that they could interplant male-sterile and pollen-producing lines and harvest hybrid seed in large quantities from the male-steril plants. Tree breeders especially attempted to follow the hybrid corn example. From 1925–1940, when the advantages of hybrid corn were becoming more evident with each passing year, tree breeding was in its infancy. The few fulltime workers all knew each other personally, at least by correspondence. Most tree breeding projects of that period, and indeed for another 15 years, emphasized the possibilities of mass-produced F_1 hybrid varieties.

Thus, much data has been gathered about crossability patterns and about the growth potentials of F_1 hybrids. Some are promising, but most are not. It now appears that greater emphasis should have been placed on the F_2 and later generations, but data about them are scanty. Consequently those subjects are treated here as possibilities rather than actualities.

Mass Produced F_1 Hybrids

Requirements for Successful Large-Scale Use

F_1 hybrids must fulfill two conditions to be considered for large-scale use. First, they must possess hybrid vigor or some combination of traits that makes them more valuable than either parental species or race. Second, they must be capable of being mass produced. Combinations fulfilling both requirements are uncommon.

Edgar Anderson (1953) postulated that pure species are usually well adapted to the habitats on which they evolved and that hybrid vigor is usually most manifest on hybrid habitats. This seems to be true in many forest trees. The natural hybrid swarms reported in oak, eucalyptus, and spruce are more or less confined to intermediate or disturbed habitats. In such cases the possibilities of putting hybrids to practical use depend on the extent of such intermediate habitats. In the case of the southwestern oak hybrids studied by Muller, the intermediate habitats on which hybrids grow best are limited in area, and it would be uneconomical to develop hybrid varieties specifically for such habitats. In Australia, where man has created relatively large areas of "hybrid habitats" by fire and other disturbances, the use of hybrid eucalypts could be more promising.

Most reports of hybrid vigor for artificially produced hybrids involve cases in which one or both parental species have been planted outside their natural ranges. Hybrid poplars (mostly crosses between European black poplar and an American species) are planted most extensively in Europe and Argentina where the American species are not native. Japanese × European larch hybrids are planted most extensively in Europe, but at lower elevations than occupied by the native European larch stands. Other examples could be quoted, indicating that in general the use of F_1 hybrids is apt to be greatest in those countries which plant exotics extensively.

The second requirement, that F_1 hybrids must be capable of being mass produced, eliminates many otherwise promising combinations. There is no real problem in poplars and willows, in which numerous combinations can be mass produced at small cost by controlled pollination or can be propagated vegetatively. At the other extreme are the oaks and maples, in which the development of a suitable mass production technique seems unlikely. The pines, spruces, and larches are intermediate, and in them some of the species×species combinations may be mass produced at reasonable cost by one of the techniques described below.

Mass Production by Interplanting Seedlings

As mentioned in the previous chapter, Austrian (♀) × Japanese red (♂) pine and Japanese (♀) × Korean (♂) larch are two species combinations that can be made so easily that protection of the female flowers against unwanted pollination is unnecessary. Red (♀) × Siberian (♂) elm is a third easily made combination. In each case it appears necessary only to interplant the two species; a large proportion of the seed produced will be hybrid.

Possible Mass Production on Young Female Flowering Trees

Macedonian white pine, eastern white pine, and white spruce are among the species that often produce large quantities of female flowers for several years before they produce much pollen. If Macedonian white pines are planted among older eastern white pines (the two cross readily), only hybrid seeds are produced for a 10–15 year period. This particular combination is of limited interest because the hybrids grow slowly. Other combinations, such as eastern white (♀) × western white (♂) pine or white (♀) × Engelmann (♂) spruce, are of greater interest. Seed orchards for their production might be (but have not) obtained by planting seedlings of the species used as the female parent in older stands of the other species.

Mass Production by Controlled Pollination

Economical mass production of pitch × loblolly pine hybrids was obtained in the Republic of Korea. This was possible because (1) the pitch pine female parents flowered when small and so could be bagged and pollinated from the ground or short ladders, (2) seed sets were high, and (3) labor costs were low. Several hundred thousand hybrid seeds were produced in some years. With gradually increasing wage rates there has been a gradual deemphasis of the mass production of these F_1 hybrids.

A "1000-bag" project was undertaken in the late 1940's at the Placerville Institute of Forest Genetics. Workers there devoted nearly 1000 pollination bags to each of several species × species combinations of pine which had given reasonably high seed sets in preliminary crosses. In several combinations the seed yield was a few thousand seeds per man day. For the most favorable combinations, the use of control-pollinated seed resulted in an approximate doubling of per seedling costs.

Birches, willow, and poplars produce several female inflorescences per branch and each female inflorescence can produce 100+ seeds. Some birches flower when young and can be control-pollinated outdoors with little difficulty. All three genera can be control-pollinated indoors on grafted flowering branches. Hence mass production by controlled pollination is feasible for some species × species combinations.

Mass Production by Vegetative Propagation

Willows and some poplars root so easily that mass production by seed is not necessary. Where these genera are planted extensively in Europe and Argentina, hybrids are most commonly used and are reproduced by cuttings.

Mass Production by Interplanting One Clone with Seedlings

If two species cross reasonably well and flower at the same time, F_1 hybrids can be mass produced by interplanting a self-sterile clone of one with seedlings of the other, as suggested by Larsen (1956). All seeds produced by the self-sterile clone will be hybrid. This scheme is used to produce Japanese × European larch hybrids, and is probably the surest way for most pine, spruce, or eucalypt hybridizations.

As a variant of this method, one self-sterile clone can be planted by itself and be hand pollinated. Costs can be relatively low, as the expenses of bagging, labeling, and debagging can be avoided.

Examples of Successful F_1 Hybrids

A corn breeder maintains hundreds of selfed lines and crosses them in many different ways. The proportion of hybrid combinations that grow well enough to supplant existing varieties is low, and many thousands of hybrid combinations must be tested for each new hybrid variety released to the public.

Thus it is no wonder that, of the hundreds of hybrid tree combinations that have been produced, relatively few are planted on a large scale. Some with growth promise, such as Japanese red × Scotch pine are ruled out because of very low seed set. Peter Garrett of the Northeastern Forest Experiment Station recently supplied me with growth data on 20- to 25-year-old pine hybrids produced and tested at the Station. Of the numerous species × species combinations, one (Austrian × Japanese red pine) might be acceptable for large-scale planting. This particular hybrid is being tested extensively in Michigan (Wright et al., 1969b). The seed orchard for its production is a 1-acre, 30-year-old Austrian pine stand situated to the lee of Japanese red pines. In years of abundant seed crops, 20,000 or more hybrid seeds can be obtained by collecting squirrel-dropped cones (probably more by climbing). Through age 10 the F_1 hybrids have outgrown Austrian pine by 30 to 40% (the tests did not include Japanese red pine) and have grown faster than any other hard pine tested in the state. There is a limitation to their use, however. Neither parental species is reliably hardy where the average January temperature is below 18°F (−8°C), but the greatest amount of forest land in the state is situated in areas where the winters are that cold. Hence, for the present in Michigan, this combination may find its greatest use as an ornamental or for planting along roadsides where damage from salt spray affects many other conifers. Vidakovic

(1974) has found this combination useful in parts of Yugoslavia close to the native range of Austrian pine.

Pitch × loblolly pine hybrid show exceptional promise in the Republic of Korea but not in their native land. Pitch pine is a rather crooked tree of moderate growth rate which thrives on poor rocky soils and is hardy where the average January temperature is as low as 22°F (−6°C); it is nearly unique among the pines in being able to sprout from the base after cutting. Loblolly pine is a much faster growing, straighter, and more valuable tree, but is not hardy in most parts of Korea. F_1 hybrids, while not superior to the parental species in any one trait, possess a desirable combination of characters and can be mass produced at relatively low cost. They grow faster and straighter than pitch pine but retain that species' ability to grow on poor sites and to sprout (important where illegal cutting of young trees for charcoal is a common practice). Hence they are planted on a large scale, but may be supplanted by F_2 hybrids in the future. Reports of their production and use are found in the annual Research Reports of the Korean Institute of Forest Genetics at Suwon.

Most pitch × loblolly pine hybrids have been produced on pitch pine females believed to be of Pennsylvania origin, using whatever loblolly pine pollen was available. Recent experiments show that further improvement can be expected by choosing the loblolly pine race more carefully (Hyun and Hong, 1969). At a number of test sites in Korea, hybrids produced with loblolly pine pollen collected near the species' northern limits in New Jersey were hardiest and grew fastest.

Where poplars and willows are planted extensively, as in Europe and Argentina, hybrids are used more than pure species. Most commercially grown poplars are hybrids, usually involving European black poplar, eastern cottonwood, or northern black cottonwood as one parent (Schreiner, 1959). Several tree species of willow have been crossed to produce the hybrids planted in Argentina (Barrett and Rial Alberti, 1972).

Since 1923 Schreiner (1972) and his co-workers have produced and tested many different poplar hybrids in northeastern United States. Several combinations had desirable growth or form characteristics, but have not been planted extensively in the region because of the scarcity of suitable fertile land. Of the many clones selected in the northeast and distributed elsewhere, some proved excellent in other regions, but many did not. This indicates that a poplar hybridization project should be conducted in a region of intended large-scale use.

Hybrids between European and Japanese larch were first noticed in 1900 when E. J. Elwes and A. Henry found exceptional fast-growing seedlings in the open-pollinated offspring of Japanese larch trees situated close to European larch on an estate near Dunkeld, Scotland (Larsen, 1937). In 1926 Danish workers obtained seed from the Dunkeld Japanese larches and found that 60% of the resulting seedlings were hybrids. The hybrids outgrew pure Japanese larch by 18% at age 10. This experience prompted the establishment of seed orchards and interest in many parts of Europe and America. There are at least 30 published reports of hybrid vigor at different localities in northern Europe. Nearly all the reports are from plantations at lower elevations than the native European larch stands.

Hybrids between Japanese and Korean larch appear to have promise in Hokkaido, the northernmost Japanese island. Neither parental species is native to the island, but Japanese larch has been planted extensively. F_1 hybrids grow at about the same rate as that species but are more cold hardy and more resistant to voles. Initial steps for their mass production were made in 1969 when thirteen 1 hectare openings were made in a pure stand of Japanese larch, each opening to be planted with a single clone of Korean larch. It was not known at the time whether the 13 Korean larch clones were sufficiently isolated to prevent cross pollination within species. If this happens, it might be necessary to eliminate all but one clone.

Elm breeding work started in the Netherlands in 1926, to find trees resistant to the Dutch elm disease. This improvement work has included hybridization among the native European species, as well as between them and exotic species. Many of the initial selections were themselves hybrids. The emphasis has been on the production of single resistant clones with desirable growth and form characteristics. Several such resistant clones have been developed and released. They are usually propagated by grafting. During the course of the work there was worry that a mutation could occur in the disease which would make the disease virulent enough to attack clones hitherto regarded as resistant. This happened, apparently in Great Britain, and an accidental introduction of the British strain of the disease organism into the Netherlands threatens some hitherto promising material. Hence it may be necessary to renew the hybridization work, using species with higher levels of resistance.

Utilization of F_2 Generation Hybrids

Let us return briefly to the concept of hybrids in terms of genes at a single locus. Assume that AA and aa trees are crossed to produce

Aa trees, which are in turn crossed to produce an F_2 generation of 1 AA:2 Aa:1 aa trees. If hybrid vigor is due to overdominance (heterozygosity as such), it is present in Aa but not in AA or aa trees. However, several loci are probably involved, of which all are heterozygous in F_1 hybrids, half are heterozygous in F_2 hybrids, and one-fourth are heterozygous in F_3 hybrids. In this case, hybrid vigor manifested in the F_1 generation should be reduced by 50% in the F_2 generation and by 75% in the F_3 generation.

There are other explanations of hybrid vigor, however. It may be due to the prevalence of dominant genes. If so, they are present at all variable loci in F_1 hybrids, at $\frac{3}{4}$ of the loci in F_2 hybrids, and $\frac{9}{16}$ of the loci in F_3 hybrids. Consequently, hybrid vigor manifested in the F_1 generation should be reduced by 25% in the F_2 and by $\frac{7}{16}$ in the F_3 generation.

With the third and fourth possible explanations (genes with additive effects and the "hybrid habitat" hypothesis, see Chapter 3), it is theoretically possible to maintain hybrid vigor almost undiminished in the F_2 and F_3 generations.

Many F_1 tree hybrids are fertile enough to produce large quantities of seed if planted together. Thus, F_2 seed could often be produced with ease, and might be as productive or nearly as productive as F_1 seed. If so, a great many more species hybrids might be put to practical use than is the case at present. There has been a general unwillingness to proceed with testing of the F_2 and F_3 generations, however, partly because of the time involved and partly because of overreliance on the overdominance and dominance explanations of hybrid vigor. One of the first mentions of the use of the F_2 generation trees is that by Rohmeder and Schönbach (1959). They found that F_2 and backcross hybrids of Japanese × European larch tested in Europe grew vigorously but did not possess as much hybrid vigor as F_1 hybrids.

Hyun *et al.* (1972) are among the few to pursue mass production of F_2 generations and seriously consider planting them on a large scale. They found that F_2 pitch × loblolly pine were as tall and a little larger in diameter than F_1 hybrids at age 7 and could be produced in almost any desired quantity by collecting open-pollinated seed from F_1 plantations. Thus, considerable numbers of F_2 hybrids are now being planted in Korea.

In most tree hybridization work is has been usual to produce F_1 hybrids by controlled pollination and to interplant those with seedlings of the parental species or with other hybrid combinations in order to test their growth potential. In such a plantation, F_2 seed can be produced only by controlled pollination, and this has not commonly been done. If an F_2 generation is the goal, it would be better to sacrifice on testing of the F_1 hybrids and instead plant the F_1 hybrids of any particular combination in a single isolated block where they could produce F_2 seed by natural crossing among themselves.

Hybridization and Subsequent Selection in Later Generations

General

Except in corn, crop plant breeders usually use hybridization in the initial stages of a breeding project to generate a variable population from which they can subsequently select plants having the best combination of traits. The usual practice is to start with two lines that between them possess the desired traits, cross these, plant the F_1 hybrids in such a way that they self or cross among themselves, plant and select among the F_2 hybrids, plant and select among the F_3 hybrids, etc.

This approach has not appealed to tree breeders because of the long generation length. However, it is frequently the best way to accomplish a given objective. It seems the only way to breed a blight-resistant timber-type chestnut or to obtain a Scotch pine resistant to several different pests and having good timber or Christmas tree characteristics. The time element has probably been overemphasized. There are enough trees which flower in 5 to 6 years that useful new varieties could be produced in the 30 to 40 year working life of a single tree breeder. In retrospect, chestnut breeding has been in progress for 80 years, since 1894, but much time was devoted to avoidance rather than implementation of a long-term program. So the long-term (i.e., 30 to 40 year) program is still to be undertaken, whereas it could have been completed.

Aside from time, the costs of such work need not be high. Presumably, a tree breeder starting on a 5-generation project would carry on several similar projects simultaneously. Production of the necessary F_1 hybrids might be time consuming, but thereafter not a great deal of effort would be required. Because of the increased genetic variability resulting from hybridization, heritabilities should be higher than in most natural populations, so the processes of selection and progeny testing should be easier than when working with nonhybrids.

A multigeneration breeding project should be planned as such from the start. Before going into the details of the planning, it is desirable to discuss the number of generations that might be needed to achieve the desired results.

Number of Generations Needed

As already noted, Hyun *et al.* (1972) in Korea obtained a satisfactory but not necessarily ideal variety with the F_2 pitch $\times$ loblolly generation. In their case the F_1 generation had a desirable combination of characteristics.

More often, multigeneration breeding will be useful where the parents have the desired combination of traits, but the F_1 and F_2 are still far from acceptable. As a general rule, the number of generations needed will depend to a large extent on the number of genes involved and the size of each generation. Assume that species A having desirable growth rate is crossed with species B having desirable disease resistance. Assume further that both growth rate and resistance are governed by 10 pairs of genes. The desired final product is to be homozygous for "plus" genes at each of 20 variable loci. The recovery rate in the F_2 generation is $\frac{1}{2}^{2n}$, where n is the number of variable loci. Thus, with 20 variable loci, the F_2 recovery rate is $(\frac{1}{2})^{40} = 1/1,099,511,627,776$. That many F_2 trees would have to be raised to find one which was homozygous at all loci. That is a practical impossibility, but it is possible to raise $2^{10} = 1024$ trees, of which one will be homozygous for "plus" genes at each of 5 loci, raise 1024 F_3 trees, and select one homozygous for "plus" genes at 5 additional loci. By continuing the process and selecting 1 of 1024 trees in the F_4 and again in the F_5, the desired final product of a tree homozygous for "plus" genes at all 20 loci can be obtained.

This example is of course oversimplified. Several trees should be selected each generation. Also, the number of variable loci is usually not known. It is clear, however, that results can be delayed by raising too few trees each generation.

Recommended Procedures

A multigeneration hybridization project should be planned as such from the start, as it differs from the usual hybridization study of the past in important ways. The essential steps are as follows.

1. Select parental types having the desired combination of traits and cross them to obtain 20 to 50 F_1 hybrids.

2. Plant the F_1 trees in an isolated location and treat them well in an attempt to induce flowering as early as possible. The F_1 generation is usually uniform, so no selection need be practiced. The isolated F_1 plantation is almost a necessity in chestnut, oak, maple, and other genera in which controlled pollination is time consuming and costly. It is not

so necessary in genera that are easier to control pollinate and in which large numbers of F_2 hybrids could be produced artificially.

3. Establish an F_2 plantation of a few thousand trees and give it intensive culture to induce good growth and early fruiting of a reasonably high percentage of the trees. Practice moderate phenotypic selection.

4. Establish a large F_3 plantation as a replicated half-sib progeny test, measure it, and start practicing family selection by removal of the poorest families and poorest trees in the best families.

5. Repeat that process for the F_4 and F_5 generations, or until the desired product is obtained in a moderately true breeding form.

In Michigan, where such a project is contemplated for Austrian $\times$ Japanese red pine hybrids, preliminary data from other experiments indicate that three different hybrid varieties will be needed to be adapted to the northern, central, and southern thirds of the state. That being the case, the development of the three varieties should be initiated with the F_2 generation. One F_2 plantation should be established in each part of the state, and selections made in any one of those plantations should be the basis of the F_3, F_4, and F_5 work in the same part of the state. In other words, three different breeding projects would be necessary.

The reason for recommending moderate phenotypic selection in the F_2 generation and practicing more intensive family selection in the later generations should be mentioned. Ordinarily, much of the superiority of the F_1 generation is due to genes with nonadditive effects. The amount of nonadditive genetic variance is still high in the F_2 generation and is reduced in each subsequent generation. If a tree's genetic superiority is due to nonadditive factors, it is not passed on to the offspring. Consequently the effectiveness of selection increases after the F_2 generation.

Possibilities in Chestnut

Chestnut is the one genus in which a multigeneration hybridizing project seems to offer more promise than any other improvement method. American chestnut, which was a common tree in the original hardwood forests of eastern United States, is nearly extinct because of the chestnut blight, a disease introduced from Asia. It is one of three species in the genus (the other two are from southern Europe and China) that become large trees. It formerly produced a valuable wood and could grow well on dry rocky soils in the southern Appalachians to which other valuable hardwood trees are not adapted. Thus, the loss of such a species is of more than sentimental importance.

Spanish chestnut of southern Europe, also a large tree, is valued for its nuts and wood. It, too, is very susceptible to the blight, but has a discontinuous range that prevented rapid spread of the disease; consequently it is not in danger of immediate extinction, but resistance nevertheless would be desirable.

In neither American nor Spanish chestnut is there evidence of enough variation in resistance to the blight to give hope of developing a resistant variety by selecting within the species. Resistance might come from the Henry chestnut, the other large tree in the genus, but that species from western China is relatively little known in the western hemisphere. Chinese and Japanese chestnuts, both small trees, are usually regarded as more likely sources of resistance.

F_1 hybrids between American or Spanish and Japanese or Chinese chestnuts have been produced. They have somewhat greater resistance than the American or Spanish parents but not enough to remain alive indefinitely. Thus, even if they could be mass produced, they would be of little practical value. Some F_2 hybrids and backcrosses have been produced but not in sufficient quantity to provide much of a basis for selection work. Also, the average level of resistance in the F_2 is too low. However, the variation in resistance and growth traits is enough to give hope of developing a satisfactory variety if the breeding work were to be continued a few more generations (Jaynes and Graves, 1963).

Most of the chestnut crossing work was done in the hope of finding one or two F_1, F_2, or backcross hybrids which would be satisfactory and which could be vegetatively propagated. Consequently such hybrids were planted in test plantations along with others of very different parentage. They were not planted in such a way as to ensure natural cross pollination such as would give sizable F_2 or F_3 populations in which to practice selection. In some ways it would be best to start over and produce new F_1 or F_2 hybrids, but that is not easy because the disease has taken such a toll of American chestnut and the older F_1 hybrids. Consequently it might be necessary to start with whatever material is now available, graft those trees, and establish clonal orchards in which natural crossing could occur, and then be sure to establish large plantations in which selection could be practiced.

Judging from experience with agricultural crops, a multigeneration hybridization and selection project could be successful in producing a new chestnut variety that is resistant to disease, capable of producing high-quality nuts, capable of producing valuable wood, and capable of growing to large size in either Europe or eastern United States. To be successful, though, the work would have to be planned for several generations in advance, and would have to be undertaken on a large scale. Some hybrids have produced small quantities of seed as early as age 3, so a generation length of 6 to 10 years could reasonably be assumed. With such a generation length, a useful new variety could be produced in 30 to 50 years.

Possibilities with Hybrid Austrian × Japanese Red Pine

A late-generation segregate variety derived from this cross can be produced much more easily than in the case of chestnut. F_1 hybrids are easily mass produced, grow rapidly, and are probably commercially valuable in themselves. They flower early, and large crops of F_2 seed have been produced on 7-year-old trees. The hybrids are inexpensive to plant. Thus a 4- to 5-generation project to be completed in 30 to 40 years can be considered realistic and easily accomplished.

The present F_1 hybrids are satisfactory on soils of moderate fertility in the southern third of Michigan, but the biggest need for improved forest planting stock is on soils of lower fertility in the northern two-thirds of the state. That is the principal reason for contemplating the multigeneration breeding work. Hopefully, enough segregation will occur to permit the development of hardier types that retain much of the growth and form superiority now evident in the F_1. The F_2 hybrids have already been produced and will soon be field planted in those places for which new tree varieties are most urgently needed.

Possibilities with Other Species Combinations

In eastern United States, eastern white pine grows well, is moderately resistant to white pine blister rust, but is very susceptible to white pine weevil; western white pine grows less well, is very susceptible to blister rust, but resistant to the weevil. There appears to be enough intraspecies variability to permit improvement in growth rate and rust resistance but not in resistance to the weevil. The two species can be crossed moderately easily but a satisfactory way to mass produce F_1 hybrids has not yet been developed. The F_1 hybrids are fertile and can flower at age 10 to 15. In general, they have not been as vigorous as eastern white pine when tested in the east.

Most of the available white pine hybrids were made many years ago without particular regard to selection of the best parents. Since they were made there have been intensive intraspecies improvement programs in both species, and it would now be possible to produce F_1 hybrids between rust-resistant western white and fast-growing eastern white pine. Thus, instead of starting with the scattered F_1 hybrids now available, it might be best to start anew. Several decades might be

FIG. 17.1. Eight-year-old alder hybrids in northern Japan. Several promising hybrids combinations involving *Alnus japonica, A. inokumae,* and other species have been obtained. The whiteness of the lower bark is caused by paint.

required, but the possibilities of success seem greater than with other methods.

A few F_1 maple hybrids have been produced which grow well and might have forestry or ornamental possibilities if they could be mass produced. That does not seem likely. So at present they remain curiosities. If they are to be used at all, it will be through multigeneration breeding or polyploidy. Several similar examples can be quoted in other genera (Fig. 17.1).

Study Questions

1. Breeding aimed at the production of F_1 hybrids that could be mass produced on a large scale was modeled after experience with what crop plant?

2. What are the requirements for a successful mass produced F_1 hybrid combination?

3. Describe each of three possible methods for mass producing F_1 hybrids.

4. What is unique about the ease with which Austrian × Japanese red pine hybrids can be produced?

5. Are F_1 species hybrids used more inside or outside their natural ranges? Why?

6. What are the advantages and disadvantages of F_2 hybrids as compared with F_1 hybrids?

7. F_1 hybrids possessing some growth promise have been produced by crossing some eucalypt species, by crossing Japanese red × Scotch pine, by crossing live × overcup oak, and by crossing Norway maple with some of its relatives. What are the possibilities of producing useful F_1 and F_2 hybrid varieties, respectively, in these combinations?

8. List four possible explanations of hybrid vigor and for each tell the relative amounts of hybrid vigor to be expected in the F_1, F_2, and F_3 generations, respectively.

9. If species A and B differ in two characteristics, each governed by 6 pairs of genes, how many F_2 trees must be raised to obtain one homozygous for "plus" genes at all loci? If each trait is controlled by 12 pairs of genes, how many F_2 trees must be raised to obtain one homozygous for "plus" genes at all loci?

10. How is it possible to obtain such a tree without growing astronomical numbers of trees each generation?

11. Why is it desirable, when planning a multigeneration hybridization project, to plant the F_1 hybrids together on a particularly favorable site?

12. In what genus is a multigeneration hybridization project particularly desirable? Why, in this genus, have F_1 and F_2 varieties not been satisfactory for large-scale forest planting?

13. In what ways is a multigeneration hybridization project similar to an intraspecies selective breeding project as discussed in Chapters 9 to 12? In what ways dissimilar?

14. Compare the amounts of empirical data available about F_1, F_2, and late-generation tree hybrids.

18

Introduction of Exotic Species

Introduction

Originally almost every food plant or domestic animal was native to a relatively small part of the world from where it has been spread to other regions by man. Name any food item and the chances are that 95 to 99% of it is produced in a country other than the country to which the plant or animal was native originally. Even most weeds in cultivated fields are not native. Ornamental trees and shrubs have also been shifted around, to the extent that a visitor to cities in the warm temperate or subtropical zones may see more foreign than native woody vegetation.

To an inhabitant of a well-forested region, such as north central United States, it is difficult to visualize that the same thing could happen in forestry—cutting down most of the native stands and replacing them with foreign trees. Even in that region it is happening to a slight extent, as the majority of plantation-grown Christmas trees are not native species. In some subtropical regions, even some with a relatively rich native flora, it has happened. The majority of the forest trees planted in the Republic of South Africa, Chile, Hawaii, Taiwan, and Australia are from other continents. In some places these have grown well enough to supply over 95% of the pulpwood and timber production. There are places in which so little native forest vegetation remains that special efforts are needed for its preservation.

Differences among species are ordinarily very large as compared with differences among races within a species or as compared with differences among individual trees within a forest. Thus the introduction of exotic tree species has been the single most important aspect of forest tree improvement for some areas. On the other hand, some regions have such excellent native species or such harsh growing conditions that trees from other lands have proved of little value. There are various intermediate regions in which tree introduction is one of several improvement methods which should be considered.

The demand for new forest trees has not been as urgent as that for new food plants or domestic animals, so the history of tree introduction is relatively short. It is a few centuries old in countries with a long maritime tradition, one to two centuries old in southern hemisphere countries colonized relatively recently by Europeans, and only 10 to 20 years old in some developing countries. Most of the history concerns sea captains who carried seed or plants on the decks of sailing vessels, botanists who exchanged seed out of curiosity, or wealthy people who wished to adorn their estates. Only in the past half-century has there been a systematic evaluation of the forestry possibilities of trees from other countries.

In spite of the empirical and often haphazard approach, considerable progress has been made. In many countries, hundreds of species have been tested. For the world as a whole several million acres have been planted to exotic tree species. In this chapter I shall discuss some of the reasons for their success, factors involved in successful introduction work, exchange relationships among regions, the methodology of introducing and testing new species, and finally short discussions of their usefulness in selected parts of the world. "Exotic Trees in the British Commonwealth" (Streets, 1962) is a good general reference on exotics. "Cultura pi valorificarea pinului strob" (Radu, 1974) is a very detailed account of the possibilities of a single species (*Pinus strobus*) in a single country (Romania).

Reasons for Expecting Success from Exotics

Through natural selection, native species become well adapted to their native environments. They can be expected to grow better than any random introduction from abroad, and in some cases grow so well that they are not surpassed by any introduction. However, there are valid reasons for expecting that some introduced species may outperform natives in some cases. These reasons are discussed below.

Natural Selection Favored Survival Rather than Economic Traits

A natural forest may be well adapted to its natural habitat and produce a near-maximum amount of dry matter per acre, but this does not guarantee that it produces large amounts of products valuable to man. The original forests of Hawaii were composed mostly of relatively low growing and crooked trees of limited value for lumber production. The original forests of the southern hemisphere contained no pines, larches, or Douglas-firs which could produce straight stems and long-fibered wood valuable for construction and pulpwood. American beech, which is a major component of the climax forests of eastern United States and Canada, is usually left after logging. The native white pines of Europe and Japan are valuable but too slow growing to be planted. This list could be expanded almost indefinitely and could be accompanied by an equal list of introduced species of value because they do produce valuable products.

Lag between Changes in the Environment and Evolutionary Response

Natural selection is a slow process. A species can be in perfect harmony with its environment only to have that environment change. Perhaps another 10 or even 100 generations would be required before the species could change sufficiently to again occupy its site fully. In the meantime a species from another region could grow better in the changed site conditions.

Evolution has been particularly conservative in the conifers, of which only about 500 species exist. Torrey pine of California, kauri of New Zealand, Brewer spruce of Oregon–California, and golden larch of the Peoples' Republic of China are examples of conifers evolved long ago and in such a precarious relation to their modern natural environments that they are in danger of extinction.

Lack of Evolutionary Response to Man-Caused Changes

As an extension of the above, consider the particularly drastic changes that have occurred in many forest regions in the past few centuries because of man's activities. Strip-coal mining in Germany and central United States, drainage of the Parana delta in Argentina, shifting agriculture and subsequent erosion in the tropics, and clearcutting and burning in many places have created environments far different from those in which natural selection operated for millions of years. Many such changes have occurred so rapidly and been so drastic that there was virtually no opportunity for an adequate evolutionary response on the part of the native species.

The amount of environmental change caused by man varies from slight for forests carefully logged by the selection system to drastic for the examples given above. The greater the change, the greater the need to shift from the culture of trees that were part of the climax forest to other trees that can endure the altered conditions. With changes as drastic as those that have occurred at low–medium elevations in Hawaii, only introduced species thrive.

Limited Evolutionary Possibilities in a Flora, Family, Genus, or Species

Once evolution has started there are constraints such that it can proceed only in certain directions within a group. This is evident in the history of the horse, which started as a 1-ft tall animal in Eocene times and evolved into a modern animal larger than a man. Once the horse ancestor became a grazing animal incapable of digging or climbing trees, increase in size became advantageous so that it could see enemies above the grass and run away rapidly. Thus there was a consistent evolutionary trend from Eocene to modern times toward increasing size of bones, feet, teeth, etc.

Australia furnishes an example of the limited evolutionary capabilities of an entire flora. The continent has been separated from the rest of the world for so long that its flora is unique. Its native gymnosperms could evolve into a few species capable of growing in the warm, moist forests of Queensland, but were not capable of changing sufficiently to compete in the drier conditions prevailing in 99% of the continent. Such evolutionary capabilities were present, however, in northern hemisphere genera, such as pine and juniper. Those and other northern hemisphere conifer genera have proved very successful when planted in Australia.

Conversely, the eucalypt genus was capable of evolving into 300 to 500 species in Australia, one or more of them adapted to practically every ecological niche in the continent. The genus is unique in being capable of evolving into species that grow rapidly and tall on almost any warm site receiving enough precipitation to support woody vegetation. This capability is lacking in the xerophytic genera of hardwoods native to other continents. Now, when eucalypts are planted in other warm dry countries they succeed very well.

If evolutionary capabilities are limited for entire floras, they are also limited for families, genera, and species. In the United States, American and slippery elms are threatened by the Dutch elm disease (introduced, fatal), American chestnut is threatened by the chestnut blight (introduced, fatal), and eastern white pine is damaged seriously by the white pine weevil (native, deforming). Experimental evidence indicates that there is no or slight possibility of developing resistant varieties to those pests by selection within those species; genes for resistance are present in other species, however.

Decimation of Native Species by Introduced Pests

The three species mentioned above represent examples in which native species cannot be recommended for planting because of an introduced pest. They may even become extinct. Their most exact replacement is probably a foreign species or a hybrid.

Lack of Plantability of Native Species

Ability to be planted economically on a large scale is an important attribute of a species destined to be used in forest improvement through stand conversion. Most tropical and subtropical forests are florestically complex, containing many tree species per acre. Relatively few are valuable. As logging proceeds, the valuable ones are eliminated, and the forest lessens in value. Successful management through natural regeneration is extremely difficult, and planting of the most valuable native species has often met with little success. Thus regardless of ultimate growth possibilities, foreign species that can be planted have been favored.

This is true to a lesser extent in United States' central hardwood region. The hardwoods grow well, are generally valuable, but are expensive to plant. Pines and spruces from other regions grow well, are less valuable, but are relatively easy to plant. Thus, most planting has involved trees moved north or south of their native ranges.

Nongrowth Factors Limiting to a Species' Natural Distribution

Islands, such as Great Britain, Ireland, Hawaii, and New Zealand, depended for their natural flora on the whims of bird, ocean currents, winds, and ancient colonizers. These agents brought trees and herbs whose propagules were capable of being transported long distances. They did not receive all the plants, or even a random sample of all the plants, that might thrive.

Even in continental areas a species' natural range may be much smaller than the area in which it can grow successfully. Eastern white pine thrives but sets very little full seed in southeastern Pennsylvania and parts of Indiana. Presumably it might have been able to spread naturally to those places if it had been able to reproduce itself. The relatively small natural ranges of black locust and osage orange in south central United States are difficult to understand. Certainly when moved outside those original ranges they grew well, became naturalized, and now seem a part of the native flora.

Factors Governing Successful Introduction

There are literally thousands of tree species in the world. Several lifetimes would be required to introduce and test them all in any one region. The percentage that can be useful in any one country is probably small. Fortunately, however, there are several generalizations that help in the choice of species and increase the possibility that any given introduction will be useful.

Performance and Characteristics in a Tree's Native Habitat

These are the most important clues as to a tree's potential usefulness in a strange country. Shrubs remain shrubs if moved around. Trees with crooked boles which produce poor quality wood continue to do so when moved to another country. Therefore in order to introduce a new timber-type tree, learn which foreign species grow straight and tall and produce good wood, and test them first. To produce a new ornamental tree, learn which foreign trees are resistant to salt spray, or produce nice flowers, or have an artistic branch arrangement, and test those species first. For planting on sterile sandy soils, concentrate on trees that grow on such soils in their homelands.

Monterey pine furnishes an excellent example of this generalization. Its hardiness, growth rate, stem form, genetic variability, and wood quality are essentially the same in New Zealand, Australia, Chile, and the Union of South Africa as in the native stands of California and Baja California. Only in maximum height (160 ft in New Zealand versus 120 ft in native stands) has it performed differently than was expected in the southern hemisphere. This is probably a consequence of the immense acreages planted in New Zealand, which has thousands of acres of Monterey pine plantations that offer protection to each other and permit greater maximum height than is possible in the few square miles of native forest on the windswept Monterey Peninsula of California. Listed below are some of the many other examples showing the correlation between native habitat and strange habitat performance.

Western hemlock, subject to woodrot in western America, has the same tendency when planted in England.

Pitch pine, tending to be large-branched and crooked in native stands, has the same characteristics when planted in Korea.

Species of eucalypt that produce poor sawtimber in Australia have produced little usable sawtimber when planted in the Republic of South Africa or Argentina.

The desirable wood properties of eastern white and slash pines have been maintained wherever these species have been grown.

Dawn redwood is fast growing and adapted to moist sites in its native China and also in the United States.

Magnolias produce large flowers before the leaves come out in its native Japan as well as in the United States. The Japanese magnolias are low-growing trees wherever grown.

Cryptomeria from Japan is a fast-growing, straight tree producing durable wood whether planted in its native Japan or in warmer conditions in Taiwan.

Climatic Similarity in Regions of Origin and Use

The most successful transfers of species are between climatically similar regions. But what is meant by climatic similarity?

Monterey pine has one of the narrowest ranges of tolerance of any tree species. On its native Monterey Peninsula in California the minimum winter temperature is about 18°F, growing season frosts are uncommon, and rainfall is moderately high in winter and scant in summer. This species grows well only when subjected to those conditions. It has suffered pronounced winter injury when exposed to 14°F and suffers damage from disease when grown in areas with high summer rainfall. A potential foreign grower of this tree will probably succeed only if the climate near his home matches that in coastal California more closely than it does any other part of North America.

At the other extreme is weeping willow. The clone which is usually grown is among the fastest in its locality whether maximum summer temperature is 80° to 110°F, whether minimum winter temperature is −20°F or 20°F, whether growing season frosts are common or rare, and whether grown under subtropical or temperate daylength conditions. Some clones of hybrid tea roses tolerate even wider climatic extremes—the same clone growing well near tropical palms at sea level close to the equator, near orange trees in Buenos Aires, or in regions receiving several feet of snow.

Minimum winter temperature was the basis for Rehder's (1940) classification of tree hardiness zones in the United States, since copied by other authors. Its effects are often difficult to separate from duration of winter cold or desiccation by winter winds. Lack of winter hardiness is especially limiting to trees introduced from the relatively mild climates of Pacific North America. It is also limiting to many species of eucalypts and to trees introduced into the most northern regions.

Growing season frost is the most important limiting factors when moving northern species southward. This is especially true of spruce and fir. Douglas-fir is also sensitive to damage from growing season frosts, whereas most pine species are not.

Low summer temperature, if accompanied by a long growing season and suitably warm winters, is usually not a limiting factor. Nearly all species from eastern United States are subjected to much warmer summers in their native habitats than in parts of northern Europe and New Zealand, where they thrive.

High summer temperature is usually accompanied by a decreased precipitation–evaporation ratio and may be limiting for that reason. There is some evidence that high temperature itself may be harmful. Even when soil moisture is maintained by irrigation, most trees from northern United States barely survive when planted along the Gulf Coast. Also, hemlock and yew from cool parts of Michigan do not grow well in the southern part of the state where summers are longer and warmer.

Low total annual precipitation is usually more limiting than is high rainfall, but may often be overcome by irrigation. Seasonal distribution is frequently just as important. In this latter respect the eucalypts are much more tolerant than the pines. Several western American pine species that are accustomed to moist winters and dry summers grow well for a few years when grown in regions with moist summers, but succumb early to diseases. Japanese and European larch apparently require abundant and frequent rainfall during the growing season and suffer if exposed to droughts of a few weeks duration.

Daylength seems to have been of relatively little importance in determining the vegetative success of a tree in a new habitat, but may explain why the flowering of some exotics has been delayed. Most transfers have involved latitudinal shifts of 15° or less. We have no evidence of what would happen if transfers were made between high elevational regions in the tropics and temperate or Arctic regions with similar temperature but vastly different daylength regimes.

The climatic tolerance of most tree species seems to be such that one need make only an approximate match between areas of origin and intended use when planning introductions. Often it is sufficient to recognize that one lives in a region with a climate which is cool-temperate with abundant growing season precipitation or moist subtropical and make introductions from other regions characterized in the same manner.

Deficiencies in Flora of Recipient Region

Exotics are most useful in regions in which the native flora is deficient. Southern hemisphere countries are deficient in native conifers. The Republic of South Africa has none. Chile, Argentina, New Zealand, and Australia have native conifers, but these are mostly slow growing and prefer deep fertile soils in high-rainfall areas. They are not adaptable to a wide range of site conditions as are some northern hemisphere conifers. Conifers are almost essential to an industrial economy because they have straight single stems, have soft and easily worked wood useful in general construction, and produce long fibers needed in paper making. Consequently the introduction of pines and to a lesser extent other genera into southern hemisphere countries fulfilled a great ecological and industrial need.

In northern Europe the deficiency is a general one. Pleistocene glaciation caused the extinction of many species, leaving the modern flora very limited in variety. Europe north of the Alps has five native coniferous tree species, in contrast to 25 for Japan. The British Isles, almost without native conifers (only Scotch pine is native there), rely mostly on introductions from continental Europe and western America.

In all subtropical regions except Australia the deficiency is in hardwoods capable of the extremely rapid growth of the eucalypts. These are capable of forming a tall forest on sites so dry as to support only grass of scrub vegetation otherwise.

However, the use of exotics is not limited to regions with impoverished floras. The island of Taiwan (12,000 square miles) has an immensely rich flora with about 800 species of woody plants. Yet *Cryptomeria* from Japan, luchu pine from the Ryukyu Islands, and China fir from mainland China are much used in reforestation work. Northeastern United States is relatively rich in native species, but scores of exotics have found a place as ornamentals, and species, such as Norway spruce,

Scotch pine, and English oak, seem capable of replacing native trees as forest species on some sites.

Differences in Adaptability

A tree planted as an exotic always encounters conditions slightly different than those to which it was naturally adapted. Species differ markedly in their adaptability to changed conditions. For example, white fir, eastern white pine, and northern red oak are three American species that grow well in other countries, whereas balsam fir, red pine, and white oak are three other American species that do not.

A review of past experience is practically the only means of detecting whether or not a given species is adaptable enough to succeed in a new place without actually testing it. There is a large amount of literature available, although not always in the language of one's choice. Books from Sweden, England, France, Germany, and Rumania contain many clues as which species might grow best in the cooler parts of the United States. "Selection and Breeding to Improve Some Tropical Conifers" (Burley and Nikles, 1972, 1973a) and "Tropical Provenance and Progeny Research and International Cooperation" (Burley and Nikles, 1973b) are books that summarize a vast amount of experience with tropical species. More often than not, species growing well in one strange country can be expected to grow well in other countries having similar sites.

The forests of northern Japan, the mountains of western China, Taiwan, and of many tropical rain forest regions contain many valuable tree species that might prove useful elsewhere but are practically unknown outside their native countries. Actual tests of every promising species seems to be the only way to determine the adaptable ones. Morphological variability, size of natural range, and tolerance of varied site conditions within the native habitat do not provide good clues. For example, eastern white pine and red pine have similar geographic ranges in northern United States and southeastern Canada. Eastern white pine tolerates a wide range of site conditions in its native habitat and grows well in several parts of Europe and Japan; red pine tolerates a narrower range of site conditions in its native habitat and is not considered a promising exotic in any other country. Almost the opposite is true of the American white and northern red oaks. White oak is extremely variable morphologically and tolerates a wide range of site conditions in its native range but is not a useful exotic; northern red oak tolerates a narrower range of site conditions in its native range but is widely planted in Europe.

The special case of northern Europe should be mentioned. As previously mentioned, Pleistocene glaciation caused the disappearance of many tree species north of the Alps. Only a few species capable of surviving in isolated pockets and recolonizing large areas in the 10,000 or so years since final retreat of the glaciers were left to contribute to the modern forest flora. In other words, the modern species have proved their adaptability by being present now. Nearly all have demonstrated this adaptability when planted in other cool temperate regions.

Monotypic Genera

Monotypic genera are those which contain only one species. They vary widely in growth and use characteristics but have one special feature that makes them of special interest in introduction work. They are nearly free from attacks by insects and diseases. This feature was recognized by J. R. Schramm. It is explained partly by the fact that most monotypic genera have small ranges. Even a minor pest can occasionally reach epidemic proportions and destroy an entire stand. If this occurred to a species that grew in a restricted area, it might become extinct. Thus the mere existence of such a species indicates that it has not been subjected to any serious pests. Those with such pests probably became extinct long ago.

The other reason that monotypes are free of insects and diseases even when transplanted abroad is that they have no close relatives supporting pests that might change slightly and become important. Ginkgo, for example, is the only living representative of an entire order and could be attacked only by an insect capable of feeding on a number of unrelated trees. Fortunately, most insects are incapable of doing this.

A monotype does not necessarily have any other desirable combination of traits. Its hardiness, adaptability, growth rate, and use characteristics must be tested as in any other type of species. However, the insurance against pest damage is enough to merit paying considerable attention to monotypic genera as a group. Below is a partial list of monotypic genera that could be used in temperate and subtropical regions.

Asiatic species: *Biota orientalis* is a small, hardy tree with durable wood. *Cercidiphyllum japonicum* attains large size and produces valuable furniture wood. *Eucommia ulmoides* is a large shrub or small tree. *Ginkgo biloba* is a large, hardy tree producing a desirable, soft, even-grained wood and is very adaptable. *Metasequoia glyptostroboides* is very adaptable and fast-growing, and produces durable but brash wood. *Platycarya strobilacea* is a fast growing, small tree. *Pseudolarix amabilis* is hardy, of medium growth rate and has a desirable growth form and autumn color. *Pteroceltis tatarinowii* is a fast growing small tree. *Sciadopitys verticillata* is slow growing.

American species: *Maclura pomifera* is adaptable, crooked, and produces hard durable wood. *Sequoia sempervirens* has limited adaptability in nearly frost-free regions, grows rapidly and to large size, has good bole form, and produces very durable wood of much commercial value. *Sequoiadendron giganteum* is more cold hardy, grows rapidly in Europe and western United States, and produces durable but brash wood. *Umbellularia californica* is of limited adaptability in nearly frost-free regions and produces hard wood with beautiful grain.

Possibilities as Parents of Hybrids

Some species deserve introduction more for their use in species hybridization programs than on their own merits. Practically all the successful hybrids mentioned in Chapters 16 and 17 involved exotics as one or both parents.

Genetic Variability within Species

Most species with large natural ranges are genetically variable. No one seedlot can be considered as representative of the entire species. Pacific Coast races of most western American species would be regarded as failures if planted in eastern United States, whereas Rocky Mountain races of the same species often succeed.

Most past introductions involved seed of unknown origin, and not necessarily the best origin. Due consideration should be given this fact when interpreting past trials. Often a species deserves further trial if any single tree or plantation performs 75% as well as comparable native species. There is a growing trend to obtain several different seedlots when introducing a new species.

Other Factors of No or Minor Importance

Some other factors are often mentioned in tree introduction work but prove to be of little importance. Among them are the following.

Size of Natural Range

Successful exotics include species with very small natural ranges (Monterey pine, Japanese larch, and Serbian spruce) as well as species with large natural ranges (slash pine, loblolly pine, Douglas-fir, and Scotch pine).

COMMERCIAL IMPORTANCE IN NATIVE RANGE

Some species with small natural ranges are little more than botanical curiosities in their native countries, but planted on many thousand (Japanese larch) or many hundred thousand (Monterey pine) acres abroad. Many other species (e.g., balsam fir and red pine) are of great commercial importance at home because of their large ranges but have not succeeded in other countries.

LATITUDE

Latitude alone is a poor indicator of a region's climate. Paris, France is at nearly the same latitude (49° N) as northern Newfoundland, but has winters more like those of Norfolk, Virginia (37° N). Many Virginia trees can grow near Paris, but few French trees can grow in Newfoundland.

Exchange Relationships among Regions

The major regional exchange relationships are summarized in Table 18.1. In some cases there has been a mutual exchange of valuable tree species, as between the various countries with Mediterranean-type climates.

Sometimes the exchange has been in one direction only. The Pacific Northwest (Oregon, Washington, and British Columbia) has a large number of valuable native timber trees that have proved useful elsewhere, but has little need for introduction from other countries. Such introductions that have been made have been used primarily as ornamentals, being unable to compete with the native species in the forest. Southeastern United States has also acted primarily as a donor region, sending loblolly and slash pines and eastern cottonwood to other countries but receiving little of value to forestry in return. Although the southeast has a generally mild climate, there are occasional very cold spells [with temperatures of 0°F (−18°C)], which cause severe damage to fast-growing species from other countries.

Exchange relationships are best known for countries with long established forestry programs or with dramatic examples of successful introductions. Relatively little is known about the moist tropics. Mexico and Central America are known to be very good sources of pines, but little is known of the possibilities of the many other species growing in those countries nor of the possibilities of exotics in those countries.

TABLE 18.1

Regions between Which Tree Exchange Has Been and Has Not Been Successful

Recipient region	Donor regions	
	Successful	Unsuccessful
Northeastern United States	Rocky Mountains, northern Europe, mountains of the Mediterranean basin, Honshu, western China, and northern India	Pacific coast of United States and Canada, Mexico, southeastern and southwestern United States, Hokkaido, Scandinavia, Siberia
British Isles	Pacific coast of United States and Canada, Northern Europe, mountains of Honshu, western Europe, and Mediterranean basin	Northern Asia, Mexico, lowlands of Japan, China, and Mediterranean basin, Himalayan mountains
Northern Italy, southern France, and Germany	Eastern and western United States, mountains of Honshu, western China	Northern Europe, Siberia
Southern Scandinavia	Siberia, mountains of central Europe and western United States	
California, New Zealand, Australia, Union of South Africa, southern Italy	California, Australia, Mediterranean basin, mountains of Mexico, southeastern United States	Mountains of Honshu, western China, northern Europe, northeastern United States
Kenya, Tanganyika	Mediterranean and California lowlands, Mexico, India	
Northern Argentina	Australia, southeastern United States, Japan, parts of Mexico	Western and northern United States

Methods of Introducing and Testing New Species

Selection of Species and Donor Regions

With literally thousands of tree species in the world, some preliminary selection work is necessary to make introduction work more efficient. This should take the form of a survey of what has already been done in one's own country, a study of the world's climates to determine suitable donor regions, and a study of the tree floras of those regions.

Mail Order versus Personally Collected Seed

Seed from Japan, Europe, United States, or Australia can usually be obtained by writing. Personal collecting expeditions may be necessary to obtain seed from countries with less developed road systems and smaller forestry programs. Most of the desirable tree species in the last-named countries occur in mountainous areas that are remote from large centers of population. In such remote areas there are relatively few people trained to recognize species or to collect seed. Most trees obtained from the mountains in the western part of the Peoples' Republic of China were sent to other countries by botanists who spent 2 to 3 years collecting. Most tree seeds obtained from Mexico and Central America were collected by special teams sponsored by groups such as FAO and the Commonwealth Forestry Institute.

The Benefits of Cooperative Approaches

Useful tree introduction work is apt to be expensive, but the benefits from any particular introduction are often shared by several states or countries. It is only natural that such states or countries cooperate to increase the effectiveness of the work. One such cooperative effort was described in a symposium held as part of the fifteenth IUFRO (International Union of Forest Research Organizations) Conference at Gainesville in 1971, and details are found in papers published in the books edited by Burley and Nikles (1972, 1973a). Collectors from the Commonwealth Forestry Institute (Oxford) spent considerable time in Central America, collecting seed from many different stands of pine species that had previously proved promising in subtropical countries. Whenever possible large quantities of seed were collected. The seed collections were assembled at Oxford, stored there, and distributed to all subtropical countries that showed an interest in the collections. Also distributed with the seed were data about the collections and suggestions as to the best testing procedures.

The NC-99 Committee of north central United States operates a little more formally. One member undertakes collection of seeds of a species or genus, assembles the seeds, and records and distributes seed or planting stock to other members. The trees are usually planted in several states in a single year using similar planting designs. After a suitable interval a cooperative report prepared by all members of the committee is prepared and published. If some of the races or species prove especially valuable for reforestation, one or more members see to it that seed or planting stock is made available to the public.

Sampling Intraspecies Diversity

Formerly, most introduction work was accomplished by obtaining seed from one or two trees in one to two localities, often with very little data about the parental trees or the localities of collection. Sometimes the names given were incorrect. This practice had three undesirable consequences. First, recollection of a promising species was often difficult. Second, little became known about the true potential of a genetically variable species. Third, the offspring of single-tree collections often exhibit low seed set, presumably because of inbreeding; thus they could not be reproduced if promising.

Thus it is an increasingly common practice to request seed from several trees in each of several widely scattered localities and to request source data about each seedlot, even when making a first-time introduction. The number of seedlots may vary from two to three for a small-range species of moderate promise to 50+ for a large-range species of great promise.

Provision for Future Seed Supplies

Golden larch, Ernest fir, and dawn redwood were introduced into the United States from the Peoples' Republic of China in 1854, 1904, and 1947, respectively. They have proved to be very desirable ornamentals and could possibly be important forest trees as well. However, that can hardly be tested because of the general unavailability of seed. Their native ranges are small, remote, and difficult for collectors to reach. The specimen trees that were planted in the United States are widely scattered, unable to cross with one another, and thus constitute a ready source of cross-fertilized seed. If there were a desire to plant these seed on a large scale, one would have to wait another 20 to 40 years until seed orchards could be developed.

Countries interested in Mexican and Central American pines faced a seed availability problem in the past. Some races and species grew so rapidly that they are considered for large-scale testing or even commercial planting after 4 to 5 years of preliminary testing. Frequently the best parental stands were of unknown location or had been cut, and the test plantations had not been designed with seed production in mind, so practical utilization of the promising imports was delayed for another 10 to 30 years. Recognition of this problem prompted several speakers at the 1971 IUFRO symposium (Burley and Nikles, 1972, 1973a) on tropical conifers to emphasize the need to plan for future seed production while planning the introduction and testing.

Thus it seems essential to raise at a very early time the question: How can seed be obtained if a new species does prove worthwhile? A partial solution is to establish a small, isolated, widely spaced "seed orchard" plantation of each race or species at the time of introduction.

Testing and Evaluation

The procedures and designs used to test exotics are the same as used in the study of individual tree inheritance and racial variation (see Chapter 8). Exotic testing should be done in two or three stages. The first preliminary tests should include several scattered plantations on different soils types and with different climates, with a few blocks per plantation and small plots. These first tests may well include a few hundred seedlots of several different species or even genera. They may be established over a period of years, with separate plantations for species with different growth rates and growth habits.

The second-stage tests should concentrate on those races or species that grew best in the first-stage tests. There should be more replication and perhaps larger plots at each test site, and increased attention can be given to individual tree variation. The second-stage trials may often be considered as semicommercial, designed for the production of wood as well as of data. The third-stage trials can consist commercial plantations designed primarily for wood production.

In working with an exotic species for the first time, one lacks the background information of site adaptability, pest problems, and silvicultural management that is usually available for a native species. This is the reason for suggesting preliminary testing on a variety of site conditions and moderate-scale second-stage testing. Efforts to bypass preliminary testing by planting large commercial acreages of an unknown species have been numerous and frequently disastrous. Efforts to economize by planting a few trees at one location only have also been numerous and have led to 75- to 300-year lags between introduction and any considerable use.

In subtropical countries it is sometimes possible to devote only 5 to 10 years to a stage and progress from preliminary tests to large commercial plantations in 10 to 20 years. This is because the species being tested most commonly are very fast growing and because there is frequently an urgent need if any considerable amount of planting is to be done. Such short test rotations are also possible in the temperate zones, although test rotations of 15 to 20 years are more desirable there.

There is always a risk of failure if a race or species is released for commercial planting on less than 50 to 100 years of experience. However, past experience indicates that the risk is usually warranted. The alternative procedure of testing as thoroughly as nature has done for native species is very expensive.

Examples of Successful and Unsuccessful Exotics

The following examples were chosen to illustrate the range of success that can be expected. They include a few species that are widely planted outside their native ranges, a few that are of moderate importance when planted abroad, and a few that might have been expected to succeed but have shown little promise as yet.

The Eucalypts

The eucalypt genus of some 300 to 500 species is native to Australia and New Guinea and is so common in parts of those islands that it is cut rather than planted. The native Australian eucalypt stands usually contain several species and belong to one of four forest types listed in the tabulation below (Metro, 1955).

Forest type	Canopy	Annual rainfall (in.)	Average tree height (ft)
Wet sclerophyllous	Closed	40–60	200–300
Dry sclerophyllous	Closed	25–40	100–130
Woodland	Open	20–25	65–80
Mallee	Open	10–20	20–25

The most valuable species are those inhabiting the wet sclerophyllous forest type. However, in other countries sites suitable for them are often limited and occupied by valuable native species. Thus the majority of eucalypts planted as exotics are from the dry sclerophyllous forest type. The most important are *E. globulus*, *E. gomphocephala*, *E. camuldulensis*, *E. saligna*, *E. grandis*, *E. gunnii*, *E. fastigiata*, *E. scabra*, and *E. tereticornis*.

As of 1955 over 2 million acres had been planted to eucalypts in Brazil ($\frac{3}{4}$ million acres), Republic of South Africa ($\frac{1}{2}$ million acres), Madagascar, Spain, Portugal, Chile, and other subtropical countries (Figs. 18.1 and 18.2). They continue to be planted on a large scale. The reason for this interest is their ability to grow rapidly (few are considered if they grow less than 10 ft per year and some can grow 20 ft per year), and to grow tall and straight on warm dry sites previously occupied by grass or low-growing scrub trees. Thus, 100-ft eucalypt trees can be seen in the highlands of Peru surrounded by short grass or bare ground, in the Pampas of Argentina surrounded by tall grass or field crops, or in Hawaii surrounded by 40- to 50-ft native forest. Pest problems have not usually been serious, but winter cold and excessive soil moisture have. The species planted most extensively are from warm, semi-arid climates and have not succeeded well when planted in areas with winter temperatures much below 20°F (e.g., most of southeastern United States) or with consistently heavy rainfall (e.g., Taiwan).

In spite of their rapid growth, most eucalypts produce a dense

Fig. 18.1. Ten-year-old gum (*Eucalyptus propinqua*) from Australia growing in northern Argentina. These trees, approximately 100 ft tall, show why eucalypts have been planted on a large scale in many subtropical countries. Growth such as this can be achieved in many areas so dry that the native vegetation was scrubby. Difficulties in seasoning have been encountered with many eucalypts and make wood quality a prime consideration in choice of species for large-scale planting.

FIG. 18.2. Lemon-scented spotted gum (*Eucalyptus citriodora*) from Australia planted in Kwantung Province in the southern part of the Peoples' Republic of China. Dr. Ching found large plantations of this and many other exotics in southern China. Despite its good qualities (termite-resistant wood, straight clean stems, and good growth in several countries) this species has not been planted on as large a scale as it should. (Photograph courtesy of Kim Ching of Oregon State University.)

wood with a pleasant grain, and the wood of some species is quite durable. It is deficient in one very important respect, however. It is subject to very great internal stress. This stress is relieved upon sawing; newly cut boards become quite crooked and are difficult to season properly. This problem is much greater in young than in old trees (internal rot seems to relieve the stress) and in some species than others.

Because of the difficulty encountered in seasoning, eucalypts have not proved as valuable as might have been expected from their growth rate and form. They have been of limited use in the making of furniture, floors, houses, etc., in spite of an urgent need for such wood in the areas in which they were planted. They have been most useful for relatively low value products in which sawing was unnecessary (firewood, mine props, fence posts, poles). In some countries paper and chipboard industries have developed to utilize the wood. Eucalypts are of value to these industries because a large mill can be supplied continuously by a relatively small acreage of plantations.

For the future, eucalypts should be selected for their wood properties as well as growth characteristics, and the development of a large planting program should be coordinated with the overall plans for development of forest industry.

Monterey Pine

Monterey pine is native to six small areas in California and Baja California. Those places are characterized by cool, dry summers, warm moist winters, and the absence of extreme cold. At Del Monte on the Monterey Peninsula the average January and July temperatures are 49° and 61°F, respectively, the lowest temperature on record is 20°F, and the average annual precipitation is 15 in., of which only 2 in. occurs from April to October. The species' natural spread to adjacent warmer areas was apparently hindered by too little precipitation, and its spread to moister areas by too low temperatures during winter.

In spite of its small natural range, Monterey pine has been planted more extensively as an exotic than any other tree species (Fig. 18.3). It has been planted extensively in Chile, New Zealand, western and southern Australia, and in the western part of the Republic of South Africa. Most of these areas are on the western part of a continental mass and therefore experience a maritime climate suited to the species. In Victoria, Australia, suitable areas have deep and well-drained soils, average winter and summer temperatures of 40° to 50°F and 60° to 68°F, respectively, 35 to 70 in. annual rainfall (60% in six winter months), minimum winter temperatures above 18°F, and vary in elevation from sea level to 3300 ft. Climates in other Monterey pine growing areas are similar. The species has met with little success on the eastern edges of continental masses, where the winter temperatures are too low (14°F was fatal in Great Britain) or where summer rainfall is plentiful (disease is a problem in Argentina and Queensland).

Extremely rapid growth has been the principal reason for planting Monterey pine. Growth rates of 3 to 6 ft per year are common. A 37-year-old plantation in New Zealand contained trees 160 ft tall and yielded 76,000 board ft per acre. In parts of Australia the yields are so high as to justify clearing 80-ft tall eucalypt forests to provide planting space.

Where it can be used, Monterey pine tends to be planted more than all other species combined; no other species compares with it in yield. For example, there are almost a million acres each in Chile and New Zealand. Local foresters worry about the possiblity of serious disease or insect outbreaks which might result from such monoculture, and they continue to search for other species that could diversify the planting program.

FIG. 18.3. Forests of exotic species on the Kaingaroa Plains near Rotorua, New Zealand. Monterey pine grows so much faster than other species, either introduced or native, that it comprises the main portion of the plantation acreage. Large areas in the picture are pure Monterey pine plantations. (Photograph courtesy of G. B. Sweet of the Forest Research Institute at Rotorua, New Zealand.)

This tree owes its prominence in southern hemisphere countries to the fact that it can produce large quantities of long-fibered, easily worked wood suitable for lumber, construction, and pulpwood in regions without native conifers or whose native conifers grow slowly and are adapted to only the best sites. At the time that large-scale planting started there was a seemingly limitless need for such wood, but in some areas growth has far exceeded local needs so that new industries and export markets have developed.

Although the wood has good technical properties, the quality of lumber produced is not high. Trees retain dead branches and produce knotty lumber unless pruned. Also, many are crooked. This may be partly the result of introgression between Monterey and the much more crooked knobcone pine (Bannister, 1958). These two species grow together and hybridize occasionally on the Monterey Peninsula, from which some of the original seed may have come. Both New Zealand and Australia have intensive breeding programs to improve form and growth rate.

Monterey pine needs mycorrhizae, which were already present or inadvertently introduced into may areas. They had to be purposely introduced into Kenya in 1910 (Pudden, 1957), and after introduction of the necessary fungi growth rates were approximately doubled.

Slash Pine

Slash pine is a native of southeastern United States, where it has a large natural range and is planted extensively because of its ability to grow rapidly (sometimes surpassing Monterey pine) on infertile sites. In its native range it experiences long hot summers, relatively warm winters but with occasional low temperatures of $0°F$, and heavy precipitation (mostly during the summer).

As an exotic it has been planted extensively in Queensland (eastern Australia), northeastern Argentina, southern Brazil, and the eastern parts of the Republic of South Africa—all regions with a warm, moist summer rainfall climate. In Argentina it was almost unknown until 1950, but by 1965 was the principal species planted to supply one large pulp mill. In that case it was able to grow as fast or faster than the native *Araucaria* (incidentally a very valuable tree) on the best sites and to grow well on sites too poor for the *Araucaria*.

Slash pine is composed of two varieties—north Florida and south Florida slash pine. The former is much the most important in the United States and has been planted most extensively abroad. Only in the last decade has there been much comparative testing of the two. A few tests indicate that the north Florida variety actually needs exposure to some cold weather (although not necessarily as low as $0°F$) for maximum growth and that the south Florida variety grows best in regions in which the temperature drops little below freezing.

Caribbean Pine

Caribbean pine is a native of Central America, Mexico, and some of the islands in the Caribbean Sea. It is a close relative of slash pine and shares with that species the ability to grow rapidly. It has not been planted much nor managed intensively within its native range, although it probably should be.

This species received relatively little attention until the last 25 years, but in the last decade has come to be one of the species considered most promising for tropical areas. Comparative species trials in many countries indicate that it can grow faster and replace slash pine in many moist regions free of frost. Preliminary provenance tests indicated that the species is geographically variable, especially with regard to form. The Honduran variety is noted for its tendency to grow like a "foxtail" (i.e., grow a central stem without branches for several years in succession) if not subject to moisture stress; this is regarded as an undesirable trait because foxtail trees are subject to storm breakage although they grow faster than normal trees. Intensive provenance tests are underway in numerous parts of Africa, Australia, South America, the South Pacific, and the Caribbean areas. This will be one of the first species to be planted on a large scale with adequate provenance information to guide in the selection of the best seed (Figs. 18.4 and 18.5).

Fig. 18.4. Seven-year-old plantation of the Cuban variety of Caribbean pine in northern Argentina. Argentina, like many other southern hemisphere countries, is devoid of native pines. Fast-growing subtropical species such as this find a ready home and are planted on a large scale.

FIG. 18.5. "Foxtail" growth in a 4-year-old tree of Caribbean pine (Honduran variety) in northern Argentina. Such foxtail growth is common in the Honduran variety when it is not subject to moisture stress. The contrast between Figs. 18.2 and 18.3 shows the importance of provenance testing in introduced species. This form of tree is subject to breakage during storms and therefore unsuitable for timber production.

Other Warm Temperate and Subtropical Pines

Pinus taeda from southeastern United States, *P. kesiya* from southeast Asia, and Mexican *P. patula* are three other species receiving major attention in moist warm temperate regions devoid of native conifers. All three grow rapidly and have been planted or tested on large areas.

Tropical America is the home of several other pine species, some native to the same climatic zones as the species mentioned above (Fig. 18.6). Nearly all have been tested in the southern hemisphere countries having large planting programs. Some have grown too slowly to be considered for large-scale use.

Ponderosa Pine

This is a native of western North America. It grows to large sizes but at relatively slow rates (1 to 2 ft per year). Because it produces valuable lumber and thrives on sites too dry for most other large trees, it is one of the two most valuable trees of western Canada and western United States.

It grows as well in New Zealand as in its native range and has been planted on over 100,000 acres in that southern hemisphere country. However, its growth rate is so slow in comparison with that of Monterey and some other pines that it is no longer in high favor.

Whereas in its native range it encounters relatively dry summers, it succeeds in parts of eastern United States with moist summers. It grows as fast there as in its native range but not enough to make it compete with the native pines; therefore its use in the east is largely limited to ornamental plantings. On moist eastern sites it suffers winter injury at less extreme low temperatures than it encounters within its native range. Its success in eastern United States depends very much on the race which is planted.

Eastern White Pine

This is a cool temperate species native to southeastern Canada and eastern United States, growing at low elevations in the north and in the mountains farther south. It was by far the most important species in the original forest, growing to large size and producing valuable, easily worked wood. It does not produce good paper, however. On good sites it outgrows practically all other native species. In its native range it suffers severely from the white pine weevil (deforming) and slightly

Fig. 18.6. Four-year-old Chiapas pine (*Pinus strobus* var. *chiapensis* or *P. chiapensis*) in northern Argentina. This Mexican species produces a soft, even-grained wood highly valued for shelves, windows, doors, etc. Both wood quality and growth rate should be considered when deciding whether to plant it on a large scale.

from the white pine blister rust (fatal). Partly because of fear of damage from those pests and partly because of inability to use thinnings for pulp it is no longer planted extensively in some parts of its native range.

It grows very well in many parts of Europe, such as northern Italy, Romania, and Great Britain, equaling its performance within its native range and far surpassing the performance of the two native European white pine species. Without the white pine weevil it develops good form and produces wood of a quality unavailable from native species. In Japan it far outgrows the native white pine species; it promises to be one of the three most planted conifers (all exotics) in the northern Japanese island of Hokkaido, even though that island is rich in native species.

This species requires too much moisture and too much winter cold to thrive in the semi-arid or subtropical parts of the southern hemisphere where most forest planting is done. In parts of New Zealand, however, it has promise, even being able to compete with Monterey pine on the basis of value of products produced per acre per year.

Norway Spruce

This is a native of northern Europe, where it is widespread and often is the most common tree. It is useful for pulp and lumber; the wood is soft but strong for its weight and is needed for some specialty products such as violins.

Aside from Sitka spruce, which can be grown only in regions with mild winters, Norway spruce seems to be the fastest growing of all spruces. It often grows twice as fast as the native spruces in parts of northeastern United States and Hokkaido. In these regions it is planted extensively (but in rather small patches) in areas south of or at lower elevations than the natural spruce–fir forest, and the extent to which it can replace native species within the natural spruce–fir forests is still unknown. It is sensitive to frosts occurring 2 to 3 weeks after the start of the growing season. Since such frosts are common in the natural spruce–fir forests of eastern America, large-scale replacement of the native spruce species by Norway spruce may not be possible.

Douglas-Fir

Douglas-fir is common in western Canada and western United States, forming the major component of the very productive forests of the Pacific Northwest and also occurring many places in the interior mountains.

It may be the single most important tree species in the world. The few other Asiatic or California species in the genus are of limited distribution and value.

As a native species, Douglas-fir owes its preeminence to its rapid growth, abundance, good form, and excellent wood quality. Also, it is a very desirable Christmas tree. These same qualities make it desirable as an exotic in several parts of the world. It thrives and is a commonly planted tree in parts of the British Isles, Andean regions of Argentina, cool moist parts of New Zealand, and parts of eastern United States.

With its very large natural range, this species is extremely variable genetically. Its success as an exotic depends very much on the race that is planted. In Nebraska, for example, the species was regarded to be no potential value until races from the southern Rocky Mountains were tested, and now it is a promising Christmas tree. In Great Britain and New Zealand fast-growing West Coast races are grown, but these fail completely in regions with cold winters.

Cryptomeria

This conifer is a native of Japan, where it grows to large sizes and produces a valuable, durable wood. It is the most commonly planted tree there. Its excellent stem form is indicated by its Japanese name *sugi* which means straight. Although native to regions in which the winter temperatures are a few degrees below freezing, this species is very adaptable. It can be grown in Singapore (minimum temperature 68°F) or in Michigan (minimum temperature −20°F). It thrives and is the most commonly planted tree at middle elevations in Taiwan, being planted there on high-rainfall sites with little or no frost (Fig. 18.7).

In the United States this species is known as an occasional ornamental but has not been tested as a forest tree. Presumably it would grow very well in moist warm temperate regions with good soils, such as the Pacific Northwest and the southern Appalachians. Those regions already have an abundance of high-value, fast-growing trees. However, they lack species with the peculiar ability of *Cryptomeria* to produce large volumes of straight, easily worked, and decay-resistant wood. Therefore it might be a valuable asset in spite of the richness of the local forest flora.

English Oak

English oak is a common and valuable tree native to the British Isles and continental Europe. There it is one of a very few native species

FIG. 18.7. *Cryptomeria* from Japan growing in central Taiwan. This species, known to the Japanese as *sugi* (or straight), is noted for its pest resistance, valuable wood, good form and rapid growth. The excellent growth of this 20-year-old plantation shows why it is one of the most commonly planted conifers in Taiwan, Republic of China, an island rich in valuable native species.

and has been managed intensively for centuries. It is important because it is common and is a high-value cabinet wood.

The United States has an abundance of valuable oak species, of which one of the most important is white oak. That species, however, is slow growing in youth and has been planted very little. English oak, which has been planted occasionally as an ornamental in northeastern United States for centuries, has recently been tested as a forest tree. At least through age 20 it grows twice as fast (2 to $2\frac{1}{2}$ ft per year) as its native counterpart, produces very large crops of acorns palatable to wildlife, and has as good form as the native American oaks. Quite possibly, this exotic might be the heart of increased oak planting in areas such as Michigan and Indiana (Fig. 18.8).

FIG. 18.8. Ten-year-old (12 from seed) English oak (large tree by author and other larger trees) and native white oak (smaller trees retaining leaves) in southern Michigan. In addition to growing nearly twice as fast as the native species under forest conditions, this exotic produces large acorn crops (up to 90 lb per tree by age 20), has good wood quality, and is suitable for city planting. It is worth planting in a region richly endowed with valuable native trees. Peculiarly, acorn production has been much heavier on young trees in Michigan than on much older native trees in some parts of Europe.

Dawn Redwood

This species, also known as metasequoia, is native to one small valley in the Peoples' Republic of China, where the few hundred living trees grow on moist soils in a warm temperate, humid climate. It was first described (1905) as a fossil species and only became known as a living species 40 years later. The first seeds were taken out of China in 1947 and distributed to many places in the temperate zone. Because of its history, it captured the imagination of many botanists and foresters and is therefore very well known considering its recent introduction to other countries. Also, it has grown very rapidly and produces a durable (but brash) wood. For these reasons and because of its adaptability to a wide range of climatic conditions, it is often considered to have great promise as an ornamental and forest tree. However, lack of seed is a serious deterrent to its widespread testing.

Ginkgo

This gymnosperm is also a rare native of the western part of the Peoples' Republic of China, but was introduced into other countries more than a century ago. It is cultivated as an ornamental in Japan, east central United States, and parts of Europe. In congested cities with air pollution problems gingko is especially valuable.

It is almost free of insect and disease pests, grows to large size, and produces a valuable even-grained wood similar to that of white pine. These facts suggest that it could become a valuable forest tree. However, for a variety of reasons it has not been tested under forest conditions. The species is dioecious, and the female trees are regarded as obnoxious because they produce large, stinking drupes in large quantities. Also, the embryos are not fertilized when the fruits drop, so a complex after-ripening period is necessary to permit fertilization and embryo development. Thus, seed is difficult to obtain and germinate in large quantities. In addition, problems encountered in planting are much more serious than are encountered with most conifers. Hence a good deal of experimentation is necessary before its forest potential can be learned.

Redwood

This unique tree, among the tallest in the world, is native to a limited area long the Pacific Coast of northern California and southern Oregon. It grows rapidly, produces durable and valuable soft wood, is free of pest problems, sprouts from the base when cut (even from very old stems), and has been the mainstay of a very large lumber industry. Nevertheless, redwood seems to have very little potential as an exotic species in other countries. The "redwood belt" of California and Oregon is characterized by mild winters, cool summers, high winter rainfall, and a large number of days so foggy that one can hardly see the lowest branches of a large tree. Only a few very small areas in other countries have this particular climate. Attempts to plant redwood under different conditions have been very numerous and have failed.

Regional Summaries of the Use of Exotics

In the previous section a few of the several hundred tree species have been discussed which have been tested or are planted to some extent outside their native ranges. Here I shall briefly discuss the present or potential use of exotics in different parts of the world. As will be seen, the role that foreign species play in the silviculture of a region depends on many factors—biological, historical, economic, and political.

Boreal Regions

The forests of boreal regions, such as Canada, Scandinavia, and Siberia, are relatively simple floristically, being composed of a few very common and widespread species. Similarly, the foreign species that can be planted in those countries are few in number. Trials of exotics have generally been most numerous and successful in the warmer, primarily agricultural parts of such countries, and forest management in the large, forested areas continues to emphasize native species almost exclusively.

Great Britain

The British Isles have only two native gymnosperms (Scotch pine and English yew), both confined to relatively small areas. Hardwood silviculture is concerned with managing naturally reproduced forests of a few native hardwoods, but softwood silviculture is concerned almost entirely with the management of plantations comprised mainly of introduced conifers (MacDonald *et al.*, 1957). The principal species are Norway spruce and European larch from continental Europe; Japanese larch from Japan; Douglas-fir, Sitka spruce, western hemlock, western redcedar, and other species from the Pacific Northwest of America; and eastern white pine from eastern America. At present the largest acreages are devoted to those species that were introduced first and have been known the longest; substantial shifts in species preference can be expected as the result of experience and further research.

Northern and Central Europe

Scandinavia and Great Britain have already been mentioned. In other parts of northern and central Europe forestry is based primarily on the few native species available. These few are adaptable to varied site conditions and therefore easy to manage. Also, through centuries of management, they have become the basis of a diversified forest industry. There are many long established arboreta and experiment stations dealing with exotics, and a large number of temperate zone species from the United States and eastern Asia have been tested. Many grow well in the cities, towns, and the agricultural areas in the warmer parts of these countries; fewer succeed in mountainous forestry regions. Among the few exotics that are planted commercially on small to moderate scales are eastern white pine and northern red oak from eastern United States, Douglas-fir from western United States, and Japanese larch. Also, European larch has been planted extensively outside its natural range. Introduced species can be expected to fill a gradually increasing role in certain ecological or industrial niches but will probably not replace natives over large areas.

Alluvial Regions of Europe

The Low Countries and the wide river valleys of southern Europe deserve special mention. They are primarily agricultural, but a limited amount of space is devoted to trees planted individually or in rows and intensively managed to produce high-value specialty products on short rotation. Poplars and elms are most common. Most of the trees planted are introduced species or hybrids between a native and an introduced species.

Northeastern United States and Adjacent Southeastern Canada

This is a region with a large number of native species capable of filling nearly all industrial needs. It is also a region with a long history of plant introduction and in which several hundred European or Asiatic species can be grown successfully. As ornamentals, European mountain ash, English and Japanese yews, Japanese magnolias, Japanese and European larches, and western American white fir have been planted commonly, whereas the native species in those same genera have been planted little. Probably over 50% of the trees and shrubs planted in large towns and cities are introduced.

Exotics also play a very large role in the Christmas tree industry, European Scotch pine, western American Douglas-fir, and western American white fir being among the most commonly planted. Altogether over 75% of the acreage is devoted to introduced species. As for forestry, a few native species (primarily red pine, eastern white pine, and white spruce) make up nearly 100% of the plantation acreage in the heavily forested northern third of the region, but foreign species or American species shifted a few hundred miles north or south are an important part of the planting program in the warmer two-thirds of the region. In these warmer areas Norway spruce has grown much faster than any of the three northeastern America spruces; Japanese and European larches have grown faster and been more suited to mesic sites than the native (swamp) American larch; any of several firs have outgrown the native balsam fir; English oak has occasionally grown twice as fast as native white oak; and European Scotch pine has shown itself to be the fastest-growing pine on certain sites. It is probable that foreign species can be employed profitably on 10 to 25% of the plantation area.

Southeastern United States

This is a region with a large number of native trees, several of which have proved of great value as exotics in other parts of the world. It is also a region of moist hot summers and winters that are warm for the most part but have occasional periods of very low temperature. These cold spells are very injurious to most introduced species that might grow fast enough to compete with the natives. Thus species introduction plays a minor role in the region's tree improvement program.

Douglas-Fir Region of Northwestern United States and Southwestern Canada

This is another region in which exotics are of little interest to applied forest tree improvement programs. The climate is moist and mild and very favorable to tree growth. The native forests contain a number of valuable, fast-growing conifers that have so far outgrown the numerous exotics that have been tested.

Hokkaido

This northernmost of the main Japanese islands is at about the same latitude and has a climate similar to that found in parts of northeastern United States. For its size it has an especially rich flora of valuable hardwoods and conifers. However, Norway spruce, eastern white pine from America, and Japanese larch from 1000 miles farther south have grown more rapidly than their native counterparts and are among the most commonly planted trees.

Hawaii

Introduced birds, mammals, crop plants, and trees dominate nearly all low elevation areas. Even at medium and high elevations introduced hardwoods and conifers have grown so much faster and taller than the small-statured native trees that tree planting is almost automatically confined to exotics. There is more debate about the desirability than about the possibility or complete replacement of the native flora in many large areas (Fig. 18.9).

Taiwan

This subtropical island of about 12,000 square miles has a remarkably rich native flora containing about 800 woody species derived in

FIG. 18.9. Exotic grasses and shrubs in Hawaii. Much of the native flora and fauna proved incapable of competing with introduced species. The vegetation in this picture, not planted and not particularly desirable, is practically 100% from foreign lands.

almost equal amounts from Japan to the north, mainland China to the west, and the Philippines and southeast Asia to the south. Many of the native species are valuable and reach large size. The climate is generally favorable for tree growth, with very high precipitation in many areas; soils in the forested mountainous areas are often thin.

Relatively few species have been introduced, and fewer still have been tested extensively. In spite of this and in spite of the abundance of valuable native species, *Cryptomeria* from Japan, China-fir from the mainland, and luchu pine from the Ryukyu Islands are planted much more extensively than native trees; they form extensive pure stands in many localities.

Australia

This continent abounds in eucalypts and has many other hardwoods, the best of which occur in rather small high-rainfall areas. It has a few valuable native conifers, all confined to such small areas as to be of limited practical importance. Introduction of northern hemisphere conifer genera, particularly the pines, provided long-fibered trees that could grow rapidly on areas previously occupied by eucalypts of limited commercial importance. This led to a very large pine planting program, with Monterey and slash pine the most common species. The pine plantations furnish a major part of the nation's forest products.

New Zealand

The climate is generally moist and varies from subtropical to cool temperate. The approximately 15 million acres of native forest are composed mainly of hardwoods, with small amounts of valuable but extremely slow-growing conifers. The native forest is generally confined to the mountainous, least accessible regions. It is to be supplemented by approximately 2 million acres of plantations nearly all composed of eucalypts from Australia, pines from the northern hemisphere, and an assortment of other northern hemisphere conifer genera. Several of these proved to be particularly well adapted to an area of well-drained pumice soils that is relatively level and accessible but unsuited to agriculture. These exotic plantations supply a major portion of the country's lumber and pulpwood (Weston, 1957).

Monterey pine has been so successful as to account for over half the plantation acreage. This is both a blessing and a problem, and a continuous search is underway for other species that might serve to diversify the planting program. Several, such as ponderosa pine, western redcedar, Douglas-fir, and Japanese larch, grow as well as in their native ranges, but Monterey pine still remains dominant.

Republic of South Africa

In contrast to New Zealand, this country was originally mostly non-forested and was devoid of native conifers. Cutting is restricted in the remaining ½ million acres of native forest, and wood production is almost solely limited to the 2 million acres of exotic plantations. These can be divided into three approximately equal-sized groups according to species, ownership pattern, management, and use. One group consists of the plantations of one acacia (wattle) species, grown by farmers and managed on 5 to 10 year rotations primarily for its tannin-producing bark. A second group of numerous small farmer-owned plantations of several eucalypt species, managed primarily for mine props, posts, and other roundwood products. Efforts are being made to find eucalypts that are better suited for lumber production and increase utilization of both acacia and eucalypts for pulp and chipboard. A third group consists of publicly owned conifer (mostly pine) plantations managed for lumber and pulp. The amount and seasonal distribution of rainfall differ appreciably in the conifer growing regions, so a number of warm temperate and subtropical species have been planted in quantity.

Tropical African Countries

Tree planting has a much shorter history and has been on a much smaller scale in the tropical African countries than in the southern hemisphere countries mentioned previously. Most has been confined to medium rainfall zones originally occupied by a savannah-type vegetation of grass and small- to medium-sized trees. Subtropical pines and eucalypts have been used most commonly. In the last decade there has been a very great increase in interest in tree planting and particularly in the testing of Caribbean, Indian, and southeast Asian pines. Much matching of species to sites needs to be done but overall results are promising.

Misiones, Argentina

This province of northeastern Argentina, between Brazil and Paraguay, was originally covered with a dense high forest composed of a few hundred hardwood species and one conifer. Most of the original forest remains, but its composition has changed as the result of continuous removal of the best trees. Tree planting started 40 years ago with the planting of a small area to the native conifer (an *Araucaria*). Planting then increased rapidly, primarily to supply the needs of a pulp mill. About 25 years ago slash and loblolly pines were found to grow more rapidly and to be adapted to a wider range of site conditions than the native conifer, which they have gradually replaced. Trials started about 15 years ago indicate that there will be further changes, as some Mexican and Caribbean pines seem most suitable for some areas. The 1966 Forestry supplement of the Argentine journal *IDIA* is devoted to exotic forestry.

Altogether the pulp plantations and agricultural lands account for less than 25% of the province's area, the remainder being covered by unmanaged natural forest of relatively low economic value. Limited experience indicates it would be very difficult to manage that native forest in a way to obtain a high proportion of valuable native trees, most of which grow slowly when young. They have not regenerated well naturally after cutting nor have they been planted successfully. On the other hand, several foreign trees in addition to the pines have been planted and have been able to compete with the lush weedy growth. Thus, regardless of the ecological desirability of a shift away from native species, such a shift seems probable if management is intensified. This problem is met in many subtropical and tropical areas having complex rain forests.

Study Questions

1. Name the 10 most common agricultural products in your region and tell the country of origin of each.

2. To what extent is forestry dependent on introduced tree species in Finland, Australia, New Zealand, Mexico, Argentina, Nigeria, Spain, Japan, Peoples' Republic of China, southeastern United States, northeastern United States, Great Britain, and northern Canada?

3. What is the average number of years required between the initial introduction and large-scale commercial use of an introduced tree species? What are some factors making the interval as long as it is?

4. Why are most native tree species more biologically successful than most introduced tree species? If this is true, give four reasons why some exotics outperform native species, and give at least two examples showing why each reason is true.

5. Why should there be a lag between an environmental change and the evolutionary response?

6. Eastern white pine, ponderosa pine, Douglas-fir, Japanese larch, European larch, Monterey pine, and eastern cottonwood have all been used to a considerable extent as exotics. Tell which are economically important within their native ranges, and for the others, tell why they have not been important as native species.

7. What things should one consider when considering whether or not to introduce a given species into a foreign country?

8. Give three examples each of small-range and large-range species that have become successful exotics.

9. Why are monotypic genera in a class by themselves? Name five.

10. Why have the best eucalypt species not been considered for large-scale use in exotic plantations? Why have so many thousands of acres of eucalypt plantations not produced as useful products as was hoped?

11. Why are the majority of north European species successful as exotics?

12. A given species has not been introduced into your region. How do you learn, short of testing it yourself, its probable adaptability to your conditions?

13. Give the required degree of similarity between recipient and donor regions with regard to summer temperature, winter temperature, precipitation, hours of sunshine, seasonal distribution of precipitation, latitude, longitude, slope, and aspect.

14. Give the successful donor regions if northeastern United States is to be the recipient region. If New Zealand is to be the recipient region. If Great Britain is to be the recipient region. If eastern Union of South Africa is to be the recipient region. If Taiwan is to be the recipient region.

15. Under what conditions can seed be obtained by mail? By personal collections?

16. What are the benefits of cooperative approaches to tree introduction? Of provision for future seed supplies? Of adequate testing before commercial release?

17. Describe the eucalypt planting areas of the world as to temperature, rainfall, need for roundwood, need for sawlogs, abundance of fast-growing native hardwoods, abundance of fast-growing native conifers. Do the same for the Monterey pine planting areas.

18. Contrast the history of exotic testing and the role of exotics in Great Britain, eastern and western United States, and Taiwan.

19

Polyploidy, Aneuploidy, and Haploidy

Definitions

Individual chromosomes are classified as homologous (having similar structure and genes) or nonhomologous (having different structure and genes). A normal diploid plant contains n pairs of homologous chromosomes, each pair being nonhomologous to the other pairs. A normal diploid also contains two complete sets of n chromosomes, all chromosomes in one set being homologous to one chromosome in the other set.

A tree's chromosome complement is termed haploid, diploid, triploid, tetraploid, pentaploid, hexaploid, or octoploid according to whether it contains $1n$, $2n$, $3n$, $4n$, $5n$, $6n$, or $8n$ chromosomes, respectively. Trees with three or more complete sets of chromosomes are collectively termed polyploids. Sometimes in evolutionary literature a distinction is made between the x-number (x is the number in a remote ancestor) and the n-number (n is the number in a modern normal plant), but this distinction will not be made here.

Trees having more or less than complete sets of chromosomes $[2n - 1)$, $(3n + 2)$, etc.] are collectively called aneuploids (meaning not true ploids).

Polyploids can be subdivided according to whether the chromosome sets are derived from the same parental species (autopolyploids) or different parental species (allopolyploids). The prefixes auto- (meaning self) and allo- (meaning other) may also be used to form words such as autotetraploid, allohexaploid, etc. In a typical autotetraploid there are four sets of homologous chromosomes. In a typical allotetraploid, chromosomes of one species are different enough from those of the other species that there are only two sets of homologous chromosomes. Numerous intermediate conditions occur.

Colchicine (derived from the autumn crocus, *Colchicum*) and podophyllin (derived from the mayapple, *Podophyllum*) are two chemicals used specifically to double chromosome numbers in plants. Both are poisonous alkaloids and owe part of their chromosome doubling ability to their poisonous nature.

Techniques of Chromosome Study

Between cell divisions, chromosomes are long, threadlike, and so intertwined that they are very difficult to study. Hence most cytological studies are made when cells are in metaphase condition (middle stage of cell division). At metaphase the chromosomes are coiled, shortened, and stain most heavily. Also, they are most apt to be in a single plane and spread out so that they can be seen separately. As a source of dividing cells, one of the following five tissues is usually used. (1) Root tips are easily obtained from seeds germinated in a laboratory and are easy to prepare for study. Moreover, cell division tends to occur more continuously in roots than in other tissues. The chromosomes are $2n$ in number, which increases the difficulty of counting if n is large. (2) Growing points in expanding vegetative buds can be used if seed is unavailable, and it is desired to study established trees. The chromosomes are $2n$ in number, and the period of active cell division may be extremely short, necessitating a knowledge of the stages of bud development. (3) Pollen mother cells divide meiotically to produce daughter cells with n chromosomes. They are a convenient source of material for the study of pairing and segregation. Preliminary studies of flower stage development and timing of collections are very important. In some trees all meiotic divisions occur within a few hours. Many pines and birches undergo meiosis 6 to 8 months before pollen shedding. (4) The daughter cells resulting from meiotic division of pollen mother cells then proceed to divide mitotically to form tetrads (groups of four pollen grains). The germinating pollen grains divide mitotically to form two (generative and vegetative) nuclei. In these divisions the cells have n chromosomes. (5) Female gametophytes also undergo mitotic division with reduced (n) chromosome number. Pederick (1970) used these in pine because the chromosomes were longer and easier to study than in other tissues.

A "squash" or "smear" technique is usually used in preference to the older techniques that involved cutting thin sections on a microtome. Tissue is treated with a hydrolyzing agent to dissolve the substances joining cells together. Then the tissue is placed on a microscope slide, macerated, stained and covered with a cover glass to which firm pressure is applied. Thus the cells can be flattened and the chromosomes spread out in one plane for easy study.

As an example of a detailed technique, I quote that used by Saylor (1972). (1) Pretreat root tips 24 to 36 hr at 12°C in a 0.03% oxyquinoline solution to cause the living chromosomes to contract and separate from each other. (2) Kill and fix the root tips 1 to 4 hr in a 3:1 ethyl alcohol–acetic acid mixture to preserve the chromosomes. (3) Soak the root tips 10 to 15 min in normal hydrochloric acid (hydrolyzing agent) to cause the cells to separate. (4) Wash with the 3:1 alcohol–acetic acid mixture. (5) Place a root tip on a microscope slide, cover with acetocarmine (a stain which dyes chromosomes much more than other cell parts) and macerate thoroughly with a sharp lancet. (6) Apply a cover glass, blot, and press heavily with the thumb to flatten the cells. (7) Apply a ring of waterproof compound around the edge of the cover glass to prevent evaporation. (8) Study within a few days. The slides can be made permanent, but this is a laborious process and most cytologists prefer to store material in a fixative solution rather than store all completed slides.

For chromosome studies, high quality, light-conducting microscopes are used. They are fitted with special light sources and light condensers. Many microscopists have a small camera that can be placed over the eyepiece to photograph good cells as they are found. Camera lucida attachments are also used. These are small prisms attached to eyepieces and designed to enable an eye to see simultaneously a chromosome and a pencil point tracing the chromosome outline on paper.

To make a chromosome count, a microscopist scans a slide with $100 \times$ magnification until he finds a cell whose chromosomes are well separated. He immediately switches to higher magnification (usually $700 \times$ to $1000 \times$) for detailed observation. If only a chromosome count is desired and the n number is small, this may be done in a few minutes for a particular cell. Counts are usually confirmed by studying several dividing cells per slide. For studies of detailed chromosome structure (karyotype analysis), a photograph and/or drawing is necessary.

The limit of resolution of a good microscope is slightly less than 1 micron (1 μm = 0.001 mm), and chromosomes in various tree species vary from about 2 to about 13 μm. The n numbers in various tree species vary from 6 to 23, with many species having $4n$ or $6n$ chromosomes.

In the most favorable material it is possible to determine chromosome number accurately and to see some details of chromosome structure. With those species having 70 to 80 very small chromosomes, exact counts are difficult to obtain and little detailed structure can be seen.

Karyotypes of Pines

A karyotype is a plant's chromosome complement and is usually pictured in such a way as to show number, length, and details of the structure of individual chromosomes. Karyotype analysis is most advanced in the conifers (especially pines) because they are economically important and have relatively large chromosomes. For that reason, pine karyotypes are chosen for detailed description, as described in papers by Pederick (1970) and Saylor (1972).

Pine root tips have $2n = 24$ chromosomes. At metaphase they are 8 to 13 μm long and 1.5 to 2 μm thick. They are somewhat longer and thinner at metaphase of the mitotic division of a female gametophyte. Each chromosome appears similar to a pair of sausages, with a near-medial constriction (centromere).

For convenience the chromosome pairs have been numbered from 1 (longest) to 12 (shortest). Chromosomes 1 to 11 are nearly the same length, varying by only about 15%. The arms (portions on either side of the centromere) in those chromosomes are nearly equal in length, the short arm generally being 5 to 15% shorter than the long arm. Chromosome 12 is much different, being about 40% shorter than chromosome 1; the ratio of the long to the short arm is about 2:1. This general description fits all pines except those belonging to the subsection *Sylvestres*, in which there are two rather than one short chromosomes. The relative lengths of the two arms of each chromosome differ slightly but consistently among species.

In most root tip chromosomes it is possible to see one to two small constrictions in each arm. In chromosomes of the female gametophyte it is possible to see one to ten small constrictions per arm. The number and distribution of such small constrictions is constant for a species.

In spite of the species differences they found, Saylor and Pederick noted that pines are more remarkable for their similarity than for their dissimilarity in chromosome structure. This indicates that pine evolution has proceeded primarily through gene substitution rather than through gross chromosomal changes. It also shows why many species can be crossed with each other to produce relatively fertile F_1 hybrids.

Rareness of Polyploidy in Gymnosperms

The gymnosperms as a group have relatively large chromosomes, rather similar karyotypes, and relatively little variability in basic chromosome number. The numbers $n = 11$ and $n = 12$ are most common. The following list (from Darlington and Wylie, 1955; Mehra and Khoshoo, 1956) shows the base numbers in most gymnosperm genera.

$n = 10$ in *Sciadopitys*

$n = 11$ in *Actinostrobus, Athrotaxis, Callitris, Chamaecyparis, Cryptomeria, Cunninghamia, Cupressus, Juniperus, Libocedrus, Metasequoia, Sequoia, Taiwania, Thuja, Torrey, Widdringtonia*

$n = 12$ *Abies, Cedrus, Cephalotaxus, Ginkgo, Keteleeria, Larix, Picea, Pinus, Taxus, Tsuga*

$n = 13$ in *Agathis, Araucaria*

$n = 11, 12$ or 19 in *Podocarpus*

$n = 12$ or 13 in *Pseudotsuga*

$n = 22$ (probably 12 originally) in *Pseudolarix*

There are only three known natural polyploids in the gymnosperms. They are redwood ($6n$), Pfitzer juniper ($4n$), and golden larch ($4n$). These three have extremely different growth rates, forms, winter hardiness, abilities to reproduce vegetatively, and uses. It is not possible to point to any one trait that is common to all three, and this might be considered due to increased chromosome number per se.

Artificial tetraploids have been made in numerous species of pines, spruces, and larches. Generally they have grown very slowly and are of interest as dwarf plants for horticultural purposes (Fig. 19.1). Kiellander (1950) made chromosome counts of the naturally occurring dwarf Norway spruces in a Swedish nursery. Not all the dwarfs were polyploids, but all the polyploids he found were dwarfs. Thus there is little current interest in the artificial production of polyploid conifers, except possibly for the production of triploids by crossing $2n$ and $4n$ trees. One triploid larch hybrid grew moderately fast. Few other triploids have been produced because of the scarcity of flowers on the dwarf tetraploids.

Aneuploidy has, however, been important in the evolution of gymnosperms. Presumably the original base number was $n = 11$ or 12, and genera evolved through addition or loss of one or two chromosomes. Thomas and Ching (1968) postulated a possible mechanism. Douglas-fir from western United States has $n = 13$ chromosomes, whereas other species in the genus have $n = 12$. Presumably two nonhomologous chromosomes in an ancestral species exchanged chromosome arms in a tree

Fig. 19.1. Dwarf eastern white pine similar to types produced by chromosome doubling. Dwarfs such as this are the common result of polyploidy breeding in several conifers. Similar dwarfs in Douglas-fir have been found to be aneuploids, containing an extra chromosome.

that was then selfed. Then, presumably there were irregularities at meiosis so that the two chromosomes could become three. If their postulated mechanism is correct, the extra chromosome in American Douglas-fir constitutes a rearrangement rather than a duplication of chromosomal material in the ancestor. A recently found aneuploid Douglas-fir ($2n = 27$) was dwarf (Ching and Doerksen, 1971).

Importance of Polyploidy in Angiosperms

Evolution

The situation is very different in angiosperms. Approximately one-third of the species of angiosperms are polyploids (Stebbins, 1950). They seem to be as frequent in trees as in herbaceous plants. Their frequency in tree genera is shown in Table 19.1.

In a few cases polyploid has been a major evolutionary process giving rise to entire genera or families. The genus *Tilia* (basswood) has $n = 41$ chromosomes, whereas all its relatives have $n = 7$ chromosomes. Presumably a hexaploid ($6n = 42$) plant was produced from an $n = 7$ ancestor and lost a chromosome to become an $n = 41$ plant. In the birch family, *Alnus* and *Betula* seem to be polyploid derivatives of an $n = 7$ species. The family Magnoliacea ($n = 19$) has relatives with $n = 7$ and $n = 12$ and presumably arose from ancient ancestors with those numbers by hybridization and subsequent chromosome doubling. Stebbins (1950) postulated that the subfamily Pomoideae ($n = 17$) of the Rosaceae arose through an ancient hybridization and subsequent chromosome doubling between a spiraea-like plant ($n = 9$) and a cherry-like plant ($n = 8$).

In more cases polyploidy seems to have led to an evolutionary dead end. A species originated and was successful but did not necessarily give rise to a great number of other species. Presumably most species with more than $2n$ chromosomes are allopolyploid derivatives of hybrids rather than autopolyploids, but this conclusion is based on inference rather than actual data. Several species (e.g., white ash, red maple, and paper birch) contain races that differ in chromosome number.

Most species contain occasional polyploids. The number of species that are entirely polpyloid or contain polyploid races varies greatly among genera and families (Table 19.1). Polyploid species are common in the cherries and birches, but have yet to be found in the oaks and eucalypts. There is no general relationship between prevalance of polyploidy and degree of evolutionary primitiveness or life form. Polyploids are as common in primitive as in advanced families and in trees as in shrubs.

Stebbins (1950) thought that polyploidy confers a degree of evolutionary variability that permits a species to adapt readily to new habitats or rigorous growth conditions. Thus he considered that plants found in Arctic and glaciated regions are apt to have high chromosome numbers.

Bowden (1940) could not find such trends. He made intragroup studies of over 100 species, comparing polyploids with diploids in the same genus or family. Polyploids occupied favorable versus rigorous haibtats and warm versus cold habitats with equal frequency. Khosla (1968) studied a large number of woody species in India. Approximately 26% of the species were polyploids whether from tropical or temperate regions.

The viewpoint on this problem depends very much on the way in which comparisons are made. *Salix* and *Betula* are two large genera containing many polyploids and are common in boreal forests. *Quercus* and *Eucalyptus* are two large genera containing few or no polyploids and are common in temperate or subtropical regions. A Canada versus

TABLE 19.1

Chromosome Numbers of Dicotyledonous Forest Tree Genera[a]

Family and genus	Base (n) number	No. of species that are			
		2n	4n	6n or 8n	Variable
Casuarinaceae					
Casuarina	11, 12, 13	5	3	—	1
Salicaceae					
Chosenia	19	1	—	—	—
Populus	19	17	8	8	—
Salix	19	27	8	8	3
Juglandaceae					
Carya	16	4	4	—	—
Juglans	16	10	—	—	—
Pterocarya	16	3	—	—	—
Fagaceae					
Castanea	12	4	—	—	—
Fagus	12	1	—	—	—
Quercus	12	44	—	—	—
Betulaceae					
Alnus	7	—	13	10	5
Betula	14	15	10	7	2
Carpinus	8	8	—	1	1
Ostrya	8	4	—	—	—
Ulmaceae					
Celtis	10, 11, 14	4	2	—	1
Ulmus	14	14	2	—	2
Zelkova	14	1	—	—	—
Moraceae					
Broussonetia	13	2	—	—	—
Cecropia	14	3	—	—	1
Morus[b]	14	12	1	2	1
Santalaceae					
Santalum	10	1	—	—	—
Trochodendraceae					
Trochodendron	19	1	—	—	—
Tetracentraceae					
Tetracentron	19	1	—	—	—
Eupteleaceae					
Euptelea	7	1	—	—	—
Cercidiphyllaceae					
Cercidiphyllum	19	1	—	—	—
Magnoliaceae					
Liriodendron	19	2	—	—	—
Magnolia	19	21	3	8	—
Lauraceae					
Cinnamomum	12	7	—	—	—
Laurus	7	—	—	—	—
Lindera	12	3	—	1	—
Persea	12	3	—	—	—
Sassafras	12	1	—	—	—

Saxifragaceae					
Deutzia	13	5	8	—	—
Dichroa	18	1	—	—	—
Hydrangea	18	11	1	—	—
Itea	11	3	—	—	—
Hamamelidaceae					
Corylopsis	12	4	—	—	—
Liquidambar	15	1	—	—	—
Symingtonia	32	1	—	—	—
Eucommiaceae					
Eucommia	17	1	—	—	—
Platanaceae					
Platanus	21	2	—	—	—
Rosaceae—Prunoideae					
Prunus[b,c]	8	35	12	7	7
Rosaceae—Pomoideae					
Amelanchier	17	5	5	—	1
Crataegus	17	5	5	—	—
Malus[c]	17	13	7	—	—
Pyrus[c]	17	15	1	—	1
Sorbus	17	8	12	—	—
Leguminosae—Mimosoideae					
Acacia	13	45	11	—	—
Albizzia	13	7	1	1	1
Prosopis	14	11	2	1	2
Leguminosae—Caesalpinioideae					
Amherstia	12	1	—	—	—
Bauhinia	14	11	1	—	—
Caesalpinia	11, 12	15	1	—	—
Cassia	8, 12, 13, 14	34	4	—	8
Cercis	6, 7	3	—	—	—

Gleditsia	14	7	—	—
Gymnocladus	14	1	—	—
Haematoxylon	12	1	—	—
Leguminosae—Lotoideae				
Cladrastis	14	1	—	—
Robinia	10	6	—	—
Sophora	9, 14	9	—	—
Oxalidaceae				
Averroha	11, 12	1	—	—
Erythroxylaceae				
Erythroxylum	12	1	—	—
Rutaceae				
Acronychia	18	—	2	—
Aegle	9	1	1	1
Evodia	18, 20	1	2	—
Micromelum	9	1	—	—
Murraya	9	2	—	—
Skimmia	15, 16	2	—	—
Toddalia	18	—	1	—
Zanthoxylum	16, 17, 18	1	14	2
Simaroubaceae				
Ailanthus	31, 40	3	—	—
Brucea	12	1	1	—
Picrasma	12	2	1	—
Burseraceae				
Canarium	13	1	—	2
Garuga	13	2	—	—
Protium	11	1	—	—
Meliaceae				
Aglaia	20	1	1	—
Amoora	20	1	—	—

TABLE 19.1 (Continued)

Family and genus	Base (n) number	No. of species that are			
		2n	4n	6n or 8n	Variable
Meliaceae (continued)					
Aphanamixis	18	1	1	—	1
Azadirachta	14	1	—	—	—
Cedrela	12, 13, 14	1	4	—	—
Chickrassia	13	1	—	—	—
Chisocheton	23	1	—	—	—
Dysoxylum	7, 9, 10	1	—	9	—
Heynea	14	1	—	—	—
Melia	14	4	—	—	—
Swietenia	24	1	—	—	—
Euphorbiaceae					
Aleurites	11	4	1	—	1
Hevea[d]	9	1	6	2	3
Coriariaceae					
Coriaria	20	1	1	—	—
Anacardiaceae					
Anacardium	12	1	—	—	1
Lannea	15	1	—	—	—
Mangifera	20	5	—	—	—
Poupartia	12	1	—	—	—
Spondias	16	2	—	—	—
Rhus	14, 15	12	—	—	—
Aquifoliaceae					
Ilex	9, 10	—	27	—	—
Celastraceae					
Celastrus	23	3	—	—	—
Elaeodendron	17	1	—	—	—
Euonymus	8	1	—	—	—
Gymnosporia	9, 12	1	2	1	—
Lophopetalum	20	1	—	—	—
Hippocrataceae					
Hippocratea	14	2	—	—	—
Staphyleaceae					
Staphylea	13	4	—	—	—
Turpinia	13	2	—	—	—
Icacinaceae					
Apodytes	12	1	—	—	—
Iodes	10	1	—	—	—
Neostachyanthus	20	1	—	—	—
Pyrenacantha	20	1	—	—	—
Aceraceae					
Acer	13	41	3	2	2
Hippocastanaceae					
Aesculus	20	7	2	1	—
Sapindaceae					
Dobinia	7	2	—	—	—
Dodonea	14, 15, 16	2	—	—	—
Erioglossum	13	1	—	—	—

Family / Genus					
Euphoria	15	1	—	—	—
Nephelium	11, 14, 15, 16	2	—	—	—
Sapindus	15	2	—	2	—
Xerospermum	16	1	—	—	—
Sabiaceae					
Meliosma	8	1	6	—	—
Rhamnaceae					
Berchemia	12	1	1	—	—
Sageretia	12	2	—	—	—
Zizyphus	10, 12, 13	4	4	—	1
Tiliaceae					
Tilia	41	7	4	—	—
Ochnaceae					
Ochna	14	1	—	—	—
Dipterocarpaceae					
Dipterocarpus	10	3	—	—	—
Lecythidaceae					
Careya	13	1	—	—	—
Nyssaceae					
Davidia	11	1	—	—	—
Nyssa	11	1	—	—	—
Combretaceae					
Combretum	13, 14, 16	6	—	—	—
Terminalia	12, 13	8	2	—	1
Myrtaceae					
Decaspermum	11	—	—	1	—
Eucalyptus	11	66	—	1	—
Eugenia	11	12	6	4	3
Melastomaceae					
Melastoma	10	1	—	—	—
Osbeckia	10	1	1	—	—
Oxyspora	10	—	1	—	—
Cornaceae					
Cornus	10, 11	16	—	—	—
Ericaceae					
Arbutus	13	5	—	—	—
Oxydendron	12	1	—	—	—
Ebenaceae					
Diospyros	15	3	1	2	—
Oleaceae					
Fraxinus	23	18	3	3	3
Olea	23	2	—	—	—

[a] From Darlington and Wylie (1955) and Mehra and Khosla (1968).

[b] One species has 22n chromosomes.

[c] Many cultivars are triploid.

[d] *Hevea brasiliensis* has 4n, 8n, and 16n races.

Australia or a Siberia versus Central America comparison based upon total numbers of species would be weighted very heavily by the relative prevalence of those four genera in each region, whereas a cool versus warm climate comparison within a genus would show no trend.

Tree Breeding

Polyploidy has been important in the development of new types of agricultural plants, among them strawberries, tobacco, New World cottons, some bananas, and wheat. Some of the best varieties of those crops are allopolyploid derivatives of ancient species hybrids. Polyploidy also promises to be important in the development of new varieties of forest trees. Ruby horsechesnut ($4n$ *Aesculus carnea* $= 2n$ *A. hippocas-tanum* $\times 2n$ *A. pavia*) is an allopolyploid believed to have originated in 1858 (Li, 1956). It is a desirable ornamental. Pumpkin ash ($2n$ *Fraxinus tomentosa* $= ?$ $4n$ *F. americana* $\times 2n$ *F. pennsylvanica*) is also a suspected allopolyploid hybrid. Interestingly, it seems to be polyphyletic, arising anew in each of several widely separated localities. Several magnolia cultivars are probable allopolyploids or aneuploids with high chromosome numbers. The total list of polyploid tree species and species containing polyploid races is long (Table 19.1).

Size attained and use characteristics of some polyploids and related diploids are summarized in Table 19.2. Note that in general there is no consistent relationship between chromosome number and those traits that might make a tree useful. Therefore, in making polyploids, a breeder is attempting to produce something different, and it may be better or worse than the diploid from which it was derived.

Autotetraploids have been produced artificially in birch, poplar, elm, alder, and black locust. They generally grew more slowly than diploids (however, much faster than $4n$ conifers). It is possible that this slowness could be overcome by a few generations of breeding.

Triploids have proved of much greater immediate value. The number of commercially valuable $3n$ clones of apples and pears is out of all proportion to the frequency with which triploidy occurs in those genera. Müntzing (1936) was the first to discover a triploid forest tree. He found a clone of European aspen characterized by unusually large leaves and exceptionally rapid growth, which proved upon examination to have $3n$ chromosomes. Several other triploid clones have since been found in European and American aspen, and nearly all are fast growing. Scandinavian and German workers have developed methods of mass producing them, and work on mass production in American aspen is underway at the Institute of Paper Chemistry.

TABLE 19.2

Growth and Use Characteristics of Diploid and Polyploid Species in Some Temperate Zone Genera

Genus	Level of ploidy	Growth and use characteristics of different species
Alnus	4n	A. incana and rubra are 50–60 ft trees, other spp. are shrubs
	6n	A. cordata and subcordata are 50-ft trees
	8n	A. japonica and spaethii are 80-ft trees
Betula	2n	B. lenta, nigra, and pendula are medium-large trees producing valuable wood; many species are shrubs
	4n and 6n	Polyploid races of B. papyrifera are large valuable trees
	8n	B. papyrifera var. occidentalis, B. grossa, allegheniensis and davurica are valuable trees; there are no 8n shrubs
Fraxinus	2n	Among the diploid species are shrubs and large trees
	6n	F. chinensis and tomentosa are large trees, the latter producing weak wood. 2n, 4n, and 6n races of F. americana differ only slightly
Tilia	2n	T. oliveri, neglecta, petiolaris, americana, europea, and platyphyllos are small to large, commercially valuable trees
	4n	T. tuan and maximowicziana are large, valuable trees
Acer	2n	A. pennsylvanicum, japonicum, and palmatum are shrubby or small trees, A. negundo is a vigorous tree with weak wood, A. saccharum is a large (130 ft) tree with valuable hard wood
	4n or 6n	A. carpinifolium is a small tree, A. pseudoplatanus, rubrum, and saccharinum are large trees producing wood of light to medium density
Magnolia	2n	M. wilsoni and obovata are small and large trees, respectively
	4n	M. liliiflora and acuminata are small and large trees, respectively
	6n	M. dawsoniana and grandiflora are small and large trees, respectively
Morus	22n	M. nigra is of medium size for the genus

Swedish and German workers have produced fast growing triploid alder and elm seedlings (Johnsson, 1956; Eifler, 1955). Triploid birches produced in the early 1950's grew more slowly than diploids, however.

Allopolyploidy, or hybridization followed by chromosome doubling, probably offers more promise than autopolyploidy. Usually diploid F_1 species hybrids do not breed true, so their usefulness depends on the extent to which they can be mass produced or can be propagated vegetatively. However if the chromosome number is doubled, such hybrids can breed true. Norway maple has been crossed with some of its Asiatic relatives to produce hybrids with desirable growth and foliage character- istics. These cannot be utilized at present because they are difficult to grow by cuttings and almost impossible to mass produce because of differences in flowering time of the parental species. It is possible that they could be made into useful new tree varieties if their chromo- some numbers were doubled. Many other similar examples could be quoted.

To date, induced polyploidy has given the greatest practical im- provement in the poplars, a genus that contains no polyploid species. This indicates that chromosome doubling may also play a role in the improvement of other genera without natural polyploid species (e.g., oak and eucalyptus) as well as in genera containing several natural polyploid species (e.g., eugenia, ash, maple, and birch).

Breeding Behavior of Polyploids and Aneuploids

Fertility of Auto- and Allotetraploids

At meiosis in a normal diploid, the chromosomes line up at the equatorial plate, and homologous chromosomes pair with each other. Then one member of each pair proceeds to each pole so that each daughter cell receives a complete set of $1n$ chromosome. The tree is fertile.

If the chromosome number of that tree is doubled to produce an autotetraploid, there are two sets of homologous chromosomes. At meiosis they associate in groups of four. The four homologous chromosomes in one group may proceed 2 and 2, 1 and 3, or 3 and 1 to opposite poles. An egg or pollen grain may receive 2 of chromosome-1, 3 of chromosone-2, 1 of chromosome-3, etc. In other words, the eggs and pollen grains receive unbalanced chromosome complements and die. The tree is apt to be sterile.

Now let us consider a diploid hybrid and its allopolyploid derivative. The diploid hybrid contains $2n$ chromosomes, $1n$ from one species and $1n$ from the other species. Chromosomes from one species may be non- homologous with chromosomes from the other species. If so, they do not pair and proceed 1 and 1 to the poles at meiosis. Instead they may proceed 2 and 0, 1 and 1, or 0 and 2, in which case the eggs and pollen grains receive unbalanced chromosome complements and die. However, if the chromosome number of that hybrid is doubled, the resulting $4n$ tree has two complete sets of $2n$ chromosomes, each set homologous to the other. At meiosis the chromosomes pair properly and proceed 1 and 1 to the daughter cells. The tree is fertile.

Thus we can formulate a general rule: if a diploid is fertile, its tetraploid is apt to be sterile, and if a diploid is sterile, its tetraploid is apt to be fertile. Acually there are various intermediate conditions such that diploids and their tetraploid derivatives can be moderately fertile.

If an F_1 diploid hybrid is sterile or produces variable offspring, it can become fertile and true breeding if its chromosome number is doubled. This is true because all members of one homologous set of chromosomes are exactly like their counterparts in the other set. This property makes polyploid induction especially useful in conjunction with hardwood hybridization projects.

Fertility of Triploids and Pentaploids

A triploid contains $3n$ chromosomes. At meiosis, they associate in groups of 3 at the equatorial plate and migrate 1 and 2 or 2 and 1 to the egg cells or pollen grains. A gamete may receive 2 of chromosome-1, 1 of chromosome-2, 2 of chromosome-3, etc. In other words, the gametes receive unbalanced chromosome complements and are nonfunctional. Thus, triploid trees are usually sterile. The same is true of pentaploid ($5n$) trees.

Triploids and pentaploids produce occasional functional pollen grains which can be detected by their nonshrunken appearance under a microscope. Nearly always these are the result of nondisjunction and are found to contain a complete complement of $3n$ or $5n$ chromosomes. Such unreduced pollen grains can be applied to $2n$ trees ($1n$ egg cells) to produce $4n$ or $6n$ trees.

Fertility of Aneuploids

Aneuploids have $2n - 1$, $2n + 1$, etc., chromosomes. Because the chromosome sets are incomplete there is no way that normal pairing and segregation can occur during meiosis. Consequently, aneuploids are usually sterile. This is the explanation for the seedless varieties of banana and pineapple. Also, some cultivars of magnolia have polyploidy combined with aneuploidy and set very little seed.

As already noted, American Douglas-fir ($n = 13$) is believed to be an aneuploid derivative of an $n = 12$ ancestor. But modern Douglas-fir is fertile. How is that possible? Presumably the ancestor was nearly sterile but could produce an occasional pollen grain with $n + 1$ chromosomes and an occasional egg with $n + 1$ chromosomes. If, through selfing, one seed with $2n + 2 = 26$ chromosomes could be produced, it could grow into a tree that could produce more seeds with $2n = 26$ chromosomes. Thereafter there was no problem with fertility.

Hybrids between Trees Differing in Chromosome Number

Differences in chromosome number do not usually prevent two trees from crossing with each other. Some $2n \times 4n$ and $2n \times 6n$ hybrids can be made rather easily. Such hybrids may be vigorous. They are, however, apt to be sterile because of abnormal pairing at meiosis. Thus a change in chromosome number, whether through aneuploidy or polyploidy, may be the start of a process leading to the formation of a new species. Returning to the Douglas-fir example, once a change had taken place, $2n = 24$ trees could cross with each other and produce fertile offspring, $2n = 26$ trees could cross with each other and produce fertile offspring, and hybrids between the two groups could be produced. Such hybrids were sterile, though, so that gene interchange between the two groups was no longer possible. Thereafter selection could produce different effects in the two populations.

Segregation Ratios in Polyploids

Segregation ratios may be quite different in polyploids and diploids. Assume a diploid (Aa) tree, with A being dominant. If it is selfed and the F_1 hybrids are crossed, the result will be an F_2 generation segregating 1 AA:2 Aa:1 aa genotypically and 3:1 phenotypically.

Assume that the tree has its chromosome number doubled to become AAaa. After selfing and crossing of the F_1 hybrids, the F_2 genotypic segregation ratio will be 1 AAAA:4 AAAa:6 AAaa:4 Aaaa:1 aaaa. If gene A is completely dominant (AAAA and Aaaa trees indistinguishable), the phenotypic segregation ratio will be 15:1. However, the genes may behave in such a manner that Aaaa and aaaa trees are distinguishable from the remainder, in which case the phenotypic segregation ratio might be 11:4:1.

Also, tetraploids can have four alleles per locus per tree, whereas diploids have only two alleles per locus per tree. Thus, for more than one reason, one should expect very different segregation ratios in a polyploid species, such as red maple, than in the diploids that have been studied most intensively to date.

Response to Inbreeding

Response to inbreeding can also differ between tetraploids and diploids. If a diploid Aa tree is selfed, $\frac{1}{2}$ of its offspring are AA or aa, and the inbreeding rate is $F = \frac{1}{2}$ (see Chapter 3). If a tetraploid AAaa tree is selfed, only $\frac{1}{8}$ of its offspring are homozygous AAAA or aaaa; the inbreeding rate is only $F = \frac{1}{8}$. Consequently there is less need to worry about loss of variability and inbreeding depression in polyploids than in diploids.

Induction and Recognition of Polyploidy

Induction by Shock

If a meristem is exposed to excessive heat, drought, cold, etc., the shock may be enough to hinder cell wall formation but not chromosome division. If this happens, chromosomes destined for two cells are included in one cell, and the chromosome number is doubled. This is presumably the mechanism by which many polyploids arose in nature.

Somatic chromosome doubling by this mechanism occurs with moderately high frequency (sometimes in 1% or more of the dividing cells). Most frequently the resulting 4n cells are outgrown by surrounding tissue, and the plant as a whole remains diploid.

Pollination by Unreduced Gametes

In normal meiosis, chromosome number is reduced from 2n to n. Sometimes the reduction division fails to occur and 2n pollen grains or egg cells result. If a 2n pollen grain fertilizes a 1n egg cell, a 3n embryo results, or it may happen that a rare unreduced pollen grain fertilizes a rare unreduced egg cell, resulting in a 4n embryo.

D. W. Einspahr uses a variation of this method to produce 4n aspen. He screens pollen from a 3n male clone through a fine cloth. The majority of the pollen is aborted, small, and passes through. The small portion retained on the cloth consists of unreduced 3n pollen grains. These are then applied to a normal 2n female (with 1n egg cells) to produce 4n seedlings.

Induction by Colchicine and Podophyllin

These chemicals are the most common means of inducing polyploidy. They do not affect chromosome division as much as they inhibit the spindle fiber mechanism by which divided chromosomes are drawn to the poles. Thus the chromosomes divide but do not pass poleward and a $2n$ cell becomes $4n$.

Of the two chemicals, colchicine is used most commonly. It is usually applied as a weak (0.1 to 0.5%) aqueous solution for a period of several hours or several days. It is effective on any actively growing tissue, such as a root tip or germinating pollen grain. However, the goal is a tree or branch capable of producing $4n$ flowers. To achieve that goal it must be applied to an actively growing apical meristem, such as in a germinating seed or an actively growing bud. Germinating seeds can be soaked in a colchicine solution. Growing buds can be surrounded by colchicine-soaked cotton. Einspahr (1965) obtained 50 $4n$ aspen seedlings by soaking entire female catkins in a 0.3% solution for 1 day at a time when the fertilized embryos consisted of a few cells. At the Hessian Forest Research Institute, good results have been obtained by treating young seedlings having only one leaf with a 0.3 to 0.4% colchicine solution for 72 hr (Anonymous, 1975). The third year after treatment approximately 40% of the seedlings were polyploid or mixoploid.

An actively growing apical meristem in a germinating seed or bud consists of two self-perpetuating tissues, the outer tunica and the inner corpus. These in turn differentiate into other meristems, such as leaf primordia, flower primordia, and cambium. The exact meristems derived from the tunica and corpus, respectively, vary among groups of trees.

Both the tunica and the corpus consist of several dividing cells. It is relatively easy to find a concentration of colchicine which can cause chromosome doubling in a few cells of the tunica, but it is difficult to obtain a concentration that will cause simultaneous chromosome doubling in all cells of both layers. Trial and error is the best method. The best concentrations and times are those which result in some mortality and in a high percentage of seedlings or shoots with gross abnormalities. The best results have been obtained with small-seeded species, such as the poplars, willows, elms, birches, and alders. In these species it has been easiest to obtain good penetration into both tunica and corpus.

Recognition and Treatment of Polyploids after Colchicine Treatment

The poisonous effects of colchicine seem to last for several cell generations, and newly induced $4n$ tissue seems to grow more slowly than surrounding $2n$ tissue. As colchicine rarely acts on all cells in a growing point, artificial induction usually results in mixtures (called mixoploids or chimeras) of $2n$ and $4n$ tissue. These mixtures may be of various types, i.e., one branch 4n and another branch 2n, top of leaf 2n and bottom of leaf 4n. If part of the top of a leaf is 4n and the bottom is faster growing 2n, the leaf curls upward; if the reverse is true the leaf curls downward. Thus, misshapen leaves, crooked branches, and general growth abnormalities are the first indications of successful colchicine treatment. The abnormalities gradually lessen and totally disappear as a branch or tree becomes wholly 4n or 2n.

After the disappearance of such abnormalities, 4n cells usually have twice the volume and 20 to 25% greater length or width than 2n cells. Such increases in cell size are detected most easily in stomatal guard cells, which can be measured at magnifications of about 100 × with a light-reflecting microscope. Stomatal measurement is a quick way to determine whether all or most of the leaves on a branch are polyploid.

Leaf size is also a convenient way to determine approximate chromosome number. In conifers generally, polyploidy is usually accompanied by a decrease in rate of cell division so that leaves are apt to be small in spite of the increase in cell size. In many hardwoods there is no change in rate of cell division so that leaves and other organs are apt to be larger than normal.

However, since the tunica and corpus give rise to different organs, the presence of 4n leaves on a branch is no guarantee that the branch will produce 4n flowers. Thus final proof must come at time of flowering. This can be done by examination of pollen grains (4n grains are usually larger than 2n grains), by actual chromosome counts in dividing gametophytic cells, or by actual chromosome counts of seedlings produced by the branch.

As already noted, there is a tendency for newly produced 4n tissue to grow more slowly than surrounding 2n tissue. Therefore complete reversion to the 2n condition is common. To overcome this tendency it is advisable to observe treated seedlings at frequent intervals, identify 2n and 4n sectors quickly, and remove 2n buds or branches.

Production of Triploids by Seed

Since the triploid condition is such a favorable one for growth rate, much effort has gone into the production of tetraploids that could in turn be used for the production of triploids. In northern Europe, 3n aspens have been mass produced by crossing a single 4n female clone with one or several 2n male clones. Much of that work has been done in a greenhouse. The same thing can be accomplished by planting a single 4n female clone in the midst of a natural 2n stand. Seed collected from that female clone will be 3n.

Haploid Breeding

Purpose of Haploid Breeding

The vegetative tissues of a seed plant are normally diploid. Occasionally, however, haploid plants arise. They are usually weak and of little interest in themselves. However, if the chromosome number of an haploid plant is doubled, the result is a completely homozygous diploid, the same as if a diploid had been selfed for several generations.

Diploids derived from haploids are homozygous but not necessarily superior. Those that survive and are moderately thrifty can be guaranteed free of the most serious deleterious recessive genes. For that reason, hybrids among them may prove vigorous. The production of haploids and chromosome doubling of those to produce diploids is now an accepted technique with several agricultural crop plants including corn, potato, tobacco, rice, petunia, cotton, alfalfa, and barley. In such crops the haploid-induced lines are treated as ordinary pure lines and are selfed, crossed, and progeny-tested in an ordinary manner. All aspects of improvement work promise to be much hastened by the fact that a breeder can work with pure lines.

Work on haploid induction in trees has been summarized by Illies (1964), Pohlheim (1968), Winton and Einspahr (1968), and Stettler and Bawa (1971). The best general reference on the subject is the book "Haploids in Higher Plants, Advances and Potential" edited by Kasha (1974). This book includes many papers on many plants (including trees) presented at a symposium held in 1974.

Possibilities in Trees

Haploids have been found in two gymnosperms. Pohlheim (1968) found a clone of western redcedar, which had been propagated for many years, that is haploid. It frequently reverts to a faster-growing, normal-appearing diploid form. Illies (1964) studied abnormal seedlings of Norway spruce and found 10 haploids (all sacrificed for chromosome counts) among many thousand normal seedlings. She postulated that they arose from spontaneous division of an unfertilized egg cell.

Most tree work has been done with poplars and has involved stimulation of unfertilized egg cells by the application of pollen that had been killed by heat, X-rays, or the chemical toluidine blue. Success has been limited. In one experiment, 5 haploids were obtained after the pollination of approximately 1.5 million ovules. Other experiments yielded 4, 5, or 6 haploids after pollinating several hundred female catkins. In still another experiment, 282 seedlings were obtained which were presumed to be haploid on the basis of marker genes but not on the basis of chromosome counts. These poplar haploids have been weak growing and have survived (if at all) only after reversion to the diploid condition. They have not yet been used in breeding work.

Much but not all of the success in crop plants has been with species that are tetraploids so that "haploid" plants in reality have $2n$ chromosomes. This may be a factor in trees, and work should possibly be concentrated in those species that are normally polyploid.

Also, several crop plant breeders have reported the greatest success when working with inbred lines that contain relatively few deleterious genes. Such lines are not available in trees. Perhaps this difficulty can be overcome by preliminary selection work (small selfing experiments) to determine which individuals carry the fewest undesirable recessive genes.

The total number of haploids produced so far in trees is too small to justify a true assessment of the possibilities of haploid breeding in forestry. Pure line breeding and subsequent hybridization have been successful in short-generation plants. Perhaps haploid breeding is a way to obtain similar results in long-generation trees.

Artificial Induction by Tissue Culture of Anthers or Pollen Grains

Unfertilized female egg cells can be cultured on artificial media or induced to develop parthenogenetically by the application of pollen that has been inactivated in some way. Such methods have been used with varying success in crop plants and are the source of all tree haploids produced so far. The success ratio is usually small (although approaching 3% in some herbs), and it is always necessary to examine a very large number of normal seedlings to detect the few haploids produced.

Greater success has attended the artificial culture of anthers or pollen grains, a technique that has developed rapidly in the past decade. Under favorable conditions, relatively large numbers of anther segments or pollen grains can be stimulated to develop into embryos, and these can develop under competition-free conditions into viable seedlings. The task of screening such seedlings for level of polyploidy is relatively easy.

The culture work must be done under sterile conditions. The culture medium is usually made of agar to which growth-promoting substances are added. Among these substances are sucrose, amino acids, plant hormones, mineral nutrients, and anther extracts. Temperature and light are important. Sometimes development can start in the dark, but the young embryos must be transferred quickly to the light in order to produce chlorophyll.

A heat or drought shock treatment, applied to the parental plant before anthers are removed, has been helpful in some plants. Apparently such shock treatment helps stimulate the male gametophytes to develop in an abnormal way (i.e., produce a whole plant) rather than a normal way (i.e., fertilize a female gametophyte). Stage of anther development is also important. If anthers are removed too early, they may be in a pregametophytic (diploid) stage. In tobacco, pollen grains cultured when the generative and vegetative nuclei were dividing simultaneously produced 3n rather than 1n embryos.

Another or pollen culture has proved successful enough in a wide variety of herbs that its success can almost be assured in some trees.

Reversion to Diploidy

Haploids must revert to the diploid condition to be useful. Of the several methods used with herbs, early treatment with colchicine seems most likely to be successful in trees. Since the 2n condition is "normal," a change from the 1n to the 2n condition is probably accomplished more easily than the change from 2n to 4n needed when producing polyploids.

Study Questions

1. Define haploid, diploid, polyploid, aneuploid, triploid, tetraploid, allopolyploid, autopolyploid, and karyotype.

2. In what tissues can chromosome counts be made? What are the advantages and disadvantages inherent in use of each tissue?

3. Describe a technique used to prepare material for chromosome measurements or counts.

4. Describe the karyotype of a typical pine species. In what respects do karyotypes differ among species of pine?

5. Compare the frequency of polyploidy in gymnosperms and angiosperms. In herbs and trees. In cool-climate and warm-climate trees.

6. How has aneuploidy played a role in evolution?

7. Which gymnosperms are natural polyploids? What are the characteristics of artificially produced polyploids in gymnosperms?

8. In which large angiosperm genera is polyploid frequent? Rare?

9. Compare ordinary diploids, diploid hybrids, aneuploids, triploids, allotetraploids, and autotetraploids as to fertility.

10. Name a crop plant and a tree believed to be of allopolyploid origin.

11. What is the most likely chromosome number of pollen produced by a 2n tree? 4n tree? 3n tree?

12. What level of polyploidy seems especially conducive to rapid growth?

13. How may a vigorous but hard to reproduce hybrid between two hardwood species be changed to become of practical value?

14. Describe how to produce a 4n tree artificially. A 3n tree.

15. Why is it difficult to produce a tree that is entirely 4n? What can be done to increase the chance of success?

16. Compare inbreeding rates in diploids and tetraploids. Why the difference?

17. What is the purpose of haploid induction?

18. How can haploids be induced? Be changed to diploids?

19. What factors may affect the success of haploid breeding in trees?

Appendix: Common and Scientific Names of Trees

acacia *Acacia*
agathis *Agathis*
ailanthus *Ailanthus altissima.* See also tree-of-Heaven
alder, black *Alnus glutinosa*
alder, Japanese *A. japonica*
alder, red *A. rubra*
alder, speckled *A. incana*
araucaria *Araucaria*
arborvitae, American *Thuja occidentalis.* See also northern white-cedar
arborvitae, Japanese *Thuja standishii*
arborvitae, giant *Thuja plicata.* See also western redcedar
arborvitae, oriental *Biota orientalis*
ash, Chinese *Fraxinus chinensis*
ash, green *F. pennsylvanica*
ash, pumpkin *F. tomentosa*
ash, white *F. americana*
aspen, European *Populus tremula*
aspen, largetooth *P. grandidentata*
aspen, trembling *P. tremuloides*
basswood *Tilia.* See also linden
basswood, American *T. americana*
beech, American *Fagus grandifolia*
beech, European *F. sylvatica*
bigtree *Sequoia gigantea = Sequoiadendron washingtoniana*
birch, European *Betula pendula*
birch, gray *B. populifolia*
birch, Maximowicz *B. maximowicziana*
birch, paper (canoe) *B. papyrifera*
birch, water *B. occidentalis*
birch, yellow *B. alleghaniensis*
boxelder *Acer negundo*
buckeye, red *Aesculus pavia.* See also horsechestnut
California laurel *Umbellularia californica.*
callitris *Callitris*
caragana *Caragana arborescens.* See also Siberian pea-tree
cedar, Atlas *Cedrus atlantica*
cedar, Lebanon *C. libani*
ceridiphyllum *Cercidiphyllum japonicum.* See also Katsura tree
cherry, black *Prunus serotina*
cherry, sour *P. cerasus*
cherry, sweet *P. avium*
chestnut, American *Castanea dentata*
chestnut, Chinese *C. mollissima*
chestnut, Henry *C. henryi*
chestnut, Japanese *C. crenata*
chestnut, Spanish *C. sativa*
China-fir *Cunninghamia lanceolata*
China-fir, Formosan *C. konishii*
coconut, coconut palm *Cocos nucifera*
cottonwood, eastern *Populus deltoides.* See also poplar

cottonwood, northern black *P. tricho-carpa*

cryptomeria *Cryptomeria japonica.* See also sugi

cypress *Cupressus*

dawn redwood *Metasequoia glyptostroboides*

Douglas-fir *Pseudotsuga menziesii*

elm, American *Ulmus americana*

elm, Chinese *U. parvifolia*

elm, red (slippery) *U. rubra*

elm, Siberian *U. pumila*

eucalypt, eucalyptus *Eucalyptus.* Various species are known as gum, messmate, stringybark, ironbark box, etc.

eucalypt, blue gum *E. bicostata*

eucalypt, cabbage gum *E. pauciflora*

eucalypt, candle bark gum *E. rubida*

eucalypt, Maiden's gum *E. maideni*

eucalypt, Murray red gum *E. camaldulensis*

eucalypt, red box *E. polyanthemos*

eucalypt, red stringybark *E. macrorryncha*

eucalypt, spotted *E. maculata*

eucommia *Eucommia ulmoides*

fir, balsam *Abies balsamea*

fir, Bornmueller *A. bornmuelleriana*

fir, grand *A. grandis*

fir, Greek *A. cephalonica*

fir, Nordmann *A. nordmanniana*

fir, white *A. concolor*

ginkgo *Ginkgo biloba*

hawthorn *Crataegus*

hemlock, western *Tsuga heterophylla*

hickory *Carya*

holly *Ilex*

honeylocust *Gleditsia triacanthos*

horsechestnut, common *Aesculus hippocastanum.* See also buckeye

horsechestnut, red *A. carnea*

juniper, Pfitzer *Juniperus chinensis* var. *pfitzeriana*

Katsura-tree *Cercidiphyllum japonicum*

larch, American *Larix laricina*

larch, European *L. decidua*

larch, golden *Pseudolarix amabilis*

larch, Japanese *Larix leptolepis*

larch, Korean *L. gmelini*

larch, Lyall *L. lyallii*

larch, western *L. occidentalis*

linden, American *Tilia americana.* See also basswood

linden, Chinese *T. tuan*

linden, common European *T. europaea*

linden, large-leaved *T. platyphyllos*

linden, Maximowicz *T. maximowicziana*

linden, Oliver *T. oliveri*

linden, pendent silver *T. petiolaris*

locust, black *Robinia pseudoacacia*

magnolia, cucumbertree *Magnolia acuminata*

magnolia, Dawson *M. dawsoniana*

magnolia, lily-flowered *M. liliiflora*

magnolia, southern *M. grandiflora*

maple, black *Acer nigrum*

maple, boxelder *A. negundo*

maple, hornbean *A. carpinifolium*

maple, Japanese *A. palmatum* and *A. japonicum*

maple, moosewood *A. pennsylvanicum*

maple, Norway *A. platanoides*

maple, red *A. rubrum*

maple, silver *A. saccharinum*

maple, sugar *A. saccharum*

maple, sycamore *A. pseudoplatanus*

metasequoia *Metasequoia glyptostroboides*

mulberry, black *Morus nigra*

oak, durmast *Quercus petraea*

oak, English *Q. robur*

oak, Gambel *Q. gambelii*

oak, Havard *Q. havardii*

oak, Mohrs *Q. mohriana*

oak, northern red *Q. rubra*

oak, post *Q. stellata*

oak, sand post *Q. margaretta* (= *Q. gambelii* × *stellata*)

oak, white *Q. alba*

Osage-orange *Maclura pomifera*

pea tree, Siberian *Caragana arborescens*

pine, Aleppo *Pinus halepensis*

pine, Apache *P. engelmannii*

pine, Arizona *P. arizonica* (*P. ponderosa* var. *arizonica*)

pine, Armand *P. armandi*

pine, Austrian *P. nigra* var. *austriaca*

pine, Bishop *P. muricata*

pine, Canary Island *P. canariensis*

pine, Caribbean *P. caribaea*
pine, Chinese *P. tabulaeformis*
pine, Corsican *Pinus nigra* var. *poiretiana*
pine, Coulter *P. coulteri*
pine, eastern white *P. strobus*
pine, Gregg *P. greggii*
pine, Himalayan white *P. griffithii*
pine, jack *P. banksiana*
pine, Japanese black *P. thunbergiana*
pine, Japanese red *P. densiflora*
pine, Japanese white *P. parviflora*
pine, Jeffrey *P. jeffreyi*
pine, Khasi *P. kesiya* (*P. insularis*)
pine, Korean *P. koraiensis*
pine, knobcone *P. attenuata*
pine, Lawson *P. lawsoni*
pine, limber *P. flexilis*
pine, loblolly *P. taeda*
pine, lodgepole *P. contorta*
pine, longleaf *P. palustris*
pine, luchu *P. luchuensis*
pine, Macedonian white *P. peuce*
pine, maritime *P. pinaster*
pine, Masson *P. massoniana*
pine, Mexican weeping *P. patula*
pine, Mexican white *P. ayacahuite*
pine, Monterey *P. radiata*
pine, Montezuma *P. montezumae*
pine, mugo *P. mugo*
pine, pinyon *P. edulis*
pine, pitch *P. rigida*
pine, ponderosa *P. ponderosa*
pine, Pringle *P. pringlei*
pine, red *P. resinosa*
pine, sand *P. clausa*
pine, Scotch *P. sylvestris*
pine, shortleaf *P. echinata*
pine, slash *P. elliottii*
pine, Sonderegger *P. sondereggeri* (= *P. palustris* × *taeda*)
pine, southern Loblolly, slash, shortleaf, longleaf and other pines of southeastern United States.
pine, southwestern white *P. strobiformis*

pine, spruce *P. glabra*
pine, sugar *P. lambertiana*
pine, Swiss stone *P. cembra*
pine, Taiwan red *P. taiwanensis*
pine, Torrey *P. torreyana*
pine, tropical *P. tropicalis*
pine, Virginia *P. virginiana*
pine, western white *P. monticola*
pine, whitebark *P. albicaulis*
pine, Yunnan *P. yunnanensis*
poplar, balsam *Populus balsamifera*
poplar, European black *P. nigra*
poplar, white *P. alba*
pteroceltis *Pteroceltis tatarinowii*
redcedar, eastern *Juniperus virginiana*
redcedar, western *Thuja plicata*
rubber tree *Hevea brasiliensis*
sassafras *Sassafras albidum*
sciadopitys *Sciadopitys verticillata*
spruce, black *Picea mariana*
spruce, blue *P. pungens*
spruce, Engelmann *P. engelmannii*
spruce, Norway *P. abies*
spruce, red *P. rubens*
spruce, Serbian *P. omorika*
spruce, Siberian *P. obovata*
spruce, Sitka *P. sitchensis*
spruce, white *P. glauca*
sugi *Cryptomeria japonica*
sweetgum *Liquidambar styraciflua*
taiwania *Taiwania cryptomerioides*
tamarack *Larix laricina*
tree-of-Heaven *Ailanthus altissima*
tupelo, black *Nyssa sylvatica*
umbrella pine *Sciadopitys verticillata*
walnut, black *Juglans nigra*
walnut, Persian *J. regia*
wattle, black *Acacia mearnsii*
wattle, green *A. decurrens*
white cedar, northern *Thuja occidentalis*
willow, weeping *Salix babylonica*
yellow poplar *Liriodendron tulipifera*
yew *Taxus*

Bibliography

Ahlgren, C. E. (1962). Some factors influencing survival, growth and flowering of intraspecific and interspecific pine grafts. *J. Forest.* **60**, 785–789.

Allen, R. M. (1967). Influence of the root system on height growth of three southern pines. *Forest Sci.* **13**, 253–257.

Anderson, E. (1953). Introgressive hybridization. *Biol. Rev. Cambridge Phil. Soc.* **28**, 280–307.

Anonymous. (1975). "Hessische Forstliche Versuchsanstalt, Jahresbericht 1974." Hann.-Münden.

Bannister, M. H. (1958). Evidence of hybridization between *Pinus attenuata* and *P. radiata* in New Zealand and variation in samples of two-year-old *Pinus attenuata, P. radiata,* and their hybrids. *Trans. Roy. Soc. N. Z.* **85**, 217–236.

Barrett, W. H. G. (1969). Adaptacion de especes de *Pinus* al Noreste Argentino. Crecimiento a los cinco anos. *IDIA (Inform. Invest. Agr.), Suppl.* Forestal No. **5**, pp. 11–26.

Barrett, W. H. G. (1970). Variacion geografica en *Pinus elliottii* Engelm. y *Pinus taeda* L. I. Caracteres de semillas y progenies en vivero. *IDIA (Inform. Invest. Agr.), Suppl.* **6**, 12–32.

Barrett, W. H. G., and Rial Alberti, F. (1972). Valor de la seleccion temprana en progenies de sauces. *IDIA (Inform. Invest. Agr.), Suppl.* **7**, 3–8.

Bateman, A. J. (1947a). Contamination in seed crops. II. Wind pollination. *Heredity* **1**, 235–246.

Bateman, A. J. (1947b). Contamination in seed crops. III. Relation with isolation distance. *Heredity* **1**, 305–336.

Bateman, A. J. (1949). Pollinating agents and population genetics. *Internat. Proc. Int. Congr. Genet., 8th, 1948* pp. 532–533.

Becker, W. A. (1964). "Manual of Procedures in Quantitative Genetics." Washington State University Bookstore, Pullman.

Bey, C. F. (1973). Growth of black walnut trees in eight midwestern states—A provenance test. *USDA Forest Serv., Res. Pap.* **NC-99**, 1–7.

Bingham, R. T. (1968). Breeding blister rust resistant western white pine. IV.

Mixed-pollen crosses for appraisal of general combining ability. *Silvae Genet.* **17,** 133–138.

Bingham, R. T., Olson, H. J., Becker, W. A., and Marsden, M. A. (1969). Breeding blister rust resistant western white pine. V. Estimates of heritability, combining ability, and genetic advance based on tester matings. *Silvae Genet.* **18,** 28–38.

Blakely, W. F. (1955). "A Key to the Eucalypts." Forestry and Timber Bureau, Canberra, Australia.

Bobrov, E. G. (1973). Introgressive Hybridisation, Sippenbildung und Vegetationsänderung. *Feddes Rep.* **84,** 273–294.

Bouvarel, P. (1960). Les vieux pins laricio greffes de la foret de Fontainebleau. *Silvae Genet.* **9,** 41–44.

Bowden, W. M. (1940). Diploidy, polyploidy and winter hardiness relationships in the flowering plants. *Amer. J. Bot.* **27,** 357–377.

Brevigliere, N. (1951). Ricerche sulla biologia florale e di fruttificazione dell *Castanea sativa* e *Castanea crenata* nel territorio di Vallombrosa. *Firenze, Cent. Stud. Castagno, Publ.* **1,** 15–50.

Brown, J. H. (1970). Seedling growth of three Scotch pine provenances with varying moisture and fertility treatments. *Forest Sci.* **16,** 43.

Burley, J. and Nikles, D. G. (ed.) (1972, 1973a). "Selection and Breeding to Improve Some Tropical Conifers," Vols. 1 and 2. Commonwealth Forestry Institute, Oxford, England, and Queensland Department of Forestry, Queensland, Australia.

Burley, J. and Nikles, D. G. (ed.) (1973b). "Tropical Provenance and Progeny Research and International Cooperation." Commonwealth Forestry Institute, Oxford, England.

Burley, J. (1966). Provenance variation in growth of seedling apices of Sitka spruce. *Forest Sci.* **12,** 170–175.

Callaham, R. Z., and Hasel, A. A. (1961). *Pinus ponderosa* height growth of wind-pollinated progenies. *Silvae Genet.* **10,** 33–42.

Callaham, R. Z., and Liddicoet, A. E. (1961). Altitudinal variation at 20 years in ponderosa and Jeffrey pines. *J. Forest.* **59,** 814–20.

Canavera, D. S. (1969). Geographic and stand variation in jack pine (*Pinus banksiana* Lamb.) Ph.D. thesis, Michigan State University, East Lansing.

Canavera, D. S. (1975). Geographic and stand variation in jack pine (*Pinus banksiana* Lamb.). *Silvae Genet.* **24,** (1).

Canavera, D. S., and Wright, J. W. (1973). A 4-year provenance test of jack pine. *Mich., Agr. Exp. Sta., Res. Rep.* **204,** 1–7.

Ching, K. K., and Doerksen, A. (1971). A natural chimera of Douglas-fir. *Silvae Genet.* **20,** 209–210.

Chmelar, J. (1967). Über die Wurzelungsfähigkeit der Weiden. *Acta Univ. Agr., Brno, Fac. Vet.* **36,** 141–151.

Conkle, M. T. (1963). The determination of experimental plot size and shape in loblolly and slash pines. *North Carolina State Coll. Sch. Forest., Tech. Rep.* No. 17, pp. 1–51.

Conkle, M. T., Libby, W. J., and Hamrick, J. L. (1967). Winter injury among white fir seedlings. *U.S., Pac. Southwest Forest Range Exp. Sta., Res. Note* **138,** 1–6.

Coyne, J. P. (1957). Control of cone insects in southern pine. *South. Conf. Forest Tree Improve. Proc.* **4,** 64–66.

Critchfield, W. B. (1963a). The Austrian × red pine hybrid. *Silvae Genet.* **12,** 187–192.

Critchfield, W. B. (1963b). Hybridization of the southern pines in California. *Proc. Forest Genet. Workshop, 1962* pp. 40–48.

Critchfield, W. B. (1967). Crossability and relationships of the closed-cone pines. *Silvae Genet.* **16**, 89–97.

Critchfield, W. B., and Little, E. L., Jr. (1966). Geographic distribution of the pines of the world. *U.S., Forest Serv., Misc. Publ.* **991**, 1–97.

Darlington, C. D., and Wylie, A. P. (1955). "Chromosome Atlas of the Flowering Plants." Allen & Unwin, London.

Dengler, A., and Scamoni, A. (1944). Über den Pollenflug der Waldbäume. *Z. Gesamte Forstwissenschaft* **76**, 136–155.

Douglass, B. S. (1974). "Final Report, Grand Fir (*Abies grandis*) Provenance Study," Mimeo. USDA Forest Serv., Portland Oregon.

Duffield, J. W. (1952). Relationships and species hybridization in the genus *Pinus*. *Z. Forstgenet. Forstpflanzenzuechtung* **1**, 93–100.

Duffield, J. W., and Righter, F. I. (1953). Annotated list of pine hybrids made at the Institute of Forest Genetics. *USDA Calif. Forest Range Exp. Sta., Res. Note* **86**, 1–9.

Duffield, J. W., and Snyder, F. B. (1958). Benefits from hybridizing American forest tree species. *J. Forest.* **56**, 809–815.

Eiche, V. (1966). "Cold Damage and Plant Mortality in Experimental Provenance Plantations with Scots Pine in Northern Sweden." Studia Forestalia Suecica, Skoghögskolan, Stockholm.

Eifler, I. (1955). Künstliche Polyploid-Erzeugung bei *Picea abies* und *Betula verrucosa*. *Z. Forstgenet. Forstpflanzenzuechtung* **4**, 162–166.

Einspahr, D. W. (1965). Colchicine treatment of newly formed embryos of quaking aspen. *Forest Sci.* **11**, 456–459.

Erdtman, G. (1954). "An Introduction to Pollen Analysis," 2nd ed. Chronica Botanica, Waltham, Massachusetts.

Falconer, D. S. (1960). "Introduction to Quantitative Genetics." Ronald Press, New York.

Farmer, R. E., Jr. (1970). Genetic variation among open-pollinated progeny of eastern cottonwood. *Silvae Genet.* **19**, 149–151.

Forbes, D. C. (1974). Black walnut controlled-pollination techniques. *Tree Planters' Notes* **25**, (3), 9–11.

Fowler, D. P. (1966). A new spruce hybrid—*Picea schrenkiana* × *P. glauca*. *Lake States Forest Tree Improve. Conf. Proc., 7th, 1965* pp. 16–25.

Fowler, D. P. (1967). Low grafting and deep planting may prevent mortality due to incompatibility in pine. *Forest Sci.* **13**, 314–315.

Fowler, D. P., and Dwight, T. W. (1964). Provenance differences in the stratification requirements of white pine. *Can. J. Bot.* **42**, 669–673.

Fowler, D. P., and Lester, D. T. (1970). Genetics of red pine. *USDA Forest Serv., Res. Pap.* **WO-8**, 1–13.

Franklin, E. C. (1968). Artificial self-pollination and natural inbreeding in *Pinus taeda* L. Ph.D. thesis, North Carolina State University, Raleigh.

Franklin, E. C. (1969). Inbreeding depresssion in metrical traits of loblolly pine (*Pinus taeda* L.) as a result of self-pollination. *North Carolina State Univ. Sch. Forest Resour. Tech. Rep.* No. 40, pp. 1–19.

Garrett, P. W., Schreiner, E. J., and Kettlewood, H. (1973). Geographic variation of eastern white pine in the Northeast. *U.S., Forest Serv., Res. Pap.* **NE-274**, 1–14.

Genys, J. B. (1968). Geographic variation in eastern white pine. Two-year results of testing range-wide collections in Maryland. *Silvae Genet.* 17, 6–12.

Genys, J. B., Wright, J. W., and Forbes, D. C. (1974). Intraspecific variation in Virginia pine, results of a provenance trial in Maryland, Michigan and Tennessee. *Silvae Genet.* 23, 99–104.

Gerhold, H. D. (1968). Mini-bags for tree breeding. *Silvae Genet.* 17, 31.

Goddard, R. E., Hollis, C. A., Kok, H. R., Rockwood, D. L., and Strickland, R. K. (1973). Cooperative forest genetics research program, 15th annual report. *Univ. Fla., Sch. Forest Resour. Conserv., Res. Rep.* No. 21, pp. 1–19.

Grigsby, H. C. (1973). South Carolina best of 36 loblolly pine seed sources for southern Arkansas. *USDA, Forest Serv., Res. Pap.* SO-89, 1–10.

Grant, V. (1949). Pollination systems as isolating mechanisms in angiosperms. *Evolution* 3, 82–97.

Gregory, P. H. (1945). The dispersion of air-borne spores. *Brit. Mycol. Soc. Trans.* 28, 26–72.

Hamrick, J. L., and Libby, W. J. (1972). Variation and selection in western U.S. montane species. I. White fir. *Silvae Genet.* 21, 29–35.

Hanover, J. W. (1966). Inheritance of 3-carene concentration in *Pinus monticola*. *Forest Sci.* 12, 447–450.

Hardin, J. (1956). Studies in the Hippocastanaceae. III. A hybrid swarm in the buckeyes. IV. Hybridization in *Aesculus. Rhodora* 59, 45–51 and 185–203.

Hare, R. C. (1973). Chemical and environmental treatments promoting rooting of pine cuttings. *Can. J. Forest Res.* 4, 101–106.

Hare, R. C., and Switzer, G. L. (1969). Introgression with shortleaf pine may explain rust resistance in western loblolly pine. *USDA Forest Serv., Res. Note* SO-88, 1–2.

Hartmann, H. T., and Kester, D. E. (1968). "Plant Propagation, Principles and Practices," 2nd ed. Prentice-Hall, Englewood, New Jersey.

Hatakeyama, S., and Adachi, Y. (1968). Geographic variation of birch species in Hokkaido. 1. Clustering of populations and correlations between progeny performance and characteristics of native habitat. (In Japanese, English summary.) *Hokkaido Forest Exp. Sta., Bull.* 6, 109–135.

Hattemer, H. (1968). Versuche zur geographischen Variation bei der japanischen Lärche. I. *Silvae Genet.* 17, 186–192.

Hattemer, H. (1969). Versuche zur geographischen Variation bei der japanischen Lärche. II. *Silvae Genet.* 18, 1–23.

Howe, G. E. (1971). Assessment of genetic variation in a multi-plantation test of half-sib families of Scotch pine. Ph.D. thesis, Michigan State University, East Lansing.

Hyun, S. K. (1956). Forest tree breeding work in Korea. *Inst. Forest Genet., Suwon Pap.* 1, 1–16.

Hyun, S. K., and Hong, S. H. (1969). The growth performance of pitch-loblolly hybrid pine produced by different races of loblolly pine in their early age. *Inst. Forest Genet., Suwon, Korea, Res. Rep.* No. 7, pp. 35–44.

Hyun, S. K., Ahn, K. Y., and Hong, S. H. (1972). Developing advanced generation breeding populations for a hybrid breeding program. *Inst. Forest Genet., Suwon, Korea, Res. Rep.* No. 9, pp. 1–8.

Illies, Z. M. (1964). Auftreten haploider Keimlinge bei *Picea abies. Naturwissenschaften* 51, 442.

Iwakawa, M., Watanabe, M., Mikami, S., Inuma, M., and Kida, S. (1967). Inheritance of some characters in open-pollinated progenies of *Pinus densiflora*. (In Japanese, English summary.) *Jap. Govt., Forest Exp. Sta., Bull.* **207**, 31–67.

Jaynes, R. A. (1964). Interspecific crosses in the genus *Castanea*. *Silvae Genet.* **13**, 146–154.

Jaynes, R. A., and Graves, A. H. (1963). Connecticut hybrid chestnuts and their culture. *Conn., Agr. Exp. Sta., New Haven, Bull.* **657**, 1–29.

Johnsson, H. (1952). Ungdomsutvecklingen hos stjälek, druvek och rödek. *Sv. Skogs-vardsfoereningen Tidskr.* **50**, 168–193.

Johnsson, H. (1956). Auto- and allotriploid *Betula*-families, derived from colchicine treatment. *Z. Forstgenet. Forstpflanzenzuecht.* **5**, 65–70.

Johnstone, R. C. B. (1973). An approach to selection for second phase clonal seed orchards. *In* "International Symposium on Genetics of Scotch Pine," IUFRO Working Party 2.2.03.5., 1973. Warsaw and Kornik, Poland.

Jokela, J. J. (1966). Incidence and heritability of *Melampsora* rust in *Populus deltoides* Bart. *In* "Breeding Pest-Resistant Trees," (H. D. Gerhold *et al.*, eds.), pp. 111–117. Pergamon, Oxford.

Karrfalt, R. P., Gerhold, H. D., and Palpant, E. H. (1975). The possibilities for interracial hybridization in Scotch pine: Geographic flowering patterns and cross-ability. *Silvae Genet.* **24**, (in press).

Kasha, K. J., ed. (1974). "Haploids in Higher Plants, Advances and Potential," Proc. 1st Int. Symp. University of Guelph, Guelph, Ontario, Canada.

Kellison, R. C. (1970). Establishment and management of clonal seed orchards of pine. *World Consult. Forest Tree Breed., 2nd, 1969* Vol. 2, pp. 1355–1366.

Khosla, P. K. (1968). Cytomorphological studies of woody elements of some families of Polypetalae of Darjeeling, and Khasia and Jaintia Hills with special reference to the trees of forestry importance. Ph.D. thesis, Panjab University, India.

Kiellander, C. L. (1950). Polyploidy in *Picea abies*. *Hereditas* **36**, 513–516.

King, J. P. (1971). Pest susceptibility variation in Lake States jack pine seed sources. *USDA Forest Serv., Res. Pap.* **NC-53**, 1–10.

King, J. P., and Nienstaedt, H. (1965). Variation in needle cast susceptibility among 29 jack pine seed sources. *Silvae Genet.* **14**, 194–198.

Kriebel, H. B. (1965). Technique and interpretation in tree seed radiography. *Lake States Forest Tree Improv. Conf. Proc.* **7**, 70–75 (USDA Forest Serv. Res. Paper No. NC-6).

Kriebel, H. B. (1972). Embryo development and hybridity barriers in the white pines (section *Strobus*). *Silvae Genet.* **21**, 39–44.

Kriebel, H. B., and Gabriel, W. J. (1970). Genetics of sugar maple. *USDA, Forest Serv., Res. Pap.* **WO-7**, 1–17.

Kriebel, H. B., Namkoong, G., and Usanis, R. A. (1972). Analysis of genetic variation in 1-, 2-, and 3-year old eastern white pine in incomplete diallel cross experiments. *Silvae Genet.* **21**, 44–48.

Kung, F. H., and Wright, J. W. (1972). Parallel and divergent evolution in Rocky Mountain trees. *Silvae Genet.* **21**, 77–85.

Kuo, P. C., Wright, J. W., Lemmien, W. A., Bright, J. N., Day, M. W., Sajdak, R. L., Skog, R. E., and Yao, N. Y. (1971). Improving Michigan's white pine. *Mich., Agr. Exp. Sta., Res. Rep.* **149**, 1–12.

La Farge, T. (1971). Inheritance and evolution of stem form in three northern pine species. Ph.D. thesis, Michigan State University.

LaFarge, T. (1974). Genetic variation among and within three loblolly pine stands in Georgia. *Forest Sci.* **20**, 272–278.

Langlet, O. (1936). Studier över tallens fysiologiska variabilitet och dess samband med klimatet. *Med. Skogsforsokanst. Stockholm* **29**, 219–470.

Langner, W. (1953). Eine Mendelspaltung bei Aurea-Formen von *Picea abies* (L.) Karst als Mittel zur Klärung der Befruchtungsverhältnisse im Walde. *Z. Forestgenet. Forstpflanzenzuecht.* **2**, 49–51.

Larsen, C. S. (1937). The employment of species types and individuals in forestry. *Royal Vet. and Agr. Coll. Yearbook* (Copenhagen) 1937, 74–154.

Larsen, C. S. (1956). "Genetics in Silviculture." Essential Books, Fairlawn, New Jersey.

Lepistö, M. (1974). Successful propagation by cuttings of *Picea abies* in Finland. *N. Z. J. Forest.* **4**, 367–370.

Lerner, I. M. (1958). "The Genetic Basis of Selection." Wiley, New York.

Li, C. C. (1955). "Population Genetics." Univ. of Chicago Press, Chicago, Illinois.

Li, C. C. (1975). "First Course in Population Genetics." Boxwood Press, Pacific Grove, California.

Li, H. L. (1956). The story of the cultivated horse-chestnuts. *Morris Arbor. Bull.* **7**, 35–39.

Libby, W. J., and Conkle, M. T. (1966). Effects of auxin treatment, tree age, tree vigor, and cold storage on rooting young Monterey pine. *Forest Sci.* **12**, 484–502.

Lines, R., and Mitchell, A. F. (1965). Differences in phenology of Sitka spruce provenances. *Forest Res. Rep.* (*Gt. Brit.*) pp. 173–184.

Lines, R., and Mitchell, A. F. (1969). Western hemlock. Provenance. *Forest Res. Rep.* (*Gt. Brit.*) pp. 45–49.

Livingston, G. G., and Ching, K. K. (1967). The longevity and fertility of freeze-dried Douglas-fir pollen. *Silvae Genet.* **16**, 98–101.

McAlpine, R. G. (1965). Vegetative propagation methods for hardwoods. *South. Forest Tree Improve. Conf. Proc., 8th, 1965* pp. 14–20.

MacDonald, J., Wood, R. F., Edwards, M. V., and Aldhous, J. R. (eds.) (1957). Exotic forest trees of Great Britain. *Forest. Comm., London, Bull.* **30**, 1–167.

McGee, C. E. (1974). Elevation of seed source and planting site affects phenology and development of red oak seedlings. *Forest Sci.* **20**, 160–164.

Maronek, D., and Flint, H. L. (1974). Cold hardiness of needles of *Pinus strobus* as a function of geographic source. *Forest Sci.* **20**, 135–141.

Mehra, P. N., and Khoshoo, T. N. (1956). Cytology of conifers. I. *J. Genet.* **54**, 165–180.

Mehra, P. N., and Khosla, P. K. (1972). Cytogenetical studies of East Indian Hammaelidaceae, Combretaceae and Myrtaceae. *Silvae Genet.* **21**, 186–190.

Metro, A. (1955). "Eucalypts for planting. *FAO Forest. and Forest Products Stud.* **11**, 1–403.

Moffett, A. A., and Nixon, K. M. (1966). (b) Plant breeding and genetics. I. Black wattle. A. Selection, progeny testing, and accessions. C. Controlled intercrossing *Wattle Res. Inst., Univ. Natal., S. Afr., Rep.* **19**, 30–34.

Muller, C. H. (1952). Ecological control of hybridization in *Quercus*: A factor in the mechanism of evolution. *Evolution* **6**, 147–161.

Munger, T. T., and Morris, W. G. (1936). Growth of Douglas fir trees of known seed source. *U.S., Dep. Agr., Tech. Bull.* **537**, 1–40.

Müntzing, A. (1936). The chromosomes of a giant *Populus tremula. Hereditas* **21**, 383–393.

Namkoong, G. (1966). Statistical analysis of introgression. *Biometrics* **22**, 2488–2502.

Namkoong, G., Snyder, E. B., and Stonecypher, R. W. (1966). Heritability and gain concepts for evaluating breeding systems such as seedling orchards. *Silvae Genet.* **15**, 76–84.

Nanson, A. (1968). La valeur des tests précoces dans la sélection des arbres forestiers, en particulier au point de vue de la croissance. Ph.D. thesis, Fac. Sci. Agron. Etat, Gembloux, Belgium.

Nanson, A. (1972). The provenance seedling seed orchard. *Silvae Genet.* **21**, 243–249.

Nanson, A. (1974a). Genetique et amélioration des arbres forestiers. *Rev. Agr. (Brussels)* **27**, 759–795.

Nanson, A. (1974b). Tables comparatives de l'éfficience de la sélection individuell, inter-familles et intra-famille par rapport à la sélection combinée. *Rev. Biometrie-Praximetrie* **14**, (1–2), 1–11.

Nienstaedt, H. (1968). White spruce seed source variation and adaptation to 14 planting sites in northeastern United States and Canada. *Can. Comm. Forest Tree Breed. Proc.* **10**, (2), 102–121.

Nienstaedt, H., Cech, F. C., Mergen, F., Wang, C. W., and Zak, B. (1958). Vegetative propagation in forest genetics research and practice. *J. Forest.* **56**, 826–839.

Nilsson, B. (1968). "Studier av nagra kvalitetsegenskapers genetiska variation hos tall (*Pinus silvestris* L.)," Rapporter och Uppsatser 3. Inst. Skogsgenet. Skogshøgskolan, Stockholm.

North Carolina State University. (1970). "School Forest Resources," 14th Annual Report. North Carolina State University, Raleigh.

Ohba, K., Iwakawa, M., Okada, Y., and Murai, M. (1971). Estimation of the degree of natural self-fertilization by the frequencies of chlorophyllous variants in Japanese red pine, *Pinus densiflora* Sieb. et Zucc., and the inheritance of the variants. *J. Jap. Forest. Soc.* **53**, 327–333.

Okada, S., and Mukaide, H. (1969). The investigation of the provenance character of Saghalien fir seedlings. III. The variation in heights and the frequency of secondary shoot occurrence of Saghalien fir seedlings from some provenance and mother trees. (In Japanese, English summary.) *J. Jap. Forest. Soc.* **51**, 6–11.

Orr-Ewing, A. (1957). A cytological study of the effects of self-pollination on *Pseudotsuga menziesii* (Mirb.) Franco. *Silvae Genet.* **6**, 147–161.

Palmer, E. J. (1948). Hybrid oaks of North America. *J. Arnold Arboretum, Harvard Univ.* **29**, 1–48.

Parks, G. C. (1974). Field grafting ponderosa pine. *Tree Planters' Notes* **25**, (3), 12–14.

Patton, R. F., and Riker, A. J. (1966). Lessons from nursery and field testing of eastern white pine selections and progenies for resistance to blister rust. *In* "Breeding Pest Resistant Trees," (H. D. Gerhold *et al.*, eds.), pp. 403–414. Pergamon, Oxford.

Pederick, L. A. (1970). Chromosome relationships between *Pinus* species. *Silvae Genet.* **19**, 171–180.

Perry, T. O., Wang, C. W., and Schmitt, D. M. (1966). Height growth of loblolly pine provenance in relation to photoperiod and growing season. *Silvae Genet.* **15**, 61–64.

Piatnitsky, S. S. (1960). Evolving new forms of oak by hybridization *In* "Questions of Forestry and Forest Management," (A. B. Zhukow, ed.), pp. 231–244. USSR Acad. Sci., Moscow.

Piccarolo, G. (1952). "Il pioppo." Ramo Edit. Agr., Rome.

Pitcher, J. A. (1971). Parental and family selection in *Prunus serotina* Ehrh. Ph.D. thesis, Michigan State University, East Lansing.

Pohlheim, F. (1968). *Thuja gigantea gracilis* Beissn. —ein Haplont unter den Gymnospermen. *Biol. Rundsch.* 6, 84–86.

Polk, R. B. (1974). Heritabilities of some branching habits in jack pine. *Cent. States Forest Tree Improve. Conf., Proc., 8th, 1972,* pp. 33–39.

Polunin, N. (1951). Arctic aerobiology: Pollen grains and other spores observed on sticky slides exposed in 1947. *Nature (London)* 168, 718–721.

Pudden, H. H. C. (1957). "Exotic Trees in the Kenya Highlands." Government Printer, Nairobi.

Pryor, L. D. (1957). Selecting and breeding for cold resistance in *Eucalyptus. Silvae Genet.* 6, 98–109.

Radu, S. (1974). "Cultura si Valorificarea Pinului Strob." Ceres, Bucharest, Romania.

Radu, S., and Blada, I. (1966). Beobachtungen über die Entwicklung der Douglasien-Pfropflinge im ersten Jahre und die Möglichkeit ihrer Ausnutzung für die Überprüfung der Plusbame. *Tagungsber., Deut. Akad. Landwirtschafeswiss. Berlin* 69, 77–84.

Read, R. A. (1971). Browsing preference by jackrabbits in a ponderosa pine provenance plantation. *U.S., Forest Serv., Res. Note* RM-186, 1–6.

Rehder, A. (1940). "Manual of Cultivated Trees and Shrubs." Macmillan, New York.

Righter, F. I. (1955). Tree improvement activities at Placerville, California, affecting southern species. *South. Conf. Forest Tree Improve. Proc., 3rd, 1955* pp. 97–99.

Riker, A. J., Kouba, T. F., Brener, W. H., and Byam, L. E. (1943). White pine selections tested for resistance to blister rust. *J. Forest.* 41, 753–760.

Robertson, A. (1957). Optimum group size in progeny testing and family selection. *Biometrics* 13, 442–450.

Rohmeder, E., and Schönbach, H. (1959). "Genetik und Züchtung der Waldbaume." Parey, Berlin.

Rowe, K. E., and Ching, K. K. (1973). Provenance study of Douglas-fir in the Pacific Northwest region. II. Field performance at age nine. *Silvae Genet.* 22, 115–119.

Ruby, J. L. (1964). The correspondence between genetic, morphological and climatic variation patterns in Scotch pine. Ph.D. thesis, Michigan State University, East Lansing.

Rudolph, T. D., Schoenike, R. E., and Schantz-Hansen, T. (1959). Results of one-parent progeny tests relating to the inheritance of open and closed cones in jack pine. *Minn. Forest. Notes* 78, 1–2.

Sakai, K., and Miyazaki, Y. (1972). Genetic studies in natural populations of forest trees. II. Family analysis: A new method for quantitative genetic studies. *Silvae Genet.* 21, 149–154.

Sarvas, R. (1962). Investigations on the flowering and seed crop of *Pinus silvestris. Comm. Inst. Forest. Fenn.* 53, (4), pp. 1–198.

Saylor, L. C. (1972). Karyotype analysis of the genus *Pinus*—subgenus *Pinus. Silvae Genet.* 21, 155–163.

Saylor, L. C., and Koenig, R. L. (1967). The slash × sand pine hybrid. *Silvae Genet.* 16, 134–138.

Schaffalitsky de Muckadell, M. (1959). Investigations on aging of apical meristems in woody plants and its importance in silviculture. *Forstl. Forsogsva. Danmark* 25, 310–455.

Schmitt, D. (1968). Performance of southern pine hybrids in south Mississippi. *USDA Forest Serv., Res. Pap.* SO-36, 1–15.

Schramm, J. R., and Schreiner, E. J. (1954). The Michaux Quercetum. *Morris Arbor. Bull.* 5, 54–57.

Schreiner, E. J. (1959). Production of poplar timber in Europe and its significance and application in the United States. *U.S. Dep. Agr., Agr. Handb.* **150**, 1–124.

Schreiner, E. J. (1966). Maximum genetic improvement of forest trees through synthetic multiclonal hybrid varieties. *Northeast. Forest Tree Improve. Conf. Proc., 13th, 1965* pp. 7–13.

Schreiner, E. J. (1972). Procedure for selection of hybrid poplar clones for commercial trials in the northeastern region. *Northeast. Forest Tree Improve. Conf. Proc.* **19**, 108–116.

Schröck, O. (1966). Der Einfluss von Reis und Unterlage auf Wuchsleistung und Wachstumsgang bei Kiefernpfropflingen (*Pinus sylvestris*). *Tagungsber., Deut. Akad. Landwirtschaftswiss. Berlin* **69**, 37–66.

Shelbourne, C. J. A. (1974). Genetic improvements from orchard seed and controlled pollinations. *N. Z. Forest Serv., Forest Res. Inst. Rep.* pp. 22–23.

Shreve, L. W. (1974). Black walnut bench-grafting procedure reduces stock suckering. *Tree Planters' Notes* **25**, (3), 3–5.

Silen, R. R. (1962). Pollen dispersal conditions for Douglas-fir. *J. Forest.* **60**, 790–795.

Silen, R. R., and Copes, D. L. (1972). Douglas-fir seed orchard problems—a progress report. *J. Forest.* **70**, 145–147.

Smouse, P. E., and Saylor, L. C. (1973). Studies of the *Pinus rigida–serotina* complex. II. Natural hybridization among the *Pinus rigida–serotina* complex, *P. taeda* and *P. echinata. Ann. Mo. Bot. Gard.* **60**, 192–203.

Solbrig, O. T. (1971). The population biology of dandelions. *Amer. Sci.* **59**, 686–694.

Sorenson, F. C. (1973). Frequency of seedlings from natural self-fertilization in coastal Douglas-fir. *Silvae Genet.* **22**, 20–24.

Squillace, A. E. (1965). Combining superior growth and timber quality with high gum yield in slash pine. *South. Forest Tree Improve. Conf. Proc., 8th, 1965* pp. 73–76.

Squillace, A. E. (1966). Geographic variation in slash pine. *Forest Sci. Monogr.* **10**, 1–56.

Squillace, A. E. (1967). Effectiveness of 400-foot isolation around a slash pine orchard. *J. Forest.* **65**, 823–824.

Squillace, A. E. (1970). Genotype-environment interactions in forest trees. *Pap., 2nd Meet. Working Group Quantitative Forest Genet. Sect. 22 IUFRO, 1969* pp. 49–61.

Squillace, A. E. (1971). Inheritance of monoterpene composition in cortical oleoresin of slash pine. *Forest Sci.* **17**, 381–387.

Staszkiewicz, J. (1970). Sosna zwyczajna, *Pinus silvestris* L., systematyka i zmiennosc. *Polskiej Akad. Nauk, Nasze Drzewa Lesne* (*Monografie Popularnora ukowe*) *Tom* **1**, 55–77.

Stebbins, G. L. (1950). "Variation and Evolution in Plants." Columbia Univ. Press, New York.

Steinhoff, R. J., and Andresen, J. W. (1971). Geographic variation in *Pinus flexilis* and *Pinus strobiformis* and its bearing on their taxonmic status. *Silvae Genet.* **20**, 159–167.

Steinhoff, R. J., and Hoff, R. J. (1971). Estimates of the heritability of height growth in western white pine based on parent-progeny relationships. *Silvae Genet.* **20**, 141–144.

Stettler, R., and Bawa, K. S. (1971). Experimental induction of haploid parthenogenesis in black cottonwood (*Populus trichocarpa* T. & G. ex Hook.). *Silvae Genet.* **20**, 15–25.

Stevens, N. E. (1949). Characteristics of some disease-free ornamental plants. *Science* 110, 218–219.

Stout, A. B., and Schreiner, E. J. (1933). Results of a project in hybridizing poplars. *J. Hered.* 24, 216–229.

Strand, L. (1957). Pollen dispersal. *Silvae Genet.* 6, 129–136.

Thomas, G., and Ching, K. K. (1968). A comparative karyotype analysis of *Pseudotsuga menziesii* (Mirb.) Franco and *Pseudotsuga wilsoniana* (Hayata). *Silvae Genet.* 17, 138–142.

Streets, R. J. (1962). "Exotic Trees in the British Commonwealth." Oxford Univ. Press (Clarendon) London and New York.

Thulin, I. J. (1970). Breeding of *Pinus radiata* through seed improvement and clonal afforestation. *World Consult. Forest Tree Breed., 2nd, 1969* Vol. 2, pp. 1109–1117.

Tobolski, J. J., Hanover, J. W., and Wright, J. W. (1971). Genetic variation in the monoterpenes of Scotch pine. *Forest Sci.* 17, 293–299.

Tucker, J. M. (1952). Evolution of the Californian oak *Quercus alvordiana. Evolution* 6, 162–180.

USDA Forest Service. (1974a). "Forest Tree Seed Orchards." USDA, Forest Service, Washington, D.C.

USDA Forest Service. (1974b). Seeds of woody plants in the United States. Prepared by the Forest Service (C. S. Schopmeyer, Technical coordinator). USDA, Hand. 450, pp. 1–883.

Van Buijtenen, J. P., Donovan, G. A., Long, E. M., Robinson, J. F., and Woessner, R. A. (1971). Introduction to practical forest tree improvement. *Tex. Forest Serv., Circ.* 207, 1–17.

Vidaković, M. (1974). Genetics of European black pine (*Pinus nigra* Arn.) *Ann. Forest.* 6(3), 57–86.

Wang, C. W., and Pattee, R. (1975). Progeny performance of 271 ponderosa pine parents. *FWR Experimental Station Paper* 21, Univ. of Idaho.

Wang, C. W., Perry, T. O., and Johnson, A. G. (1960). Pollen dispersion of slash pine (*Pinus elliottii* Engelm.) with special reference to seed orchard management. *Silvae Genet.* 9, 65–92.

Webb, C. D. (1961). Field grafting loblolly pine. *N.C. State Forest Tree Improve., Tech. Rep.* No. 10, pp. 1–33.

Weisgerber, H., and Kohnert, H. (1974). Untersuchungen zur heterovegetativen Vermehrung bei Fichte, Douglasie und Kiefer im Freiland. *Allg. Forst-Jagdztg.* 145, 205–211.

Wells, O. O. (1964). Geographic variation in ponderosa pine. I. The ecotypes and their distribution. II. Correlations between progeny performance and characteristics of the native habitat. *Silvae Genet.* 13, 89–103 and 125–132.

Wells, O. O. (1973). Variation among shortleaf pines in a Mississippi seed source planting. *USDA Forest Serv., Res. Note* SO-162, 1–8.

Wells, O. O., and Switzer, G. L. (1971). Variation in rust resistance in Mississippi loblolly pine. *South. Forest Tree Improve. Conf. Proc. 11th, 1971* pp. 25–30.

Wells, O. O., and Wakeley, P. C. (1966). Geographic variation in survival, growth and fusiform-rust infection of planted loblolly pine. *Forest Sci. Monogr.* 11, 1–40.

Wells, O. O., and Wakeley, P. C. (1970). Variation in shortleaf pine from several geographic sources. *Forest Sci.* 16, 415–423.

Weston, G. C. (1957). Exotic forest trees in New Zealand. *N.Z. Forest Service Bull.* **13**, 1–103.

Williams, R. D., Funk, D. T., Lemmien, W. A., and Russell, T. E. (1974). Apparent freeze damage to black walnut seedlings related to both seed source and fertilizer treatment. *Tree Planters' Notes* **25**, (3), 6–8.

Winton, L. L., and Einspahr, D. W. (1968). The use of heat-treated pollen for aspen haploid production. *Forest Sci.* **14**, 406–407.

Woessner, R. A. (1965). Growth, form and disease resistance in four-year-old control- and five-year-old open-pollinated progeny of loblolly pine selected for use in seed orchards. *North Carolina State Univ. Sch. Forest. Resour., Tech. Rep.* No. 28, pp. 1–67.

Wright, J. W (1952). Pollen dispersion of some forest trees. *Northeast. Forest Exp. Sta., Sta., Pap.* **46**, 1–42.

Wright, J. W. (1953). Summary of tree-breeding experiments by the Northeastern Forest Experiment Station, 1947–1950. *Northeast. Forest Exp. Sta., Sta. Pap.* **56**, 1–47.

Wright, J. W. (1955). Species crossability in spruce in relation to distribution and taxonomy. *Forest Sci.* **1**, 319–349.

Wright, J. W. (1959). Species hybridization in the white pines. *Forest Sci.* **5**, 210–222.

Wright, J. W. (1962). "The Genetics of Forest Tree Improvement." FAO, Rome.

Wright, J. W. (1970a). Genetics of eastern white pine. *USDA Forest Serv., Res. Pap.* **WO-9**, 1–16.

Wright, J. W. (1970b). An improved record system for forest genetic nursery and plantation studies. *Silvae Genet.* **19**, 64–68.

Wright, J. W. (1972). Genetic variation patterns in Michigan's native trees. *Mich. Acad.* **5**, 61–69.

Wright, J. W. (1973). Genotype-environment interaction in north central United States. *Forest Sci.* **19**, 113–123.

Wright, J. W., and Freeland, F. D. (1960). Plot size and experimental efficiency in forest genetic research. *Mich., Agr. Exp. Sta., Tech. Bull.* **280**, 1–28.

Wright, J. W., and Gabriel, W. J. (1958). Species hybridization in the hard pines, series *Sylvestres*. *Silvae Genet.* **7**, 109–115.

Wright, J. W., and Wilson, L. F. (1972). Genetic differences in Scotch pine resistance to pine root collar weevil. *Mich., Agr. Exp. Sta., Res. Rep.* **159**, 1–5.

Wright, J. W., Pauley, S. S., Polk, R. B., Jokela, J. J., and Read, R. A. (1966). Performance of Scotch pine varieties in the north central region. *Silvae Genet.* **15**, 101–110.

Wright, J. W., Wilson, L. F., and Randall, W. K. (1967). Differences among Scotch pine varieties in susceptibility to European pine sawfly. *Forest Sci.* **13**, 175–181.

Wright, J. W., Lemmien, W. A., and Bright, J. N. (1969a). Early growth of ponderosa pine ecotypes in Michigan. *Forest Sci.* **15**, 121–129.

Wright, J. W., Lemmien, W. A., and Canavera, D. S. (1969b). Abundant natural hybridization between Austrian and Japanese red pines in southern Michigan. *Forest Sci.* **15**, 269–274.

Wright, J. W., Kung, F. H., Read, R. A., Lemmien, W. A, and Bright, J. N. (1971a). Genetic variation in Rocky Mountain Douglas-fir. *Silvae Genet.* **20**, 54–60.

Wright, J. W., Lemmien, W. A., and Bright, J. N. (1971b). Genetic variation in southern Rocky Mountain white fir. *Silvae Genet.* **20**, 148–150.

Wright, J. W., Read, R. A., Lester, D. T., Merritt, C., and Mohn, C. (1972). Geographic variation in red pine, 11-year data from the North Central states. *Silvae Genet.* **21,** 205–209.

Wright, J. W., Lemmien, W. A., Bright, J. N., Day, M. W., and Sajdak, R. L. (1975). Scotch pine varieties for Christmas tree and forest planting in Michigan. *Michigan Agr. Exp. Sta. Res. Rep.* (in press).

Wright, S. (1943). Isolation by distance. *Genetics* **28,** 114–138.

Wright, S. (1946). Isolation by distance under diverse systems of mating. *Genetics* **31,** 39–59.

Yao, Y. N., Pitcher, J. A., Wright, J. W., and Kuo, P. C. (1971). Improved red pine for Michigan. *Mich., Agr. Exp. Sta., Res. Rep.* **146,** 1–12.

Ying, C.-C. (1974). Genetic variation of eastern cottonwood (*Populus deltoides* Bartr.). *Neb., Agr. Exp. Sta., Dep. Forest. Prog. Rep.* No. 1, pp. 1–148.

Zeijlemaker, F. C. J., and Mackenzie, A. M. (1966). A note on the occurrence and inheritance of a new flavonoid constituent of black wattle (*Acacia mearnsii*). *Wattle Res. Inst., Univ. Natal, S. Afr., Rep.* 19 pp. 57–58.

Zobel, B. J. (1971). The genetic improvement of southern pines. *Sci. Amer.* **225,** (5), 94–103.

Zobel, B. J., Blair, R., Kellison, R. C., and O'Gwynn, C. (1973). An operational breeding program—theory and practice. *In* "Selection and Breeding to Improve Some Tropical Conifers," (J. Burley and D. K. Nikles, eds.), Vol. 2, pp. 187–202. Commonwealth Forestry Inst., Oxford, England and Queensland Dept. of Forestry, Queensland, Australia.

Glossary

acclimatization Adaptation to climatic change on the part of a population, usually as the result of genetic change.

acquired character A trait, not inherited, developed as the result of environmental influence during the lifetime of an individual.

adaptation A change on the part of an individual or population resulting in better survival or growth. The process of change. The changed structure or function.

additive genes Factors affecting the same character in such a way that 1, 2, 3, 4, etc., genes have 1, 2, 3, 4, etc., units of effect on the phenotype.

air layer An undetached aerial portion of a plant on which roots develop, commonly as the result of wounding or other stimulation.

albino A part of a plant or a plant lacking chlorophyll. Albinism is usually lethal.

allele, allelomorph One of a pair (in a diploid individual) or series (in a population or a polyploid individual) of genes located at the same locus in homologous chromosomes and controlling the same character. One of a pair of characters controlled by such genes and alternative to each other in inheritance.

allopatric Inhabiting separate areas.

allopolyploid An organism with more than two sets of chromosomes, the sets being different and derived from different species by hybridization.

analysis of variance A statistical analysis by which F values (variance ratios) are compared in such a manner as to determine the probability that differences among populations or treatments are too large to be due to chance.

andro Prefix referring to male.

anemophily Wind pollination.

aneuploid An organism having other than an exact multiple of the haploid number of chromosomes. An organism having $2n - 1$, $3n + 1$, $3n - 1$, etc., chromosomes.

anther The part of a stamen, usually borne on a stalk, that bears pollen.

anthesis The shedding of pollen from anthers. The period of full bloom during which pollen is shed and female flower parts are receptive.

apomixis   Reproduction without fertilization by seed or seed-like organs. Presumably present but not yet demonstrated in forest trees.

archegonium   The flask-shaped female sex organ of some gymnosperms which contains the egg cell.

asexual reproduction   Reproduction without fertilization. It includes apomixis and various forms of vegetative reproduction.

autopolyploid   An organism having more than two homologous sets of chromosomes, derived from a single species.

B

backcross   A cross between a hybrid and either of its parents. Backcrossing is usually used to incorporate a single desirable trait from one species or variety with several desirable traits from the species or variety used as the recurrent parent.

biotype   A group of individuals having the same genotype.

bisexual   Hermaphroditic, having both male and female reproductive organs. Usually used to indicate a departure from normality.

bivalent   A pair of homologous chromosomes joined together at metaphase stage of meiosis.

block   A part of an experiment containing one plot of each of several seedlots or clones. A complete block includes one plot each of every seedlot or clone in the experiment.

breeding   The science or art of changing the genetic constitution of a population of plants or animals.

breeding arboretum   A plantation or group of trees used chiefly as parental stock in forest genetic experimentation.

breeding rotation   The shortest period (from seed to seed) required for trees to reproduce by seed.

budding   Grafting by insertion of a bud with a small amount of attached bark or stem tissue into a cut in the bark of the rootstock.

bud-sport   A somatic mutation (in the broadest sense) arising in a bud and producing an abnormal branch. Includes change due to gene mutation, somatic reduction, chromosome deletion, and polyploidy.

bulk progeny test   A test of the offspring of parents which have been grouped according to phenotypic similarity and in which the identity of the offspring is maintained only for groups of parents.

C

cambium   A thin layer of longitudinally dividing cells between the xylem and phloem which gives rise to secondary growth.

catkin   A unisexual inflorescence consisting of a central stem, scaly bracts, and sessile flowers without petals.

centromere   The central, spindle attachment region of a chromosome, often constricted.

character, characteristic   The detectable phenotypic expression of the action of a gene or group of genes. A trait. A feature used to distinguish among individuals or populations.

chimera   A plant composed of two or more genetically different tissues. Includes "periclinal chimera," in which one tissue lies over another as a glove fits a hand, and "sectorial chimera," in which the tissues lie side by side.

chi square   A statistic (equals sum of squares of deviations divided by sum of squares of expected values) useful in determining the statistical significance of differences in a qualitative trait.

chromosome A small, elongated, deeply staining body found within the nucleus, consisting primarily of DNA and a protein sheath, and containing the genes or factors responsible for most hereditary traits.

chromosome number The number of chromosomes characteristic of a race or species, designated either by the n (gametic) or $2n$ (somatic or vegetative) number.

chromosome set The chromosomes inherited as a unit from one parent, and usually consisting of nonhomologous chromosomes.

chromosomes, homologous Chromosomes in which the same gene loci occur in the same sequence. A diploid cell contains n pairs of homologous chromosomes.

chromosomes, nonhomologous Chromosomes having different gene contents.

cline A geographic gradient, usually assumed to be genetically controlled.

clonal test A plantation or group of plantations consisting of clones and established primarily to determine genetic differences among clones. See also seed orchard, clonal.

clone All plants (ramets) reproduced asexually from a common ancestor (ortet) and having identical genotypes. Named clones are given non-Latin names preceded by the abbreviation "cl."

colchicine An alkaloid, derived from the autumn crocus, used specifically to inhibit the spindle mechanism during cell division and thus to cause a doubling of chromosome number.

combining ability The relative ability of an organism to transmit genetic superiority to its offspring.

combining ability, general The relative ability of an organism to transmit genetic superiority to its offspring when crossed with other organisms in general. High general combining ability usually implies the presence of genes with additive effects.

combining ability, specific The relative ability of an organism to transmit genetic superiority to its offspring when crossed with specific other organisms. High specific combining ability usually implies the presence of dominance, overdominance, or epistasis.

cone The woody, usually elongated seed-bearing organ of a conifer, consisting of a central stem, woody scales and bracts (often not visible), and seeds.

constriction A lightly staining and seemingly narrow (probably because of lack of spiralization) portion of a chromosome. The primary or central constriction is called the "centromere."

corpus The inner (or body) portion of an apical meristem consisting of a layer of several dividing cells.

correlation analysis A statistical analysis employing variance ratios and showing the strength of the relation among two or more variables. The strength of a correlation is measured by the correlation coefficient (r).

cross A hybrid (noun) or to hybridize (verb).

crossability The relative ease with which hybrids can be produced between races or species.

cross-pollination The transfer of pollen from one tree to a female flower of another tree or clone.

crossover, crossing-over Simultaneous breakage at the same locus and subsequent exchange of segments between homologous chromosomes at the metaphase stage of meiosis.

cultivar. A "cultivated variety," given a non-Latin name, and designated by cv. A cultivar is any clone, race, or product of breeding deemed worthy of cultivation and of a separate name.

cutting A detached leaf, stem, or piece of root which is encouraged to form roots. A "greenwood cutting" is made during the period of active growth, and a "hardwood cutting" is made during the dormant season.

cyclophysis A form of topophysis in which differences due to age are transmitted through a cutting or graft.

cytology The science relating to the study of cells or cellular inclusions, especially chromosomes.

cytoplasm The living matter (protoplasm) within a cell exclusive of the nucleus.

cytoplasmic inheritance The transmission of hereditary traits, from the maternal parent only, by means of factors located in the cytoplasm.

D

deficiency, chromosome The absence of a chromosome segment.

deletion, chromosome The absence of an internal chromosome segment.

deoxyribonucleic acid See DNA

diallel crossing Mating each tree with every other tree in a group. In a group of n trees, a complete diallel involves $n(n-1)/2$ different parental combinations.

dichogamy Maturation of male and female flowers or parts of flowers at different times fostering cross-pollination. Most common in maple.

dioecious Producing male and female flowers on different plants.

dioecism Being dioecious.

diploid (1) Having $2n$ or two sets of homologous chromosomes. (2) Pertaining to the chromosome number in the vegetative rather than the gametic tissue. In the latter sense it is proper to speak of the vegetative tissue of a $4n$ or $6n$ plant as "diploid," as opposed to the "haploid" gametes.

DNA Deoxyribonucleic acid. A long, double-stranded, self-replicating acid of very large molecular weight which is the genetically active portion of a chromosome.

dominance Masking effects of one allele by another so that a heterozygote (Aa) and a homozygous dominant (AA) organism are phenotypically indistinguishable. Dominance may be complete or partial.

dominant gene A gene (denoted usually by a^+ or a capital letter) which prevents its allele from having a phenotypic effect. Antonym of "recessive gene."

drift Change in gene frequency and population characteristics due to chance rather than selection, and usually most pronounced in small populations.

dysgenic Detrimental to the genetic quality of a population. Antonym of "eugenic."

E

ecotype A genetic subdivision of a species resulting from the selective action of a particular environment and showing adaptation to that environment. The word ecotype carries connotations of difference and of adaptation, whereas the word race carries a connotation of difference but not necessarily of adaptation. Ecotypes may be geographic, climatic, elevational, or edaphic.

efficiency, statistical The degree to which an experiment gives the most useful information per unit of experimental material or of cost.

egg cell The $1n$ female reproductive cell which when fertilized results in an embryo.

elite tree A tree of proven good combining ability.

emasculation Removal of anthers or male flowers prior to pollination.

embryo That portion of the seed resulting from union of male and female gametes and developing into a mature plant.

endemic Native to a small region.

endosperm Food storage tissue contained in the seed and surrounding the embryo. The endosperm is small or aborted in most hardwood tree seeds. Endosperm has $3n$ chromosomes ($2n$ from female and $1n$ from male) in angiosperms and $1n$ chromosomes (all from female) in gymnosperms.

entomophilous Insect pollinated.

epistasis Dominance exerted by nonallelic genes. An "epistatic" gene is dominant to a "hypostatic" gene.

equatorial plate The group of metaphase chromosomes lying in one plane perpendicular to and at the center of the spindle during cell division.

eugenic Favorable to the genetic quality of a population. Antonym of "dysgenic."

euploid Having an exact multiple of the typical haploid chromosome number. Antonym of "aneuploid."

evolution Long-time changes in gene frequency and phenotypic characteristics of a population or group of populations.

exotic An introduction from a foreign land.

expressivity The degree to which a gene expresses itself in the phenotype.

F

F Inbreeding coefficient.

F ratio A ratio between two variances used to determine the probability that a given difference between means is due to chance.

F_1, F_2, F_3 The first, second and third generations, respectively, following a cross. The F stands for filial.

F_2 ratio The ratio between phenotypes (or less commonly among genotypes) in an F_2 population resulting from self-pollination or like-to-like mating in an F_1 population.

factor A gene or unit of inheritance. Also frequently used to designate any agent affecting performance.

family The offspring of a single tree after open pollination or of a single pair of trees after controlled pollination.

family, full-sib The offspring of a single pair of trees, usually resulting from controlled pollination.

family, half-sib The offspring of a single (usually female) tree and having different parents of the other sex. A half-sib family may result from open pollination or from controlled pollination using a mixture of pollens.

fertile Capable of producing viable seeds.

fertilization The union of male and female gametes to produce a fertilized egg cell.

fitness The relative ability of organisms of a particular genotype to survive and produce offspring, or the contribution of one of a pair of alleles to the general vigor of an organism. Symbolized by W.

flower induction Stimulation of trees to produce flowers.

forest genetics The study of heredity in forest trees.

forest tree improvement The application of genetic principles to the betterment of forest trees.

fruit The mature ovary or cone and seeds contained therein.

fruiting rotation The shortest period (seed to seed) required for trees to produce seed. Breeding rotation.

full-sibs Trees with both parents in common.

G

gain, genetic The increase in productivity following a change in gene frequency.

gamete An haploid generation male sperm cell or female egg cell capable of developing into a zygote (embryo) after fusion with a germ cell of the opposite sex.

gametophyte The few-celled, haploid generation portion of a seed plant arising from a meiotic division and giving rise through mitosis to the male or female gametes.

gene The basic unit of most types of inheritance, occupying a fixed position on a chromosome and consisting of a portion of a DNA molecule.

gene complex The balanced system of genes constituting the internal environment within which each gene acts, and so balanced that a change in one affects the operation of other genes.

gene flow The spread of genes through crossing.

gene interaction The action of nonallelic genes upon each other in such a way that the expression of a gene depends on the presence of other genes.

genecology A combination of ecology and genetics.

genetic drift Change in gene frequency and population characteristics due to chance rather than selection, and usually most pronounced in small populations.

genetics The science dealing with resemblances and differences among organisms related by descent.

genotype (1) The entire genetic constitution, expressed or latent, of an organism. (2) The genetic constitution of an individual with respect to a few genes under consideration. (3) A group of organisms having similar genetic constitution.

germplasm The sum total of the genes and cytoplasmic factors governing inheritance.

graft To place a detached branch (scion) in close cambial contact with a rooted stem (understock) in such a manner that scion and understock unite. A plant which has been grafted.

gyn-, gyno- Prefix meaning female.

H

half-sibs Trees with one parent (usually the female) in common.

haploid Having one set ($1n$) chromosomes, as in gametes.

heritability That portion of the total variance due to genetic factors. In a broad sense, that portion of the total variance due to all genetic factors. In a narrow sense, that portion of the total variance due to genes with additive effects and most indicative of the superiority that can be transmitted by seed.

heritability, family That portion of the total variance due to differences among families, and applicable only to family means.

heritability, single tree That portion of the total variance due to genetic differences among individuals, and applicable to data from single trees.

heterosis Hybrid vigor.

heterozygous (1) Possessing different alleles at a particular locus. (2) Derived from the union of gametes of dissimilar genotype. Antonym of "homozygous." A heterozygous tree is called a "heterozygote."

hexaploid Having $6n$ chromosomes.

homologous chromosomes Chromosomes in which the same gene loci occur in the same sequence.

homozygous (1) Possessing the same allele at a particular locus. (2) Derived from the union of genetically similar gametes. Antonym of "heterozygous." A homozygous tree is called a "homozygote."

hybrid Offspring of organisms of dissimilar genotype, often the offspring of a cross between different species.

hybrid habitat A habitat of intermediate characteristics, particularly well suited to hybrids.

hybrid index A numerical rating, derived from measurement of several diagnostic traits, used to denote the probable contribution of two parental races or species to individuals in a hybrid population.

hybrid swarm A population, usually of limited extent, consisting of parental species or races and hybrids between them.

hybridization The crossing of different races or species. Also sometimes used to denote the crossing of individuals.

hybridization, introgressive Long-continued hybridization leading to the infiltration of genes from one species into another.

I

inbreeding Mating between close relatives. Mating in a population consisting of a few individuals.

inbreeding coefficient (F) A number, between 0 and 1, calculated from the number of mating individuals and their relationship to each other, showing the homozygosity remaining in an individual or population as a fraction of the loci which were originally heterozygous. For most trees, $F = 1/2N$ where N is the number of breeding individuals.

inbreeding depression The reduction in vigor which often accompanies inbreeding, usually attributed to an accumulation of deleterious recessive genes in a homozygous condition.

incompatibility The inability of pollen to effect fertilization because growth of the pollen tube is arrested in the style.

infertility Sterility. General inability to produce seed due to lack of pollen tube growth, lack of fertilization, or embryo abortion.

individual tree selective breeding Selection and crossing of individual trees as opposed to working with populations.

inter- Prefix meaning between, as in interspecific, interracial.

interaction, genotype–environment The situation in which genotype A performs better than genotype B at one place, but poorer than B at another place.

intra- Prefix meaning within, as in intraspecific, intraracial.

introgression Long continued hybridization leading to the spread of genes from one race or species into another.

inversion The reversal of one chromosome segment relative to another.

isolation The prevention of crossing among populations because of distance or geographic barriers (geographic isolation), because of growth on different sites (ecological isolation), because of differences in flowering time (phenological isolation), or because of genic or chromosomal differences preventing normal seed set (genetic isolation).

K

karyotype The characteristic chromosome complement of a race or species. Karyotype analysis usually considers the number, relative total length, relative length of the two arms, and any visible structural details such as secondary constrictions or fragments.

L

Lamarckism The theory that evolution proceeds as the result of inheritance of acquired characters.

Lammas shoot Abnormal midsummer growth from a previously dormant bud, commonly caused by excess moisture. Named after Lammas Day, which is August 1.

layering A form of vegetative reproduction in which an intact branch develops roots as the result of contact with the soil.

lethal gene A gene which in the homozygous state causes death; usually a recessive gene.

like to like mating A system of mating in which similar trees are crossed with each other regardless of relationship.

line A population consisting of one (selfed line), a few (inbred line), or many (bulked line) parents and their offspring through several generations.

line breeding A system of mating in which closely related individuals are crossed with each other.

linkage The association of characters from one generation to the next due to the fact that the genes controlling the characters are located on the same chromosome.

linkage group The genes located on a single chromosome or the characters controlled by such genes.

locus The fixed position of a gene in its chromosome.

M

male sterile Incapable of producing functional pollen.

marker gene A gene, usually dominant, with a large effect and serving to identify one parent in offspring resulting from open pollination or mixed pollination.

mean Average. The sum of a group of observations divided by the number of observations.

meiosis A form of cell division in which chromosome number is reduced from $2n$ to $1n$. The form of cell division giving rise to gametophytes from diploid tissue.

Mendel's laws or principles A set of three laws first formulated by Gregor Mendel; each is generally true but there are numerous exceptions. (1) Characters exhibit alternative inheritance, being either dominant or recessive. (2) Each gamete receives one member of each pair of factors present in a mature individual. (3) Reproductive cells combine at random.

meristem A plant tissue composed of dividing cells and giving rise to organs such as leaves, flowers, xylem, phloem, roots, etc.

metaphase The middle stage of mitosis or meiosis during which the chromosomes are contracted and lie in one plane in the middle of and perpendicular to the spindle.

metaxenia Influence exerted by pollen on the maternal tissues of the fruit. Reported only in oak (among trees), where some hybrid acorns are distinguishable from nonhybrid acorns by acorn color.

microspore Haploid male cell which will ripen into a pollen grain. Immature pollen grain.

mitosis The form of cell division in which chromosome number is not reduced, each daughter cell receiving exactly the same complement as the mother cell. Mitosis can involve $2n$ vegetative cells or can involve $1n$ gametophytic cells.

modification Nongenetic variation attributable to the environment.

monoecious Bearing separate male and female flowers on the same tree.

monoecism Being monoecious.

monotypic genus A genus having but a single species.

mutagen A mutation-inducing agent such as a chemical or ionizing radiation.

mutant gene A changed gene.

mutation A sudden change in genotype. Usually a "gene mutation" (change in a single gene) is meant, but the term is sometimes used in a broader sense to include changes due to polyploidy, chromosome deletions, chromosome inversions, etc.

mutation rate The frequency (usually denoted by u or v) with which a gene changes or mutates to its allele.

N

n, 2n The haploid (gametophytic) and diploid (somatic or vegetative) numbers of chromosomes respectively.

N Frequently used to denote numbers of individuals.

neighborhood Specifically, the largest population in which mating occurs at random. Generally, a population within which individuals can interbreed.

nondisjunction Failure of a pair of homologous chromosomes to separate at meiosis so that both members of the pair are carried to the same daughter cell.

normal curve A curve derived from the equation $y = e^{-x^2}$ applicable to the distribution of many continuous variables.

nucleotide The fundamental building block of DNA. One nucleotide consists of a 5-carbon sugar (deoxyribose) linked to phosphoric acid and to one of four nitrogenous bases (A = adenine, C = cytosine, G = guanine, T = thymine). For convenience, nucleotides are usually designated as A, C, G, or T according to their nitrogenous bases.

nucleus The central part of a living cell containing the chromosomes.

O

octoploid Having $8n$ chromosomes.

ontogeny The developmental history of an organism during its lifetime.

open pollination Pollination effected by wind or insects.

ortet The original ancestor of a vegetatively propagated clone.

outbreeding, outcrossing Mating unrelated parents.

ovary The part of an angiosperm flower which contains the ovules and which develops into the fruit.

overdominance The condition in which the heterozygote is superior to the best homozygote.

ovule A part of an ovary containing an egg cell and developing into a seed after fertilization.

P

pairing, chromosome The coming together of homologous chromosomes prior to and during the metaphase stage of meiosis.

panmictic unit A local population in which there is completely random mating.

panmixia Random mating with no selection.

parthenocarpy The formation of fruit without fertilization and without fertile seeds, as in a banana. It includes fruit development stimulated by application of pollen incapable of causing fertilization. Common in some trees.

parthenogenesis, diploid A type of apomixis in which a seed develops from an unreduced egg cell.

parthenogenesis, haploid A type of apomixis in which a seed develops from a $1n$ egg cell; the seed may be $1n$ or $2n$ according to whether or not chromosome doubling occurs. Of most frequent occurrence in seeds bearing two or more embryos or showing other germination abnormalities.

pedigree Record of ancestry.

penetrance The percentage frequency with which a gene produces an effect.

pentaploid Having $5n$ chromosomes.

periphysis A form of topophysis in which effects due to position in tree are transmitted through a cutting or graft.

phenology The study of the timing of periodic phenomena such as flowering, growth initiation, growth cessation, etc., especially as related to seasonal changes in temperature, photoperiod, etc.

phenotype The visible characters of a plant. The product of a plant's genotype and its environment.

phylogeny The evolutionary history of a race, species, or higher category.

pistil The female portion of an angiosperm flower consisting of an ovary, stigma, and style.

plasticity, genetic The capacity for adaptation due to genetic changes, as in natural selection.

plasticity, physiological The capacity for adaptation due to internal physiological changes within individual plants.

pleiotropy Control of more than one character by a single gene.

plot The largest part of a test plantation or nursery experiment consisting of a single seedlot or clone under test. Trees of a plot are planted adjacent to each other in a row or rectangle. Plots are aggregated into blocks.

plus tree A phenotypically superior tree.

polar nuclei Two of eight nuclei developing in the female gametophyte of an angiosperm, the others being the egg cell and five nuclei which disintegrate. The polar nuclei fuse with a $1n$ nucleus from a pollen grain to form $3n$ endosperm.

pole The point at either end of the spindle from which spindle fibers radiate and toward which chromosomes migrate after metaphase. Also, in a totally different sense, a long stick used to measure tree heights.

pollen A mass of pollen grains.

pollen grain A microscopic (usually between 0.01 and 0.1 mm in diameter), usually yellow particle which is the male gametophyte and which results in seed formation after application to a female flower. A pollen grain germinates to produce a tube nucleus (which later degenerates) and a generative nucleus. The latter may effect fertilization (gymnosperms) or divide into two more nuclei (angiosperms), one of which fertilizes the egg to produce an embryo and the other of which fertilizes the polar nuclei to produce $3n$ endosperm.

pollen mother cell A $2n$ cell which divides twice (once by meiosis and once by mitosis) to form a tetrad of four pollen grains.

pollen tube An outgrowth of a germinated pollen grain through which the generative nucleus passes to effect fertilization.

pollination The placing of pollen on the receptive part of a female flower.

polycross test An experiment to determine the general combining ability of a group of trees by crossing each with the same mixture of pollen from several males.

polyembryony The presence of more than one embryo in a developing seed.

polygene A gene determining quantitative inheritance and having a small additive effect.

polyploid Having more than two complete sets of chromosomes.

population (1) Genetically, a group of individuals related by common descent and treated as a unit for convenience. There is no definite limit to the size or amount of variability contained within a population, nor is there necessarily a connotation that populations differ by any set amount. (2) Statistically, a group of homogenous observations or the individuals on which such observations were made.

population genetics The study of genetic changes in groups of individuals (populations), particularly over periods of several generations.

precision, statistical The fineness with which differences among genotypes or treatments can be shown to be significant statistically.

progeny The offspring of a particular tree or a particular combination of one female and one male tree.

progeny test An experiment, usually replicated, to compare the offspring of different parents, or to compare performance of offspring and parents. Usually confined to seedling offspring.

progeny test, full-sib An experiment consisting of full-sib families.

progeny test, half-sib An experiment consisting of half-sib families.

progeny test, open pollinated An experiment consisting of half-sib families resulting from open pollination.

propagule A seedling, cutting, or graft, especially when small.

protandry Shedding of pollen prior to time of maximum receptivity of the female flowers on the same tree.

protogyny Having female flowers or flower parts receptive prior to pollen shedding on the same tree.

provenance The ultimate natural origin of a tree or group of trees. In forestry literature the term is usually considered synonymous with "geographic origin," and preferred to "origin," which could mean "nursery of origin," "seedhouse of origin," or "method of propagation." Sometimes used to denote the trees having a given place of origin.

provenance test An experiment, usually replicated, comparing trees grown from seed or cuttings collected in many parts of a species' natural range.

Q

quantitative inheritance Inheritance of characters which vary continuously and which are controlled by three to many genes.

R

r The correlation coefficient, varying from -1 to $+1$, and showing the degree to which one variable is associated with another. The level of significance for a particular value of r varies with the number of sets of observations, 25 or more sets often being considered desirable. In genetic work, values of r between -0.4 and $+0.4$ are usually considered weak, and values of r less than -0.8 or greater than $+0.8$ are usually considered strong.

race A genetic subdivision of a species, more or less distinct, having distinctive characteristics when grown in a particular environment. Sometimes used to denote a portion of a cline. Differences among races may or may not be adaptive; differences among ecotypes are presumed to be adaptive. Hence "race" is a more general term than "ecotype." A taxonomic variety or a subspecies is a genetic subdivision of a species described and named according to the strict rules found in the "International Code of Botanical Nomenclature," and usually sufficiently distinctive to be recognizable in the wild.

ramet An individual member of a clone, descended through vegetative propagation from the ortet.

random breeding unit The largest local population within which mating occurs at random. A neighborhood.

randomized complete block design An experimental design in which every seedlot or clone is represented once and only once in every block, and in which the plots are randomized within blocks.

recessive gene A gene, member of an allelomorphic series, without phenotypic effect when present in the heterozygous state.

receptive Referring to the female flower or stigma, capable of supporting a germinating pollen grain which can then effect fertilization.

reciprocal cross Repetition of a cross in the reverse direction, as female A × male B and female B × male A.

recombination Obtaining new combinations of genes through crossovers and independent segregation of chromosomes at meiosis.

reduction Halving the chromosome number at meiosis.

regression A statistical measure (usually indicated by the regression coefficient b) showing the amount of increase or decrease in one variable for a given amount of increase in another variable. Originally intended, and still frequently used, to show the degree to which the offspring of exceptional parents regress toward or revert to the average of the parental generation.

repeatability The degree to which characteristics measured at differnt times in an organism's life remains constant. Expressed as a correlation coefficient r.

replicate A portion of an experiment containing one plot of each seedlot, clone, or treatment. Synonymous with block in a randomized complete block experiment, but not in some other designs.

replication The practice of repeating experiments in different places or at different times in order to learn which differences are due to the seedlots or treatments under test as opposed to uncontrolled sources of variation.

resistance The ability to survive or thrive when grown under adverse conditions, such as the presence of insects, disease, and cold. Resistance may be partial or complete, and may be due to ability to avoid, throw off, repel, or recover from damage. The term "immunity" is usually used to denote a high degree of resistance to a disease.

ribonucleic acid RNA. A single-strand acid, formed on a DNA template, found in the protoplasm, and controlling cellular chemical activities. Whereas DNA transmits genetic information from one cell generation to the next, ribonucleic acid is an intermediate chemical translating genetic information into action.

ribosome A protoplasmic granule containing ribonucleic acid and believed to be the seat of protein synthesis.

RNA See ribonucleic acid.

roguing Systematic removal of individuals or families of undesirable genotype, especially from a seed orchard.

rootstock A rooted plant on which a scion is grafted. An understock.

S

s (1) The standard deviation of a sample, as computed from actual measurements. (2) Selection pressure.

S_1, S_2, S_3 (1) The first, second, and third generations, respectively, of selfed offspring.

S_1, S_2, S_3 Members of a series of alleles causing self-incompatibility.

satellite A short segment of a chromosome, separated from the main body by a constriction.

scion A detached plant part grafted or budded on a rootstock.

scion–stock interaction The phenotypic effect of the rootstock on the scion (or occasionally vice versa) because of which a scion on one type of rootstock performs differently than if on its own roots or on another rootstock.

seed The fertilized, ripened ovule of a flowering plant containing an embryo, an endosperm (sometimes), and a seed coat, and capable of developing into a mature plant.

seed orchard A plantation established primarily for the production of seed of proven genetic quality.

seed orchard, clonal A seed orchard composed of vegetatively propagated (usually grafted) trees established primarily for the production of seed of proven genetic quality. Primary interest centers on the genetic quality of the seed produced, whereas in a clonal test primary interest centers on the genetic quality of the clones themselves.

seed orchard, progeny test A progeny test so managed and thinned as to produce seed of proven genetic quality.

seed orchard, seedling A plantation of seedlings established primarily for the production of seed of proven genetic quality.

seed production area A phenotypically superior stand or plantation rogued and treated in such a manner as to produce large quantities of seed.

seedlot A convenience term denoting a group of seeds or their offspring which will be considered as a unit in an experiment.

segregation The separation of genes or chromosomes of maternal and paternal origin at meiosis.

selection The choosing of individuals or populations with desirable characteristics to obtain genetic improvement.

selection, clonal Choosing the best clones in a clonal test for further asexual propagation.

selection, combined Choosing the best individuals in the best families as parents.

selection, direct Choosing trees on the basis of desirability in the traits under consideration, as choosing the tallest trees to improve growth rate.

selection, family Choosing parents on the basis of average family performance in a progeny test.

selection, genotypic Choosing parents on the basis of known information about their genotypes.

selection, index Choosing parents on the basis of several desirable traits. The desirability of an individual, family or clone is judged according to an index value calculated by considering the heritability and importance of each trait.

selection, indirect Choosing individuals or populations on the basis of a trait (usually of no economic importance in itself) believed to be associated with a desired trait of economic importance, rather than on the basis of the desired trait.

selection, mass Phenotypic selection. Choosing trees on the basis of their phenotypic performance.

selection, natural Natural elimination of trees on the basis of their phenotypic inability to survive or produce seed under a particular set of environmental conditions.

selection, phenotypic Choosing trees on the basis of their phenotypic performance.

selection coefficient s. The term $1 - s$ is so defined as to show the survival rate of one gene relative to that of another gene.

selection differential The difference between a selected tree, family or clone and the average of the population from which it is taken.

selection pressure The strength of the tendency to eliminate undesirable genotypes or phenotypes, usually expressed in terms of a selection differential or as the proportion of total trees which are selected.

self To place pollen from a male flower on a female flower on the same tree (verb), or a tree resulting from such pollination (noun).

self-pollination Pollination of a female flower by pollen taken from the same tree or clone.

self-incompatibility Inability to produce seed following self-pollination. Sometimes limited specifically to cases in which the inability is due to a pollen-borne gene that prevents pollen tube growth on a stigma having the same gene.

self-sterility Inability to produce seed following self-pollination.

semilethal gene A gene (usually recessive) causing greatly reduced viability if present in the homozygous state.

sib, sibling A sexless term meaning brother or sister, i.e., an individual belonging to the same family. Half-sibs have one parent (usually the seed parent) in common. Full-sibs have seed and pollen parents in common.

sigma (σ) The standard deviation. When squared, the variance. Often, σ is used to denote a theoretical or true value, whereas s is used to denote a calculated value applicable to a particular finite population. Such a distinction is not made as much in the genetic as in the statistical literature.

smear Small amount of fixed, stained and macerated tissue, placed on a microscope slide and compressed by pressure on a cover glass, used in the study of chromosomes.

somatic Pertaining to the vegetative rather than the reproductive cells, and having $2n$ chromosomes.

species A group of similar organisms, capable of interbreeding, and more or less distinctly different in geographic range and/or morphological characteristics from other species in the same genus.

sperm cell Male gamete. The $1n$ cell derived from a pollen grain that actually fertilizes the egg cell.

spindle A diamond-shaped network of fibers formed between the two poles and the equatorial plate and visible during the metaphase stage of cell division. The spindle is the mechanism by which chromosomes migrate to the poles during cell division.

sporophyte   The normal plant with 2n chromosomes.

stamen   The male, pollen-producing organ, comprised of a stalk and an anther.

staminate   Male or pertaining to stamens.

standard deviation   Square root of the sum of the deviations squared divided by one less than the number of observations. That is, $s = [\Sigma d^2/(n-1)]^{1/2}$. A standard deviation is a measure of the variability among individual observations.

standard error   A standard deviation divided by the number of observations. A standard error is a measure of the reliability of the means of a group of observations.

sterility   Inability to produce sound seed.

stigma   The apical portion of the pistil of an angiosperm flower, on which pollen germinates.

stock   Understock, rootstock. A rooted plant on which a scion is grafted. Also, a general term to denote trees to be planted.

stoma (plural stomata)   A small opening in a leaf, surrounded by two guard cells, controlling gas and water exchange between the leaf and the atmosphere. Relative stomata size is sometimes indicative of chromosome number.

strain   A group of trees, related by common descent, differing in some respect from the main body of the species. In forestry the exact content of a strain is less well defined than is that of most other subspecific categories.

strobilus (plural strobili)   The male (catkin) or female (cone) inflorescence of a conifer, composed of a central axis, scales (sporophylls), bracts, and seeds or anthers.

subspecies   A geographically localized subdivision of a species, genetically and morphologically distinguishable from other subspecies, described according to taxonomic rules, and given a Latin name. A subspecies tends to be larger than a taxonomic variety but there is no clear distinction between the two.

sympatric   Inhabiting the same area.

synthetic variety   In agronomy, a cultivar produced by the combination of selected lines and thereafter propagated as one variety.

T

taxon   A taxonomic unit of any size.

taxonomy, descriptive   Classification and naming of groups of organisms having common ancestry and common phenotypic (usually morphological) characteristics.

taxonomy, experimental   Descriptive taxonomy supplemented by genetic studies such as provenance tests. The distinction between genetics and experimental taxonomy is often slight.

taxonomy, physiological   Classification based partly on physiological characteristics.

test plantation   A plantation, usually replicated, established to determine the extent to which differences among seedlots or clones are hereditary.

test rotation   The numbers of years from seed to observation required to obtain valid estimates of the probable performance at maturity of a group of seedlots or clones.

tester   A parent (usually male) crossed with a large number of trees to determine the general combining ability of the latter.

tetrad   A group of four pollen grains descended from a pollen mother cell and behaving as a group until the pollen matures.

tetraploid   Having 4n chromosomes.

tetravalent   A group of four chromosomes at meiotic metaphase, usually indicative of sterility.

topcross test   An experiment in which one tree (usually male) is crossed with each of a large number of trees (usually female) to determine general combining ability of the latter.

topophysis   The long persistence of nongenetic position or age effects after grafting or rooting. Examples are the horizontal growth of grafted trees produced with scions taken from lower branches and the early coning of grafted trees produced with scions taken from flower-producing branches.

topwork   To graft a scion into the top of a tree several feet tall.

translocation   Exchange of chromosome segments between nonhomologous chromosomes.

triplet   A sequence of three nucleotides in DNA or their counterparts in RNA.

triploid   Having $3n$ chromosomes.

trisomic   Having one chromosome present three times and the others present twice, that is $2n + 1$.

trivalent   A group of three chromosomes at meiotic metaphase, usually indicative of sterility.

tube nucleus   The nucleus occurring in and regulating growth of the pollen tube, but not taking part in fertilization.

tunica   The outer sheath of an apical growing point, containing several dividing cells, and giving rise to different tissues than the inner or corpus layer.

U

u   Frequently used to denote mutation rate.

understock   Rootstock. A rooted plant on which a scion is grafted.

V

v   Frequently used to denote mutation rate in a reverse direction.

V   Frequently used to denote variance.

variable, continuous   A measured characteristic in which no distinct classes are recognizable, any class limits being arbitrarily drawn.

variable, discontinuous   A measured characteristic in which distinct classes (e.g., dead versus alive) are easily recognizable.

variability, variation   Absence of uniformity. Quality of differing from the average value. These terms are usually used in a general sense and qualified by such words as low, moderate, and high. For quantitative comparisons, the concept of "variance" is used.

variance   A statistical measure of variability, defined in an exact manner. In the simplest sense, variance $= V = $ (standard deviation)^2 = (sum of deviations^2)/$(n - 1)$ where n is the number of observations and the deviations are calculated by subtracting the average or mean value from each observation. Many types of variances can be calculated and the ratios between these provide a way of determining whether observed differences or relationships are large enough to be due to factors other than chance.

variance, analysis of   A statistical procedure designed to separate variability into parts due to seedlot, replication, chance (error), etc., and thus to determine whether observed differences among seedlots, blocks, etc., are large enough to be taken seriously.

variance components   Once an analysis of variance is completed, the mean squares can be broken down into variance components by formulas such as given in Chapter 12. Variance components, while difficult to calculate, provide the most convenient way of expressing the relative amounts of variability due to the different sources of variation.

variance ratio A ratio of one variance to another which is basic to most tests of statistical significance. The F value, the correlation coefficient (r), and the regression coefficient (b) are among the most commonly used variance ratios.

variety, agronomic A distinctive seedling population or clone with enough favorable characteristics to warrant cultivation. Agronomic varieties are given non-Latin names according to priority rules which vary among different crops.

variety, cultivated (cultivar) A clone or seedling population with enough favorable characteristics to warrant cultivation and given a non-Latin name. There are no formal rules governing the description and naming of cultivated varieties. Cultivar names are fairly constant for the most commonly planted horticultural crops. They are much less constant for forest trees, a cultivar often being known by several names.

variety, geographic or taxonomic A subdivision of a species with distinct morphological characters and a distinctive geographic range, and given a Latin name according to the strict rules of the International Code of Botanical Nomenclature. A taxonomic variety is known by the first validly published name applied to it so that nomenclature tends to be stable.

vegetative nucleus Nucleus in the pollen tube.

vegetative reproduction, vegetative propagation Reproduction by other than sexually produced seed. Includes grafting, reproduction by cuttings and some types of apomixis.

W

wind pollination Pollination by windborne pollen.

Z

zygote The cell (usually diploid) resulting from fertilization of one gamete by another, or the tree resulting from such fertilization.

Index

A

Acacia, 32, 108
Adaptability, 368
Adaptation, 32, 273, 361
Additive genes, 15, 42, 53.
Africa, 396
Age effect
    on flowering, 193, 197, 201
    on rooting, 125
Air layering, 128
Air turbulence, 70, 74
Alder, 32, 38, 115, 356, 405,
    411
Allopolyploids, 412
Aneuploidy, 399, 404, 413
Angiosperms, 9, 25
    polyploidy in, 404–412
Anther culture, 419
Apoinixis, 128
Apple, 405
Araucaria, 38
Arborvitae, Japanese, 60
Argentina, 372, 396
Ash, 76, 108, 115, 125, 410, 411
Aspen, 115, 122, 127, 416
Australia, 376, 395
Autopolyploids, 410, 412

B

Bacteria, 9
Bark thickness, 218
Basswood, 93, 411
Beech, European, 115, 125, 132, 202, 206
Birch, 38, 115, 122, 124, 166, 168, 192,
    194, 222–223, 296, 346, 405, 411
Black locust, 124, 127
Blister rust, resistance to, 132, 165,
    178–179, 214
Blocks, number and design, 136,
    146–150
Branch angle, 214
Breeding arboreta, 146, 175, 262
Buckeye, hybrid, 335
Budding, 112, 114

C

Cambium, 22, 112, 114
Canada, 392–393
Cedar, true, 72, 76
Cell division, 7, 22
Centromere, 9
Chemical differences, 206, 278
Cherry, 141, 202, 217
Chestnut, 32, 38, 108, 125, 316, 335, 336,
    353

Chromosomes, 7–13
  homologous, 13, 22
  mapping, 13
  numbers of, 403, 406–409
  pairing, 317
  techniques of study, 400–401
Climate of origin, importance in exotics, 365
Climbing, technique of, 86
Cline, 214, 259, 273, 291, 301
Clonal seed orchards, *see* Seed orchards
Clonal selection, 181–184
Clonal tests, 215, 250
Clone, 111
Clone replacement in orchards, 229
Colchicine, 415
Cold hardiness, 61, 255, 275, 299, 302, 365, 379
Combining ability, 191–192
  general, 169, 178
  specific, 178
Cone
  collection, 228
  opening, 214
  production, 228
Corn, 39, 43, 313
Correlation
  age–age, 268
  coefficient, 163
  parent–offspring, 169
Cottonwood, eastern, 126, 132, 202, 218, 221, 235, 302
Crossability patterns, factors affecting, 318–320
Crossovers, 14, 23
Cryptomeria, 20, 121, 202, 206, 387–388
Cuttings
  leaf, 127
  root, 127
  stem, 121–126
Cyclophysis, 130
Cytoplasm, 7, 20, 25

D

Dalbergia, 124
Daylength, 366, *see also* Photoperiod
Deficiencies, floral, 367
Deoxyribonucleic acid, *see* DNA

Diallele crosses, 176
Dioecism, 37
Discontinuous variation, 294, 302
Disease resistance, *see* Blister rust, Dutch elm disease, Fusiform rust, *Melampsora* rust
DNA, 9, 12, 17, 130
Dominance, 15, 42, 204
Double fertilization, 25
Douglas-fir, 8, 31–32, 38–39, 72, 76, 82, 99, 115, 120, 159, 174, 189, 202, 283–287, 290–292, 386–387, 404, 413
Dutch elm disease, resistance to, 182

E

Elevational trends, 257, 291
Elm, 8, 32, 76, 93, 103, 181, 348, 411
Emasculation, 93
Embryo, 25
Endosperm, 25
Epistasis, 15
Equilibrium frequency, 48–49, 53–54
Equipment, climbing, 225
Eucalyptus, 9, 32, 93, 108, 115, 124, 202, 235, 319, 337, 362, 376–379
Europe, 267–281, 372, 392
Evaluation of exotics, 375
Evolution, 316, 320–322, 333, 362, 404–410
Exchange relationships for exotics, 371
Exotics, 359–398
  factors affecting introduction, 365–370
Experimental design, 261–263

F

F_1 and F_2 generation, 42, 44, *see also* Hybrids
Family, see *Selection*
Family size, optimum, 180
Fertility of polyploids, 413
Fertilization, 24–25
Fiber length, 221
Fir, 38, 320–321
  grand, 287
  Sakhalin, 221–222, 263
  white, 82, 287–289
Fitness, 48, 55

Flower buds, recognition of, 93
Flower primordia, 129
Flowering
    age, 193, 197, 201
    inheritance of, 217
    time as an isolating mechanism, 78
Frost, resistance to, 366
Full-sib, *see* Selection, Progeny tests
Fusiform rust, resistance to, 165, 189,
        307

G

Gain
    estimation of, 239–252
    methods of increasing, 160–164,
        207–224
Gametes, 24
Gene frequency, calculation methods,
        45–59
Generation length, 192
Genes, 12–17
    arrangement, 13
    chance extinction of, 65
    classification, 14
    deleterious recessive, 33, 38
    function, 12, 17
    marker, 72
    structure, 12
Genetic drift, 36, 258
Genetic gain, *see* Gain
Genetic information, 7, 11
Genetic load, 30
Genotype, 21
Genotype–environment interaction, *see*
        Interaction
Geographic races, 259, 301
Geographic variation, 2, 61, 77, 81, 208,
        217, 221–223, 253–312
    factors influencing, 253–255
    general trends, 255–258
    random patterns, 258, 309
Ginkgo, 390
Golden larch, 403
Grafting, 86, 111–121, 193, 226
Great Britain, 372, 391
Growth rate, 160, 207–210, 216–217, 221,
        223, 275
Gymnosperms, 25
    polyploidy in, 403

H

Half-sib, *see* Selection, Progeny tests
Haploidy, 418–420
Hardy–Weinberg equilibrium, 48
Hawaii, 394
Hemlock, 122, 285
Heritability, 162–164, 174, 189, 211, 218
    clonal, 251
    estimates, reliability of, 241, 248
    estimation of, 239–252
    family, 170–174, 243, 247, 249, 251
    single-tree, 170, 240, 249
Hermaphrodism, 92
Heterosis, *see* Hybrid vigor
Heterozygosity, 32, 44
    selection against, 58–59
    selection for, 56–58
Homozygosity, 32
Honeylocust, 132
Horsechestnut, 335, 410
Hybrid corn, 343
Hybrid habitats, 42, 315
Hybrid vigor, 41, 315, 343, 350
Hybridization, 39, 41–44, 56–59, 85,
        312–359
    between genera, 318, 322
    between species, 312–359, 410
    between varieties, 313, 317
Hybrids, 312–359
    breeding behavior, 350
    F_1, 314–316, 344–349
    F_2, 314–316, 349–351
    mass production, 344–347
Hybrid swarms, 57, 59, 320

I

Inbreeding, 28–41, 164, 193, 415
    adaptation, lack of, 32
    coefficient, 33
    consequences of, 28–33
    depression, 30
    mechanisms for and against, 37
    plus outcrossing, 40
    tropical and temperate trees, in, 37
Incompatibility
    genetic, 317
    graft, 118–121, 210

Inheritance, 12–22
  genic, 12–19
  maternal, 20
  paternal, 20
Insects
  cone, 228
  resistance to, 213, 215, 276
Interaction, genotype–environment, 149,
    252, 271, 285, 293–294, 297, 302
Introgression, 289, 294, 320–322, 333,
    341
Isolation, 77, 226, 316
  by distance, 80
Isozymes, 60

J

Japan, 372, 393
Juniper, 38, 133, 403

K

Karyotype of pine, 402

L

Labels, pollination, 102
Ladders, 87
Larch, 32, 115, 124, 322
  European, 202
  hybrid, 322, 324, 348–350
  Japanese, 198, 202, 254
Late generation breeding, 315, 351–357
Layering, 128
Leaf color, 206, 213, 275, 289–290,
    301–302
Leaf moisture, 271
Leaf nutrients, 287, 299
Leaf primordia, 22
Leaf size, 275, 306
Line breeding, 39
Linkage, 13
Locus, 15, 44

M

Magnolia, 93, 100, 411
Maize, *see* Corn
Maple, 38, 100, 108, 124, 334, 357, 411
  Norway, 206

red, 67, 122
  sugar, 116, 202, 300
Mating designs, 167, 176–177
Meiosis, 22, 412, 415
Melampsora rust, resistance to, 178–179,
    221
Metaphase, 7, 9
Metasequoia, 236, 390
Migration, 66
Mitosis, 22, 415
Monoecism, 38
Monoterpenes, 209, 278
Monotypic genera, 369
Mulberry, 411
Mutagens, 19
Mutation rate, 19, 66
Mutations, 17, 65

N

Neighborhood, 80
New Zealand, 284, 379–381, 395
Normal curve, 75
Nucleotides, 9
Nucleus, 7
Nurseries, 139, 151–152

O

Oak, 38, 42, 59, 108, 116, 125, 187, 320,
    339
  English, 206, 387–389
  northern red, 82
Oleoresin, 209
Operator gene, 17
Operon, 17
Optimizing family size, 180
Overdominance, 16, 42, 57

P

p, 47
Pea tree, 32, 49, 206
Periphysis, 130
Pest control, need for in plantations,
    141–142
Phages, 12
Phenology, 217, 221, 285, 293, 295–296,
    303–304, 317
Phenotype, 21, 268–269

Photoperiod, 286, 306
Pine, 8, 23, 32, 58–59, 95, 100, 116, 124,
    132, 187, 318, 321, 323–325, 346,
    402
  Austrian, 327–328, 347, 355
  Caribbean, 264, 381–384
  eastern white, 33, 60, 97, 116, 118,
    132, 201, 214, 256, 296–299, 325,
    355, 384–386, 404
  Himalayan white, 325
  hybrid, 323–331
  jack, 58, 120, 165, 201, 206, 214, 294
  Japanese black, 202, 326
  Japanese red, 39, 201, 216, 326, 328,
    347, 355
  knobcone, 210, 331
  limber, 82
  loblolly, 30–31, 38–39, 43, 94, 116–117,
    120, 165, 173, 175, 188, 201, 207,
    223–232, 262, 305–309, 321, 348,
    350, 384
  lodgepole, 58, 284
  longleaf, 165
  Monterey, 121, 184, 202, 210–213, 254,
    331, 364, 379–381
  pinyon, 76
  pitch, 43, 321, 348, 350
  ponderosa, 81–82, 89, 189, 192, 216,
    254, 287–288, 292, 330, 384
  red, 118, 165, 201, 206, 232–234, 296
  Scotch, 39, 76, 82, 90, 96, 117–118,
    159, 167, 199, 201, 210, 254, 259,
    267–281, 327
  shortleaf, 117, 201, 306
  slash, 39, 76, 117, 188–189, 191–192,
    197, 201, 206–210, 223–232, 254,
    305–306, 381
  southern, 40, 108, 328–329
  southwestern white, 82
  Virginia, 120, 174, 201, 307
  western white, 165, 178–179, 192, 206,
    217, 355
Pinus
  subgenus *strobus*, 323–324
  subsection *Australes*, 328–329
  subsection *Contortae*, 330
  subsection *Oocarpae*, 330
  subsection *Ponderosae*, 330
  subsection *Sylvestres*, 324–328

Plantations
  alignment in, 144
  care of, 140–144
  establishment methods, 152–156
  spacing in, 142–143
Plots, 135
  design of, 146–147
  missing, 149
Plus trees, *see* Selection, mass
Podophyllin, 415
Pollen, 67, 85
  application of, 101
  collection of, 98
  dispersion distance, 69, 71–76
  forcing of, 95
  shipment and storage of, 98–99
  testing of, 100
Pollination
  artificial, 85–109, 346
  bags, 91
  birds, by, 70
  equipment, 86–105
  indoor, 103
  insect, 37, 70, 79
  wind, 69
Polyploids, 77, 335, 399–415
  breeding behavior, 412
  induction of, 415–417
Poplar, 9, 38, 58, 121, 181, 183, 188, 192,
    202, 319, 321, 337, 339–340, 346,
    348, 418
Populations
  continuous, 80
  density of, 81
  genetics of, 47–82
  isolated, 79
  size of, 33, 44, 59
Precipitation, 366
Progeny tests
  bulk, 166
  data useful in thinning clonal orchards,
    229
  field, 135–157
  full-sib, 173–179, 191–194, 199–201,
    218, 246–249
  half-sib, 167–172, 190–194, 197, 207,
    212–213, 215–216, 232, 242–245,
    263
  nonstatistical considerations, 138–144
  nursery, 151–152

statistical considerations, 136–138, 146–149
Provenance tests, 253–265
limited range, 262–263, 299
range wide, 261–262, 284–311

Q

q, 47

R

Races, elevational and geographic, 81, *see also* Geographic variation, Elevational trends
Radiation, 73
Range discontinuities, 79, 255, 274, 290, 294
Receptivity of flowers, 100
Recessive genes, 15, 205–206
selection against, 48–53
selection for, 53
Recombination, 24
Records
plantation, 144–145, 156–157
pollination, 102
Recovery ratios, 44
Redcedar, western, 418
Redwood, 8, 390, 403
Regression, parent-offspring, 207–212, 217, 218, 240–242
Replication, 135, 146–150
Republic of South Africa, 396
Resistance to pests, 132, 289, 355, 363, *see also* Blister rust, Dutch elm disease, Chestnut blight, Fusiform rust, *Melampsora* rust, White-pine weevil
Ribonucleic acid, *see* RNA
RNA, 12
Root suckers, 127
Root tips, 22
Rooting of cuttings, 121–127
Rootstocks, effects on scions, 116
Rubber tree, 117

S

Safety precautions, 91
Sassafras, 127

Scion collection, 130–132, 226
Scions, 111
Second generation, 218, 230, *see also* Hybrids
Seed
age of production, 299
collection, 106, 228
development, 24
dispersion, 68
dormancy, 60, 299
protection of, 105
sorting, 106–107
yields, 3, 33, 108, 131, 231, 413
Seed orchards, 40, 77, 132, 187–203, 204–237, 345–346, 374
clonal, 172, 187, 191, 194, 197–201, 206–213, 223–232, 347
management, 223–236
seedling, 173, 179, 187, 194, 232–234, 264
Seed production areas, 188
Segregation ratios, 44, 352, 414
Selection, 47–62, 65–67, 86, 159–185
clonal, 180–184
coefficient, 48
combined, 179, 190
cost comparisons among methods, 171
differential, 160–162, 188
direct and indirect, 163
full-sib family, 173–179, 187–190, 194, 208
fundamental theorem, 55
half-sib family, 187–190
indirect, 163
mass, 165–166, 169–170, 188–190, 194, 204–225
natural, 60–61, 259–260, 360–361
phenotypic, *see* Selection, mass
pressure, 60
Selective breeding, 159–185
Selfing, 28, 32–33, 38–39
frequency of natural, 38
Self-replication of DNA, 9
Spacing, 227
Spruce, 32, 36, 38, 58, 100, 116, 119, 124, 187, 192, 321, 331–334
blue, 202
Engelmann, 58, 294
Norway, 58, 76, 87–88, 122, 124, 202, 206, 332, 386, 418

Serbian, 332
Siberian, 58
Sitka, 285, 294
white, 58, 108, 113, 236, 256, 293
Statistical designs
compact family, 136
incomplete block, 138, 150
Latin square, 137
randomized complete block, 136
Statistical procedures
analysis of variance, 242, 246, 251
efficiency, 136, 146–149
precision, 136
Stem form, 208, 210, 217, 219, 277, 298, 306
Sterility, 31, 350
Sweetgum, 38

T

Taiwan, 394
Taiwania, 235
Tamarack, 2, 202, 294
Tannin content, 218
Taxonomic varieties, 269, 274–280
Techniques
chromosome study, 400–401
measurement, 154–157
pollination, 100–102
Test plantations
design of, 136–138
exotics, 372–376
measurement of, 154–157
Tester crosses, 176, 191, 208
Thinning, 229, 233
Tissue culture, 419
Tissue differentiation, 129
Topophysis, 129, 181
Tracer elements, 73

Triploids, 410, 413, 417
Tupelo, black, 127

U

Understock, 111
Uniformity, 182
United States, 283–311, 372, 392–393

V

Variance, analysis of, 242–251
Variance components, 222–223, 242, 246, 251
Variation, *see also* Geographic variation, Selection
continuous, *see* Clines
discontinuous, 274, 290
Vegetative propagation, 111–133, 346
Viruses, 12

W

Walnut, black, 108, 116, 120, 125, 142, 195–196, 301, 303
Wattle, black, 202, 206, 217–220
Weed control, 140–142
White-pine weevil, resistance to, 215
White cedar, northern, 296
Willow, 9, 38, 92, 121, 126, 184, 192, 346, 348
Wood density, 159, 208, 216, 221, 310
Wood properties, 378

Y

Yeast, 9
Yellow-poplar, 32, 38, 93, 100, 108